Small Gas Engines

by

ALFRED C. ROTH
Assistant Professor, Industrial Technology
Eastern Michigan University
Ypsilanti, Michigan

RONALD J. BAIRD
Professor, Industrial Technology
Eastern Michigan University
Ypsilanti, Michigan

WITHDRAWN

South Holland, Illinois

THE GOODHEART-WILLCOX COMPANY, INC.

Publishers

Library of Congress Catalog Card Number 91-23101.
International Standard Book Number 0-87006-919-5.

56789-92-98765

Library of Congress Cataloging-in-Publication Data
Roth, Alfred C.
  Small gas engines: fundamentals, service, trouble-shooting, repair / by Alfred C. Roth, Ronald J. Baird.
     p.   cm.
  Includes index.
  ISBN 0-87006-919-5
  1. Internal combustion engines, Spark ignition.
I. Baird, Ronald J.   II. Title.
TJ789.R58   1992
621.43'4--dc20   91-23101
                                                      CIP

INTRODUCTION

SMALL GAS ENGINES provides students, do-it-yourselfers, and aspiring mechanics with practical information about small engine construction, operation, lubrication, maintenance, troubleshooting, service, rebuilding, and repair. Small engine users at all levels are finding it beneficial to have a thorough understanding of engine fundamentals and service procedures.

Since the introduction of SMALL GAS ENGINES, many revolutionary developments in automotive design have found their way into the design of small gasoline engines. Some of the results include higher horsepower-to-weight ratios, fuel injection, electronic ignition, pressure lubrication, and the use space age materials.

Today's engines last longer, use less fuel, run quieter, start easier, and are more reliable than their predecessors. The high level of technology associated with the design and manufacture of modern engines calls for closer tolerances and precise airflow and lubrication requirements. As engines have become more sophisticated, they have also become more complex.

Today, the small gas engine serves as a labor-saving power source for more products than ever before. Lawn mowers, lawn edgers, riding tractors, garden tillers, snow throwers, leaf blowers, and string trimmers are used around the home. On the farm, small gas engines power chain saws, portable pumps, post hole drills, brush cutters, log splitters, conveyor systems, and cement mixers.

Industrial applications of small gas engines include portable electric generators, hydraulic pumps, lifting mechanisms, portable winches, welding generators, drilling equipment, cement finishers, tampers, horizontal boring units, and trenchers. Other industrial uses include portable paint sprayers, flame blowers (dryers), road stripe erasers, air compressors, sweepers, and portable road warning signs.

Recreational vehicle applications represent a large segment of the small gas engine market. Small gas engines are typically used in water jet skis, motorcycles, mini-bikes, go-carts, gyrocopters, all-terrain vehicles (ATV's), ultra-light aircraft, and hot-air balloon inflators. Outboard engines are available in a wide range of horsepower ratings for a variety of watersport applications.

SMALL GAS ENGINES has been written to provide detailed technical information about one- and two-cylinder, two and four cycle gasoline engines. Diesel and LP-Gas engines are also covered in the text. Learning Objectives, Review Questions, and Suggested Activities are featured in each chapter. A WORKBOOK FOR SMALL GAS ENGINES is available as a supplement to the SMALL GAS ENGINES textbook.

Alfred C. Roth

THE MANY APPLICATIONS OF SMALL GAS ENGINES

Two and four cycle gasoline engines of low horsepower and small size play a familiar and important role in almost everyone's life. When power needs take us out of reach of the electric cord or beyond the application of an electric motor, we depend on small gas engines to ease our chores and power our recreational vehicles.

We use small gas engine-powered equipment to cut grass, remove snow, cultivate gardens, cut wood, drive generators, pump water, and sweep factory floors. We also use small engines in snowmobiles, go-carts, motorcycles, and boats. Typical small engine applications are pictured below and on pages 6, 7, and 8. Chapter 17 covers small gas engine applications in detail.

Snowmobiles are generally powered by small gas engines. In colder regions of the country, they are very common and popular.

IMPORTANT SAFETY NOTICE

Proper service and repair is important to the safe, reliable operation of small gas engines and related equipment. The procedures recommended in this book are effective methods of performing service operations. This book also contains various safety procedures and cautions that must be followed to minimize the risk of personal injury and part damage. These notices and cautions are not exhaustive. Those performing a given service procedure or using a particular tool must first satisfy themselves that safety is not being jeopardized.

SMALL GAS ENGINES contains the most complete and accurate information available at the time of publication. Goodheart-Willcox cannot assume responsibility for any changes, errors, or omissions in this text. Always refer to the appropriate service manual for specific repair recommendations.

CONTENTS

A small two cycle engine drives this hedge trimmer.

A small four cycle engine powers the rotor and driving wheels of this snow blower.

A small gas engine-propelled riding mower can be used to pull a sprayer. Pressure to sprayer is supplied by an engine-driven compressor.

Left. Rotary lawn mowers account for a large share of small gas engines in use. Center. Gas engine-powered mudjack lifts sunken sections of sidewalk. Right. Small gas engine powers this garden cultivator.

A powerful small gas engine is used to drive this amphibious all-terrain vehicle.

Riding mowers are used where lawns are spacious. Property owners often require the services and talents of small gas engine mechanics to provide maintenance and repair services for engines and vehicles.

CHAPTER 1

SAFETY IN THE SMALL GAS ENGINE SHOP

After studying this chapter, you will be able to:
- ☐ Explain why a clean, well-organized shop is extremely important.
- ☐ List several dangers associated with working in a small engine shop.
- ☐ Explain the importance of maintaining and using tools properly.
- ☐ Describe methods for minimizing the risks involved in working with small engines.
- ☐ Explain the function of OSHA.

Small gas engine work can be rewarding and exciting. However, you will encounter dangerous situations whenever you work in a small engine shop. Special precautions should be taken when working with small engines. *It is very important to recognize potential hazards and make sure that your work area is safe.*

Safety is the responsibility of everyone in the small engine shop. If you notice dangerous shop conditions or unsafe work practices, notify your instructor immediately. Never take unnecessary risks to complete a job. Safe shop practices can prevent serious injury or save a life.

The warning symbol illustrated in Fig. 1-1 appears throughout the text to signal potentially dangerous situations. The warnings that accompany these symbols should be read carefully and followed closely. Failure to follow these warnings can result in serious injury or death.

KEEP WORK AREA CLEAN

A clean, well-organized work area is very important to everyone in the shop. Floors should be free from oil and dirt. An oily floor is slippery and can cause serious falls. Always clean up after working on a project. Pick up all tools and store them properly in a toolbox or workbench. Return all unused supplies to the proper storage area and discard all waste in appropriate containers. Aisles and doorways should be free from obstructions.

Keeping the shop area clean can also eliminate fire hazards. When combustible materials are allowed to accumulate in the shop, the possibility of fire increases. Never store used rags in a closet or corner. Rags saturated with gasoline or solvent are highly flammable and can be easily ignited. An approved container for storing flammable waste is shown in Fig. 1-2. A clean work area will increase safety and productivity.

HANDLE HAZARDOUS MATERIALS PROPERLY

There are many dangerous chemicals used in the small engine shop. Always store chemicals in a safe place. Flammable liquids should be kept in closed containers when not in use, Fig. 1-3. Gasoline is an extremely flammable liquid and its vapors can explode if exposed to sparks or flames. Never fill fuel tank while the engine is running or hot. Heat from engine could ignite the gasoline.

Fig. 1-1. There are many safety hazards in the small engine shop. This warning symbol will be used throughout the text to signal potential dangers.

Fig. 1-2. Oil- or solvent-saturated waste is extremely flammable and should be stored in a proper container.

Some small gas engines are equipped with battery-operated ignition systems. The batteries used in these systems are similar to those used in automobiles. Handle batteries carefully to avoid splashing acid on clothes, skin, or in eyes. *Hydrogen gas* is produced when the battery is charged or discharged. If the hydrogen gas is ignited, the battery can explode, throwing acid and fragments from its case and in every direction. Always keep sparks and flames away from the battery.

Only use chemicals for their intended purpose. Gasoline should never be used as a cleaning solvent. Gasoline has a low *flashpoint* and can be ignited easily. Many of the chemicals encountered in a small engine shop can cause serious burns. Avoid contact with skin. Wear rubber gloves and safety goggles when working with cleaning solvents.

WEAR PROPER CLOTHING

Proper clothing should be worn when working with small gas engines. Avoid loose-fitting clothing, which can get caught in moving machine parts. Neckties and jewelry should never be worn when working near rotating machinery. To avoid serious injury, keep hands, feet, hair, clothing,

Fig. 1-3. Flammable liquids should be stored in closed containers. These safety cans are equipped with flame arrestors, which prevent flames or sparks from entering containers.

and jewelry away from rotating engine parts. Never operate machinery with safety shrouds removed.

Safety glasses should be worn to protect eyes when using drills, grinders, hammers, chisels, or compressed air, Fig. 1-4. *Goggles* should always be worn when working with chemicals. A pair of safety shoes is recommended to prevent foot and toe injury.

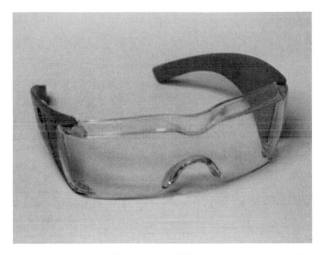

Fig. 1-4. Safety glasses should be worn when working in the small engine shop.

MAINTAIN ADEQUATE VENTILATION

The exhaust gases produced by gasoline engines contain carbon monoxide. *Carbon monoxide* is colorless and odorless. Breathing small amounts of carbon monoxide can cause drowsiness and headaches. Large amounts of carbon monoxide can cause death.

If an engine must be operated in the shop, make sure that exhaust fans are running and that large doors and windows are open. A lethal amount of carbon monoxide can accumulate in a closed, one-car garage in three minutes. Adequate ventilation is extremely important in the small gas engine shop.

Solvents used to clean engine parts can release toxic fumes. Check warnings on solvent labels and follow instructions carefully. When working with any solvent for an extended period of time, make sure that there is plenty of fresh air.

USE HAND TOOLS PROPERLY

The safe use of hand tools is often taken for granted in the small engine shop. Many accidents, however, are caused by the improper use of common hand tools.

Keep tools clean. Greasy or oily tools are likely to slip from your hand and may fall into rotating engine parts. The rotating parts can throw the tool, causing serious injury.

Tools should only be used for the job they were designed for. Never use screwdrivers or files to pry items loose. Most are made from hardened steel and may crack or shatter if improperly used.

Keep tools in top shape. Sharpen tools periodically. Dull tools require greater effort. Make sure all tools are equipped with appropriate handles. When using a wrench, always pull the handle toward your body. This will help to prevent injury if the tool slips, Fig. 1-5. Hammer heads must be securely attached to the handle. If the head is loose, it could fly off during use.

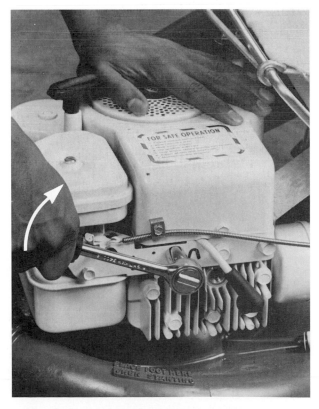

Fig. 1-5. To prevent injury, always pull wrench toward your body.

USE POWER TOOLS PROPERLY

Before using a power tool, make sure all guards and shields are in place. Wear safety goggles when operating power tools. If you are not familiar with a tool, read the operating instructions carefully or ask for help before attempting to operate the unit.

Never make adjustments on a power tool when it is running. Shut the tool off and wait for it to stop completely before attempting to service the device in any way.

All power tools should be equipped with a *dead man switch.* This type of switch automatically shuts the tool off when the operator releases the control button.

USE COMPRESSED AIR CAREFULLY

Compressed air is used in the small engine shop to accomplish various tasks. Wear safety goggles when using compressed air. Never use compressed air to clean your clothing or your hair. Flying particles can be blown into your eyes, causing serious injury or blindness. These particles can also penetrate your skin.

Check all connections before turning on a compressed air system. Always hold the hose nozzle tightly when using compressed air. Never set hose down without shutting off the air nozzle. Pressure in the hose can cause it to whip violently.

LIFT PROPERLY

Always lift heavy objects carefully. If necessary, ask for help when moving heavy items. Many shops are equipped with small overhead cranes to help move large objects.

To avoid unnecessary back strain, always lift with your legs; not with your back. Keep your back as straight as possible when lifting heavy objects, Fig. 1-6.

Never carry items that will obstruct your view. Make several trips if necessary. When carrying long items, use two people so that item is held level and both ends are attended. Never reach for heavy overhead items. The item may accidentally fall, causing severe head injury. Always use a ladder.

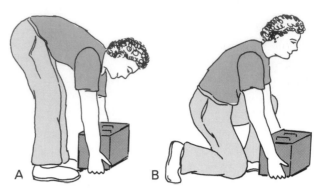

Fig. 1-6. When lifting heavy objects, keep your back straight and use your legs to lift weight. A–Improper lifting procedure. B–Proper way to lift heavy objects.

USE PROPER ELECTRICAL WIRING/GROUNDING PROCEDURES

Electrical hazards can be found in every small engine shop. Electricity is the most common cause of shop fires. Before using electrical equipment, check wires for fraying or cracking. Make sure all electrical equipment is properly grounded or double insulated. If equipment is not grounded, electrical shock can occur.

All outlets, switches and junction boxes should be covered. Label circuit breaker (fuse box) clearly so that it can be located in case of an emergency. Breaker switches should be labeled also.

Never use extension cords as permanent substitutes for fixed wiring. Extension cords should never run through holes in walls or floors.

Do not overload outlets. Too many components on one circuit can cause excess current to flow in the circuit. Overloaded circuits are a frequent cause of electrical fires.

OPERATE ENGINES SAFELY

Never operate a small engine at speeds greater than those recommended by the manufacturer. Excessive engine speed can cause parts to break loose from the engine. Severe personal injury can result from flying parts. Never tamper with the governor setting to increase maximum engine speed.

Keep hands and feet away from rotating engine parts. Small engines develop considerable torque and can cause serious injuries. Never operate engine with guards or shrouds removed.

Small engine components get extremely hot. Avoid touching the engine when it is running. Let engine cool before attempting repairs. In addition to causing burns, a hot engine can cause a fire if gasoline is accidentally spilled on hot surfaces.

Avoid touching electrical wires while the engine is running. The high voltage produced by some ignition systems can cause electrical shock. Some systems produce more than 30,000 volts.

Never operate an engine without a muffler. Wear ear protection when working on a running engine for a long period of time.

BE PREPARED FOR EMERGENCIES

In the event of an emergency, it is very important to be prepared. Emergency equipment should be stored in a highly visible place. List emergency numbers next to each telephone in the shop.

All shop areas should be equipped with fire extinguishers. These extinguishers should be mounted in highly visible, unobstructed areas. All extinguishers should be inspected monthly. Always keep the area around the extinguisher free from obstructions.

Fire extinguishers are categorized according to the type of fire that each is designed to suppress, Fig. 1-7. Class A fires involve ordinary combustibles such as wood, cloth, and paper. Class B fires involve flammable liquids such as gasoline and solvents. Class C fires are electrical fires. Be sure to use the proper type of extinguisher. Using the wrong extinguisher can be dangerous. Some fire extinguishers can be used for all types of fires, Fig. 1-8.

FIRES	TYPE	USE		OPERATION
A CLASS A FIRES ORDINARY COMBUSTIBLE MATERIALS SUCH AS WOOD, PAPER, TEXTILES, AND SO FORTH. REQUIRES . . . COOLING–QUENCHING	**FOAM** SOLUTION OF ALUMINUM SULPHATE AND BICARBONATE OF SODA	OK FOR A B		FOAM: DIRECT STREAM INTO THE BURNING LIQUID. ALLOW FOAM TO FALL LIGHTLY ON FIRE
		NOT FOR C		
	CARBON DIOXIDE CARBON DIOXIDE GAS UNDER PRESSURE	NOT FOR A		CARBON DIOXIDE: DIRECT DISCHARGE AS CLOSE TO FIRE AS POSSIBLE. FIRST AT EDGE OF FLAMES AND GRADUALLY FORWARD AND UPWARD
B CLASS B FIRES FLAMMABLE LIQUIDS, GREASES, GASOLINE, OILS, PAINTS, AND SO FORTH. REQUIRES . . . BLANKETING OR SMOTHERING		OK FOR B C		
	DRY CHEMICAL	MULTI-PURPOSE TYPE	ORDINARY BC TYPE	DRY CHEMICAL: DIRECT STREAM AT BASE OF FLAMES, USE RAPID LEFT-TO-RIGHT MOTION TOWARD FLAMES
C CLASS C FIRES ELECTRICAL EQUIPMENT, MOTORS, SWITCHES, AND SO FORTH. REQUIRES . . . A NONCONDUCTING AGENT		OK FOR A B C	NOT FOR A OK FOR B C	
	SODA–ACID BICARBONATE OF SODA SOLUTION AND SULPHURIC ACID	OK FOR A		SODA–ACID: DIRECT STREAM AT BASE OF FLAME
		NOT FOR B C		

Fig. 1-7. Chart illustrates various fire extinguisher types and fire classifications. In the small engine shop, always use an extinguisher designed for use on electrical and chemical fires.

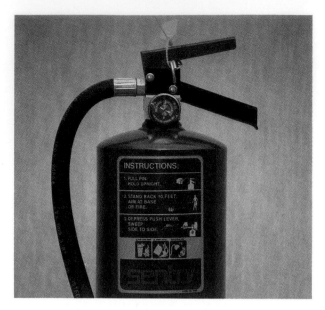

Fig. 1-8. One type of fire extinguisher. This particular extinguisher can be used for all types of fires.

First aid kits should be properly stocked and placed in prominent locations. If someone gets hurt, notify your instructor or supervisor immediately. Always seek professional help for serious injuries.

FOLLOW OSHA REQUIREMENTS

OSHA (Occupational Safety and Health Administration) is a governmental organization that establishes rules for safe work practices. All businesses and industries are required to follow OSHA regulations. It is very important to be familiar with OSHA rules and recommendations.

SUMMARY

Certain precautions must be taken when working on small gas engines. Keeping the work area clean will increase safety and productivity in the shop. Hazardous materials must be handled with care to avoid fires or chemical burns. Dispose of rags saturated with solvents in a proper container.

Proper clothing should be worn when working on small engines. Avoid loose-fitting clothing, which can get caught in rotating engine parts. Adequate ventilation is imperative when working in an enclosed area.

Use all tools properly. Read all instructions before using power tools. Never use compressed air to clean clothing.

Electrical malfunctions are the most common cause of shop fires. Do not overload electrical circuits.

Do not operate engine at speeds greater than those recommended by the manufacturer. Keep hands and feet away from rotating engine parts. Avoid touching hot engine parts.

Be prepared for emergencies. List emergency phone numbers above each telephone. Keep fire extinguishers in highly visible areas. Make sure first aid kit is properly stocked and easily accessible.

KNOW THESE TERMS

Hazards, Flammable, Solvent, Flash point, Hydrogen gas, Safety glasses, Goggles, Carbon monoxide, Toxic, Dead man switch, Fire extinguisher, First aid kit, OSHA.

REVIEW QUESTIONS–CHAPTER 1

1. There are many potential hazards in the small gas engine shop. True or False?
2. Oily or dirty floors can cause people to _____.
3. A flammable liquid frequently used in the small engine shop is _____.
4. Gasoline should always be stored in a closed container. True or False?
5. To prevent injury when working around small engines, avoid wearing:
   a. Loose clothing.
   b. Jewelry.
   c. Neckties.
   d. All of the above.
6. Carbon monoxide is _____ and _____.
7. Power tools should not be operated without proper safety shrouds. True or False?
8. Never use compressed air to clean _____ or _____.
9. Always lift with your:
   a. Arms
   b. Legs
   c. Back
   d. All of the above.

10. Electrical malfunction is the most common cause of shop fires. True or False?
11. Small engine ignition systems can produce more than 30,000 volts. True or False?
12. Batteries produce _____ _____ when charging or discharging.
13. Fire extinguishers are categorized by the type of fire they are designed to suppress. True or False?
14. OSHA establishes regulations for _____.

SUGGESTED ACTIVITIES

1. Make several safety posters warning of the potential dangers in a small engine shop and place them throughout your work area. Emphasize good housekeeping and proper storage of hazardous materials.
2. Check guards on all power tools and equipment and discuss purpose for each guard. Make sure that all guards are correctly mounted and in proper condition.
3. Walk through shop area and identify potential hazards. Discuss ways to minimize these hazards with your instructor.
4. Locate emergency equipment throughout your work area. Check fire extinguishers for sufficient charge. Make sure that they are designed for use on flammable liquids and electrical equipment. Make sure the first aid kit is properly stocked. Familiarize yourself with all the items in the first aid kit.

Cutaway view of a one-cylinder, overhead valve, four cycle engine. Note the position of the piston and the crankshaft. (Tecumseh Products Co.)

CHAPTER 2

ENGINE CONSTRUCTION AND PRINCIPLES OF OPERATION

After studying this chapter, you will be able to:
☐ Explain simple engine operation.
☐ List the qualities of gasoline that make it an efficient fuel for small engines.
☐ Explain why gasoline is atomized in the small engine.
☐ Identify the basic components of a small engine and describe the function of each part.

A gasoline-fueled engine is a mechanism designed to transform the chemical energy of burning fuel into mechanical energy. In operation, it controls and applies this energy to mow lawns, cut trees, propel tractors, and perform many other laborsaving jobs.

A gasoline engine is an internal combustion engine. Gasoline is combined with air and burned inside the engine. In its simplest form, an engine consists of a ported cylinder, piston, connecting rod, and crankshaft, Fig. 2-1.

The piston is a close "fit" inside the cylinder, yet it is free to slide on the lubricated walls. One end of the connecting rod is attached to the piston; the other end is fastened to an offset crankpin, or journal, of the crankshaft. As the piston moves up and down, the connecting rod forces the journal to follow a circular path, rotating the crankshaft.

SIMPLE ENGINE IN OPERATION

When the engine is cranked, gasoline is atomized (reduced to minute particles) and mixed with air. This mixture is forced through an intake port and into the cylinder, where it is compressed by the piston on the upstroke and ignited by an electrical spark.

Burning rapidly, the heated gases trapped within the cylinder (combustion chamber) expand and apply pressure to the walls of the cylinder and to the top of the piston. This pressure drives the piston downward on the power stroke, causing the crankshaft to turn, Fig. 2-1.

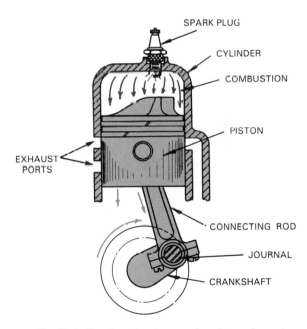

Fig. 2-1. Combustion forces the piston down to rotate crankshaft.

As the piston and connecting rod push the crankshaft journal to the bottom of the stroke, the pressure of the burned gases is released through an exhaust port. Meanwhile, a fresh air-fuel charge enters the cylinder and the momentum of the power stroke turns the crankshaft journal through bottom dead center (BDC) and into the upstroke on another power cycle.

GASOLINE

Gasoline is a hydrocarbon fuel (mixture of hydrogen and carbon), refined from petroleum. *Petroleum* is a dark, thick liquid that is extracted from the earth by oil wells. Luckily, petroleum is the second most plentiful liquid in the world; only water is available in greater quantity. Gasoline, however, cannot be recycled as water can. Therefore, it is imperative that we conserve gasoline and use it wisely.

Gasoline contains a great amount of energy. For engine use, it should:
- Ignite readily, burn cleanly, and resist detonation (violent explosion).
- Vaporize easily, without being subject to vapor lock (vaporizing in fuel lines, impeding flow of liquid fuel to carburetor).
- Be free of dirt, water, and abrasives.

Gasoline is assigned an *octane number* that corresponds to its ability to resist detonation. Premium grade gasoline ("high test" or "ethyl") burns slower than regular gasoline. It has a high octane number and is used in engines with high compression. Regular grade gasoline has a lower octane number and burns relatively fast. Generally, regular gasoline is used in small, low-compression, one- and two-cylinder, gasoline engines.

Gasoline is available in both leaded and unleaded varieties. The use of lead compounds is the most economical way to increase gasoline's octane number. For many years, most gasolines contained tetraethyl lead.

Since the mid 1970's, however, unleaded gasoline has become widely available. Instead of lead compounds, methyl tertiary butyl ether is commonly added to these fuels to increase octane levels. The main reason that unleaded gas was introduced was to provide fuel for automobiles equipped with catalytic converters. These vehicles will not operate properly on leaded fuel.

Modern unleaded gasoline is a complex substance. Ongoing research is necessary to seek ways to produce fuels that offer efficient engine performance and meet air pollution standards.

In the 1970's, a 10% ethanol blend of gasoline, known as gasohol, was introduced. Today, this product is often sold as "super unleaded" or "premium unleaded" gasoline, depending on the octane level.

The main drawback to these gasoline blends is their ability to absorb moisture, which can pass through the fuel filter and into the combustion chamber. These fuels should never be stored in high humidity areas or used in engines that set idle for long periods of time. Gasoline containing alcohol can also corrode fuel tank linings, shrink carburetor floats and seals, increase carbon deposits, and pit metal parts. For maximum performance and engine life, only use the gasoline recommended by the engine manufacturer.

GASOLINE MUST BURN QUICKLY

Gasoline placed in a container and ignited will produce a hot flame, yet it will not burn fast enough to produce the rapid release of heat necessary to run an engine. Even though a considerable quantity of fuel may be involved, a large flame will not necessarily result, Fig. 2-2.

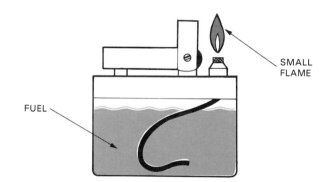

Fig. 2-2. A small flame is produced, due to small area of exposed fuel.

WARNING: Under no circumstances should experiments illustrated in this chapter be performed. Gasoline can be a very dangerous fuel and must be handled with caution. Illustrations and examples discussed here are meant to demonstrate how gasoline is prepared and used in an engine.

In Fig. 2-2, the surface area of the wick in the lighter is small. Vapor from the surface of the liquid, combined with oxygen, is what burns

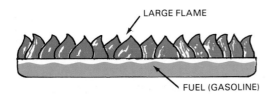

Fig. 2-3. A large flame is produced by a large area of exposed fuel.

readily. If the surface of the liquid is small, relatively little vapor will be given off to provide combustion. Since the liquid must change to vapor before it is burned, it would take considerable time to use up the fuel at this rate.

By placing the same amount of fuel in a shallow, wide container, much more surface area will contact the air and the fuel will burn rapidly, Fig. 2-3.

FUEL IS ATOMIZED

The more surface area of gasoline exposed to the air, the faster a given amount will burn. To produce the rapid burning required in an engine, gasoline must be broken up into tiny droplets and mixed with air. This is called *atomizing*.

Once the entire surface of each droplet of the air-fuel mixture is exposed to the surrounding air, a huge burning area becomes available. Given a spark, the entire amount of gasoline will flash into flame almost instantly. *In effect, atomization causes a sudden, explosive release of heat energy, Fig. 2-4.*

EXPLOSION MUST BE CONTAINED

To perform useful work, the explosive force caused by the burning gas must be contained and controlled. To illustrate this point, imagine that a metal lid is suspended on a string and held several inches from the ground. If a mixture of gasoline and air (atomized) were sprayed under it and ignited, the lid would be raised a short distance by the force of the explosion. See Fig. 2-5.

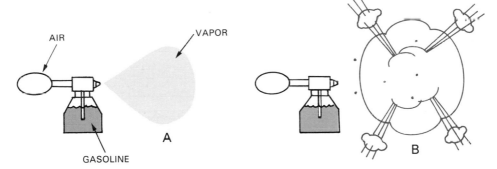

Fig. 2-4. Atomized fuel exposes a large area of fuel, which, when ignited, releases heat energy with an explosive force.

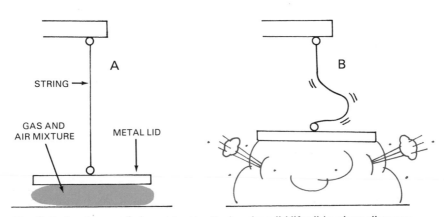

Fig. 2-5. A mixture of air and fuel ignited under a lid lifts lid a short distance.

The reason the lid hardly moved is because the explosion was not confined and directed toward the lid. Instead, the explosion exerted force in all directions, and much of the force was lost. If the gasoline and air mixture is sprayed inside a metal container with a lid, the full force of the explosion will be directed against the lid when the mixture is ignited. This will blow the lid high into the air, Fig. 2-6.

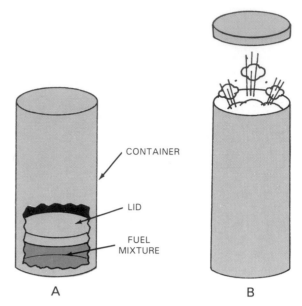

Fig. 2-7. A—Lid is placed in a long container. B—Most of energy of burning fuel is absorbed by lid, imparting greater speed to lid when explosion occurs.

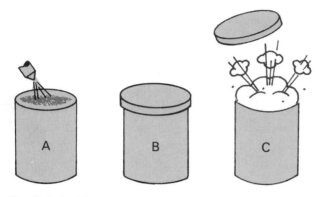

Fig. 2-6. A—Mixture of fuel and air is sprayed into a container. B—Lid is placed on top. C—Full force of explosion is directed toward base of lid when mixture is ignited, and lid is driven high into air.

FURTHER IMPROVEMENT

Even though the burning air-fuel mixture is confined by the container, once the lid starts to lift, a large amount of the force escapes to the sides. To eliminate this loss, a long, cylindrical container may be used with the lid having a close, sliding fit, Fig. 2-7. With the fuel mixture slightly compressed in the bottom of the container by the weight of the lid, the fuel will burn and direct most of the pressure against the lid as it travels up through the container. When the lid reaches the top, it will be traveling at a high rate of speed. The expansion of the gas will be nearly complete and little force will be lost, even after the lid clears the container.

BASIS FOR AN ENGINE

An elementary engine can be formed by attaching a crankshaft and a connecting rod to the setup illustrated in Fig. 2-7. The lid will serve as a piston and the container will act as a cylinder, Fig. 2-8. When the air-fuel mixture in the cylinder is ignited, it will drive the piston upward, causing the crankshaft to turn.

Although it is crude, this elementary engine illustrates the operating principles of a modern gasoline engine. Study the names of the various parts shown in Fig. 2-8 and become acquainted with their application to engine design.

There are many faults with the engine pictured in Fig. 2-8. How will a fresh air-fuel charge be admitted to the cylinder? How will the charge be ignited? What holds the various parts in alignment? How will the engine be cooled and lubricated? What will "time" the firing of the air-fuel mixture so that the piston will push on the crankshaft when the journal is in the correct position? How will the burned charge be removed (exhausted) from the cylinder? What will keep the crankshaft rotating after the charge is fired and until another charge can be admitted and fired? These questions can be categorized into five basic areas:

- MECHANICAL (engine design and construction).
- CARBURETION (mixing gasoline and air, and admitting it to cylinder).

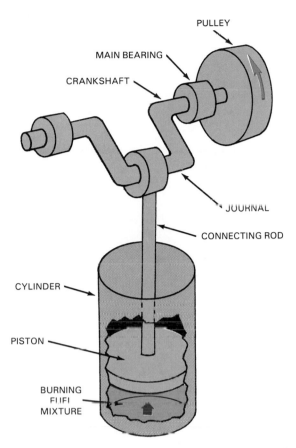

Fig. 2-8. Principles of operation illustrated here are the same as used in a modern gasoline engine. Note how burning fuel mixture forces lid (piston) upward to turn crankshaft and pulley.

- IGNITION (firing the fuel charge).
- COOLING (heat dissipation).
- LUBRICATION (oiling of moving parts).

In this chapter, emphasis will be placed on the mechanical aspects of engine design and construction. It will provide you with an opportunity to develop a workable engine. We will assume that the gasoline and air are being mixed correctly, the fuel charge is being fired at the right time, and the engine is properly cooled and lubricated.

PARTS ALIGNMENT–CYLINDER BLOCK

The *cylinder block* keeps all engine parts in alignment, Fig. 2-9. This critical engine component is usually a casting of iron or an aluminum alloy. The cylinder formed in the block can be produced accurately by modern methods. It may

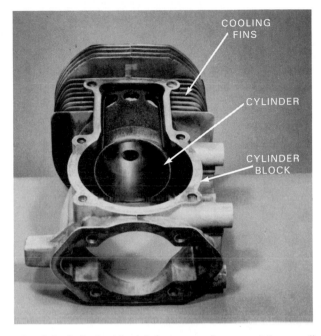

Fig. 2-9. Cylinder block is important because it keeps all moving parts in alignment. (Jacobsen Mfg. Co.)

be bored directly into the casting, or a steel sleeve may be inserted into an oversize hole bored in the block.

Aluminum cylinder blocks are cast around a steel sleeve. Aluminum, being a soft metal, would wear out quickly due to the friction of the piston. Advantages of aluminum are its light weight and ability to dissipate heat rapidly.

All air-cooled engines have *cooling fins* on the outside of the cylinder and cylinder head. *The size, thickness, spacing, and direction of the cooling fins is carefully engineered for efficient air circulation and heat control.*

The cylinder block must be rigid and strong enough to contain the power developed by the expanding gases. In some cases, the block is a separate unit; in others, it is cast as part of the crankcase. Similarly, the cylinder head may be bolted to the block or it may be cast as one complete unit. The method employed depends on the intended application of the engine and the manufacturer's preference.

Fig. 2-10 shows a combined cylinder block and crankcase with a separate, bolted cylinder head. Note the gasket that seals the unit. A sleeved, aluminum, die-cast cylinder is shown in Fig. 2-11.

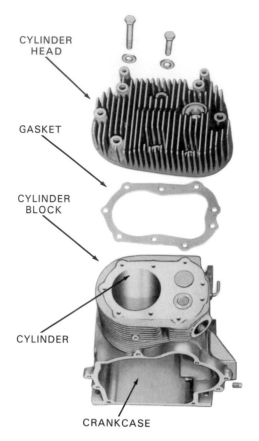

CYLINDER HEAD

GASKET

CYLINDER BLOCK

CYLINDER

CRANKCASE

Fig. 2-10. A combined cylinder block and crankcase. Cylinder head and sealing gasket are bolted to cylinder block. (Wisconsin Motor Corp.)

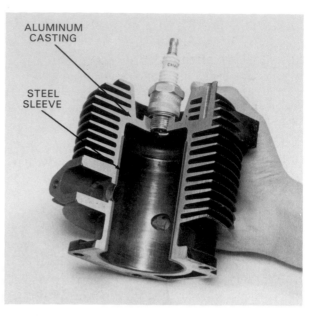

ALUMINUM CASTING

STEEL SLEEVE

Fig. 2-11. An aluminum cylinder block die cast around a steel sleeve. Note fins for air cooling.

CRANKSHAFT AND CRANKCASE

The *crankshaft* is the major rotating part of the engine, Fig. 2-12. Generally, it is forged steel, with all bearing surfaces carefully machined and precision ground. Counterweights are used to balance the weight of the connecting rod, which

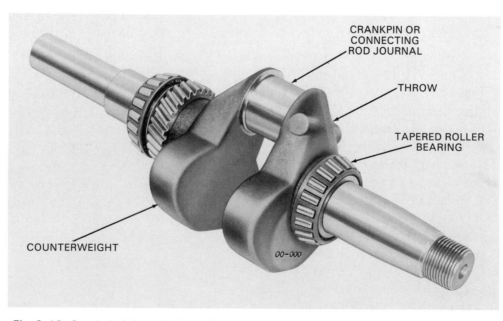

CRANKPIN OR CONNECTING ROD JOURNAL

THROW

TAPERED ROLLER BEARING

COUNTERWEIGHT

00-000

Fig. 2-12. Crankshaft for a single cylinder engine. Large counterweights opposite the crank journal balance rotational forces.

is fastened to the journal. Since connecting rods are cast or forged from different weight materials, holes are often drilled in counterweights to balance the crankshaft and prevent vibration

Fig. 2-13 shows a crankshaft being installed in crankcase. Note the tapered roller bearings. The flywheel is keyed to the end of the shaft with a Woodruff key. This type of key cannot slip out during operation. A lock washer and nut hold the flywheel in place.

Fig. 2-13. Crankshaft must be clean and carefully installed in crankcase. Tapered end fits into flywheel, which is secured with a key, lockwasher, and nut. (Tecumseh)

The end of the crankshaft and the hole through the flywheel have matched tapers that provide good holding power. When roller bearings are used to support the crankshaft, highly polished, hardened alloy steel bearing races are pressed into the crankcase to reduce friction and provide good wearability.

The *crankcase* must be rigid and strong enough to withstand the rotational forces of the crankshaft, while keeping all parts in proper alignment. Oil for lubrication is contained in the crankcase on some engines. On others, a valve system is used that allows a fuel, air, and oil mixture to enter. *The crankcase must be designed to*

protect the internal parts. Gaskets and oil seals are used to keep dirt out and clean oil in.

The crankcase and the cylinder block may be cast (metal melted and poured into a form of the desired shape) as a unit or fastened together by bolts. Fig. 2-14 shows a two cylinder engine with the cylinder block being placed on the crankcase with the crankshaft already installed. Note the tapered end on the crankshaft, which receives the flywheel.

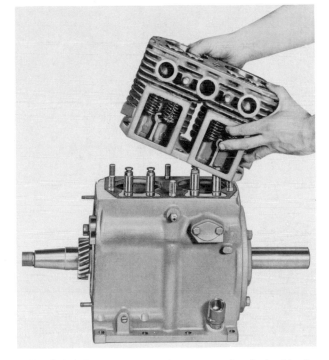

Fig. 2-14. Two cylinder crankcase and cylinder block being assembled. Stud bolts in crankcase will hold cylinder block in place when nuts are tightened.

In certain engine applications, the crankcase is not only an important part of the engine but also an integral part of the apparatus being driven. The engine parts illustrated in Fig. 2-15 are for a chain saw. Fig. 2-16 shows a similar saw completely assembled.

PISTONS

The *piston* is the straight line driving member of the engine. It is subjected to the direct heat of combustion and must have adequate clearance in

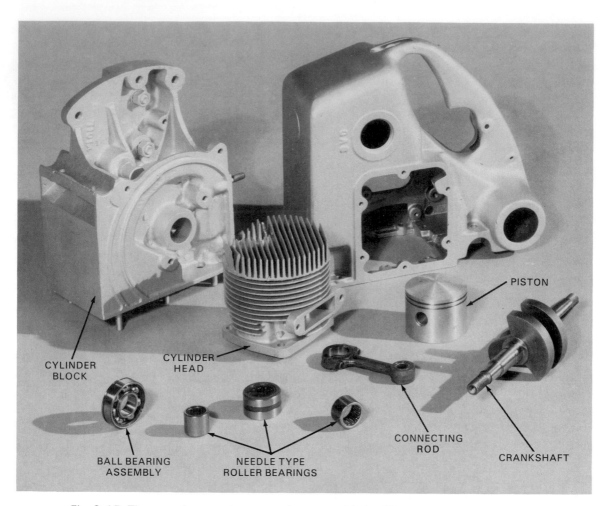

Fig. 2-15. These engine components, when assembled, will become driving members of a lightweight chain saw. (Beaird-Poulan, Inc.)

Fig. 2-16. A typical gasoline engine-driven saw.

the cylinder to allow for expansion. *The piston provides a seal between the combustion chamber and the crankcase.* This is accomplished by cutting grooves near the top of the piston and installing **piston rings**. The piston rings fit the grooves with a slight side clearance and exert tension on the cylinder wall. Properly installed, piston rings prevent blow-by of exhaust gases into the crankcase and leakage of oil into the combustion chamber.

The number of piston rings per piston depends upon the type of engine and its design. Note the two piston rings in Fig. 2-17. The piston is hollow to reduce weight. The top may be flat, domed, or contoured to provide efficient flow of gases entering and leaving the combustion chamber.

There is a hole in each side of the piston through which a piston pin, or wrist pin, is placed. This pin acts as a hinge between the connecting rod and piston and holds the two together. Generally, spring retainers hold the piston pin in place.

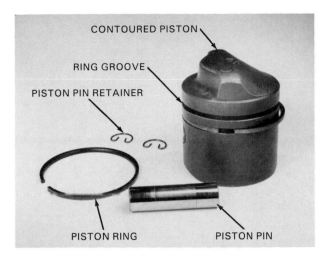

Fig. 2-1 7. The piston is the largest sliding-reciprocating part in engine. Piston rings seal combustion chamber from crankcase and must fit properly. (Jacobsen Mfg. Co.)

CONNECTING PISTON TO CRANKSHAFT

The sliding piston is connected to the rotating crankshaft with a metal link called a *connecting rod*. The big end of the connecting rod encircles the crankshaft journal and contains a bearing to permit free movement. The upper, or small end, of the connecting rod also must be movable. Note how the metal piston pin is passed through the connecting rod and piston, Fig. 2-18.

Also note the needle-type roller bearings, bearing race shells, and retainers, which must be in-

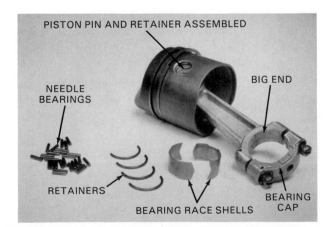

Fig. 2-18. Big end of connecting rod is ''capped'' to fit around crankshaft journal. Needle bearings are inserted to help reduce friction. Piston pin and retainers are shown assembled. (Jacobsen Mfg. Co.)

Fig. 2-19. The relative positions of connecting rod parts are illustrated with roller bearings in place. (Kiekhaefer Mercury)

stalled in the large end of the connecting rod when it is placed on the crank journal. The bearing cap holds the assembly together with connecting rod bolts or screws. Fig. 2-19 shows the relative position of the connecting rod and cap with the bearings in place.

Expanding gases push the piston toward the crankshaft, causing the connecting rod to turn the shaft. Fig. 2-20 shows how the reciprocating (up and down) movement of the piston is changed to rotary (revolving) motion by the crankshaft. Notice that the upper connecting rod bearing allows the connecting rod to swing back and forth while the lower bearing permits the crankshaft journal to rotate within the rod.

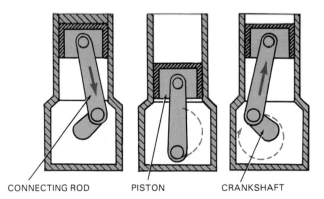

Fig. 2-20. Connecting rod must be free to pivot on piston pin, while crank journal follows a rotary path. Connecting rod must withstand severe stress in operation.

INTAKE AND EXHAUST

In developing an engine, we need to provide a way in which a fresh air-fuel mixture can be admitted to the engine and, once burned, the waste products exhausted. This can be done by using *ports* (openings) that are alternately covered and exposed by the piston (two stroke cycle design) or by using *poppet valves* to open and close the port openings (four-stroke cycle design). Both two-stroke and four-stroke designs are commonly used. Each has definite advantages and disadvantages. The engine being discussed here will be of the four-stroke cycle type.

ADMITTING FUEL MIXTURE TO ENGINE

For a four-stroke cycle engine (see Chapter 3 for additional information on fundamentals), passages leading to and from the cylinder area must be constructed. Fig. 2-21 shows two ports cast into cylinder block, one intake and one exhaust. The cylinder head is recessed to provide a passage from the ports to the cylinder.

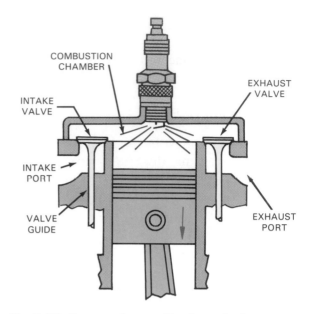

Fig. 2-22. Poppet valves seal intake and exhaust ports during power stroke. Valve guides keep valves aligned with valve seats.

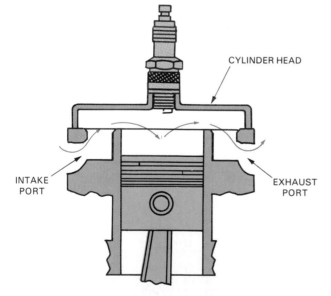

Fig. 2-21. Intake port permits air-fuel mixture to enter. Exhaust port allows burned gases to escape.

By installing a valve in each port, it is possible to control the flow of fresh fuel mixture into the cylinder and provide a means of exhausting the burned gases. During the period of expansion of the burning gases that drive the piston downward, both valves are tightly closed, Fig. 2-22.

The angled face of each valve will close tightly against a smooth seat cut around each port open-

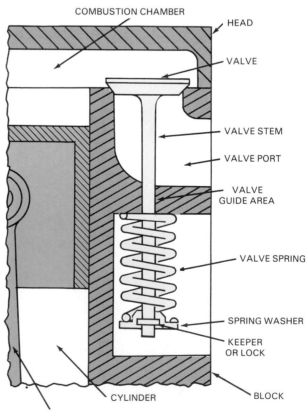

Fig. 2-23. Valve spring keeps tension on valve to ensure proper seating. Valve spring keeper and washer hold spring in place and permit removal when necessary.

ing. To align the valve and assure accurate raising and lowering in relation to the seat, the valve stem passes through a machined hole in block. This hole is called a valve guide.

VALVE SPRING ASSEMBLY

A *valve spring* must be used on each valve to hold it firmly against the seat. Placed over the valve stem, the spring is compressed to provide tension. It is connected to the valve stem by means of a washer and keeper (lock).

The spring allows the valve to be opened when necessary and will close it when pressure is removed from the valve stem. Fig. 2-23 shows the location of the spring and keeper assembled on

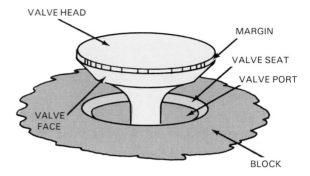

Fig. 2-25. Valve face and valve seat must be ground to correct angles and concentric to centerline of guide to seal properly.

the valve. An enlarged view of the "horseshoe" valve lock system is shown in Fig. 2-24.

A valve in the open position is illustrated in Fig. 2-25. When pressure is removed from the end of the valve stem, the spring will draw the valve down against the seat and seal off the port from the combustion chamber. *For the engine to function properly, the valves must be opened the right amount at the right time. They must remain open for a specific period and close at the correct instant.*

By using a shaft with two thick sections spaced to align with the valve stems, a basic device for opening and closing the valves is provided. By grinding the thick sections into a cam shape, the *camshaft* is formed. When the shaft is revolved, the cam lobe will cause the valve to rise and fall, opening and closing the ports. Study Fig. 2-26.

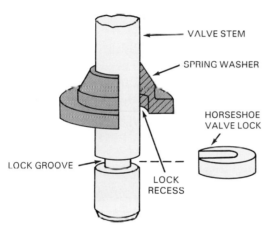

Fig. 2-24. Typical method of retaining valve spring on valve stem. Special tool generally is used to compress spring prior to removing horseshoe valve lock.

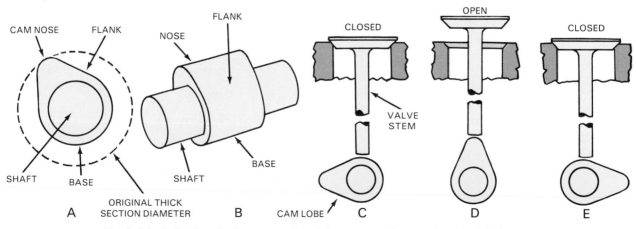

Fig. 2-26. A, B–By grinding a round shaft into a cam shape, a camshaft is formed.
C, D, E–When camshaft is revolved, cam lobe will open valve.

VALVE LIFTER OR TAPPET

In actual practice, the cam lobe does not contact the valve stem directly. By locating the camshaft some distance below the valve stem end, it is possible to insert a *valve lifter* between the lobe and stem. The valve lifter may have an adjust-

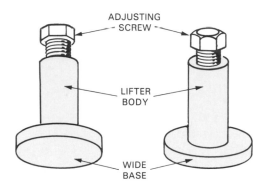

Fig. 2-27. Valve lifter may be called a tappet or cam follower. Adjustment screw allows setting of proper valve clearance. Wide base provides a larger contact area.

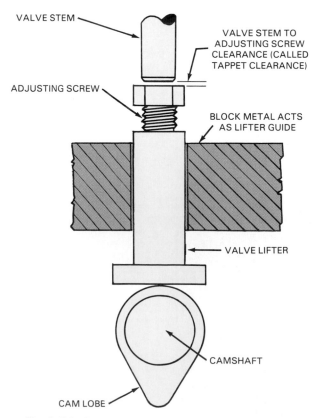

Fig. 2-28. As camshaft turns, cam lobe will operate valve lifter to open valve, then allow it to close.

ment screw in the upper end to provide a means of adjusting valve stem-to-lifter clearance. Without this adjustment, proper clearance must be obtained by grinding the end of the lifter or valve stem. The base of the lifter may be made wider than the body to provide a larger cam lobe-to-lifter contact area, Fig. 2-27.

By drilling a hole in the block above the camshaft, a guide is formed in which the lifter can operate, Fig. 2-28. The base of the lifter rides on the cam and the adjusting screw almost touches the end of the valve stem. As the camshaft revolves, the lifter will rise and fall, opening and closing the valve.

LOCATING THE CAMSHAFT

Generally, the camshaft is located in the crankcase, directly below the valve stems and valve lifters. The ends of the camshaft are supported in bearings in the block. The camshaft is shown being fitted into place in Fig. 2-29. One type of valve assembly is illustrated in Fig. 2-30. Study the relationship of the parts. Fig. 2-31 shows the location of each valve part in relation to the rest of the engine block.

The camshaft is driven by the crankshaft through gears. Fig. 2-32 shows the large camshaft

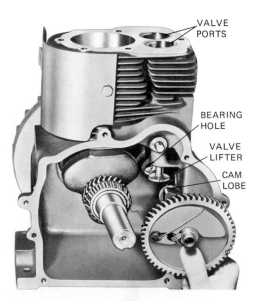

Fig. 2-29. Cam lobes are located directly under valve lifters. Camshaft turns in lubricated bearing holes. (Wisconsin Motors Corp.)

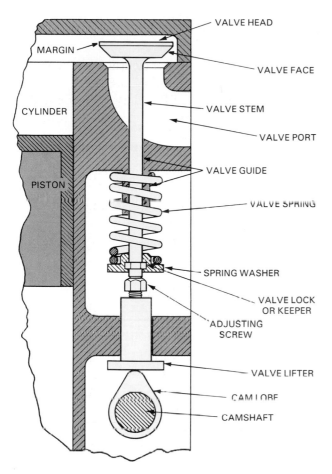

VALVE HEAD

MARGIN

VALVE FACE

CYLINDER

VALVE STEM

VALVE PORT

PISTON

VALVE GUIDE

VALVE SPRING

SPRING WASHER

VALVE LOCK
OR KEEPER

ADJUSTING
SCREW

VALVE LIFTER

CAM LOBE

CAMSHAFT

Fig. 2-30. Complete valve train. Study part names and
their relationship to each other.

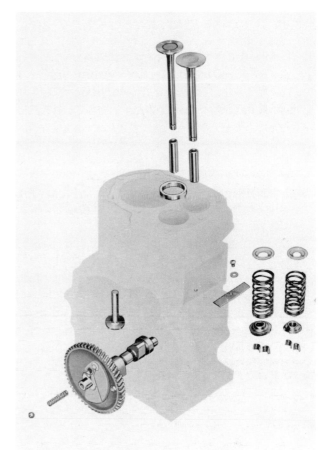

Fig. 2-31. Valve parts and their positions relative to
cylinder block and crankcase.

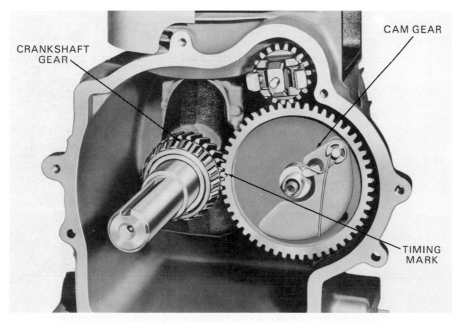

CAM GEAR

CRANKSHAFT
GEAR

TIMING
MARK

Fig. 2-32. Camshaft is gear driven from crankshaft. Camshaft gear is always twice
as large as crankshaft gear for proper timing. During assembly, timing marks must
be matched.

gear meshed with the smaller crankshaft gear. *The camshaft gear is always twice as large as the crankshaft gear. This gear ratio will be explained in Chapter 3 under four-stroke cycle engine.*

FLYWHEEL

Even though the crankshaft moves fast during the power stroke, it is relatively light and tends to slow down or stop before the next power stroke. This periodic application of power, followed by coasting, would cause the engine to speed up, slow down, and/or run roughly.

To improve the running quality of the engine, an additional weight in the form of a round *flywheel* is fastened to one end of the crankshaft, Fig. 2-33. During the non-power strokes, the inertia of the heavy flywheel keeps the crankshaft spinning and smooths engine operation. Metal fins on the flywheel act as a fan that forces air over the cylinder to cool the engine. Magnets cast into the flywheel produce electrical current for the ignition system.

SUMMARY

A gasoline engine is designed to transform the chemical energy of burning fuel into mechanical energy. For engine use, gasoline should ignite readily, burn cleanly, and vaporize easily. It should also be free from dirt and oil and resist detonation.

For efficient small engine use, gasoline must be broken into small particles and mixed with air. This process is called atomizing.

To perform useful work, the explosive force caused by burning gasoline must be contained and controlled by a piston and a cylinder.

The cylinder block keeps all engine parts in alignment. Air-cooled engines have cooling fins on the outside of the cylinder block. The crankshaft is the major rotating part in the engine. The crankcase is designed to protect internal engine parts and must be rigid enough to withstand the rotational forces of the crankshaft.

The piston is the straight line driving member of the engine. It provides a seal between the

Fig. 2-33. Flywheel is fastened to crankshaft. When rotating, its weight smooths engine operation. (Briggs and Stratton Corp.)

combustion chamber and the crankcase. The piston is connected to the crankshaft by the connecting rod.

Air-fuel mixture is admitted and burned gases are exhausted through ports in the engine. In some engines, valves are installed in the ports to help control intake and exhaust. A camshaft and lifters are used to open and close the valves.

KNOW THESE TERMS

Gasoline, Petroleum, Octane number, Atomizing, Cylinder block, Cooling fins, Crankshaft, Crankcase, Piston, Piston rings, Connecting rod, Ports, Poppet valves, Valve spring, Camshaft, Valve lifter, Flywheel.

REVIEW QUESTIONS–CHAPTER 2

1. Fuel must be atomized in the engine for the purpose of _____.
2. Force on the piston is transmitted to the crankshaft by the _____.
3. Name five desirable characteristics of gasoline for use in small engines.
4. Small gasoline engines generally use:
   a. High octane fuel.
   b. Low octane fuel.
5. What effect did lead (tetraethyl lead) have when introduced into gasoline?
6. What was the main reason that unleaded gasoline was introduced?
7. What adverse effects can result from using gasoline with alcohol blends?
8. The bearings that support the crankshaft are called _____ bearings.
9. The cylinder block is generally made of _____ or _____.
10. When are sleeved cylinders used?
11. Why are sleeved cylinders used?
12. Why are metal fins made as a part of the cylinder?
13. _____ are designed into the crankshaft to provide for engine balancing.
14. Why do some pistons have a contoured face?
15. The _____ will cause the valve to rise and fall, opening and closing the ports.
16. The angled face of each valve will close tightly against a smooth _____ cut around each port opening.
17. Which is larger, the camshaft gear or the crankshaft gear? How much larger?
18. What are three tasks performed by the flywheel?

SUGGESTED ACTIVITIES

1. Visit a local gasoline station and find out the following:
   a. Various fuel prices.
   b. Octane ratings of the fuels sold.
   c. Kind of containers fuel may be sold in.
   d. Quantities that legally may be stored at home.
2. Disassemble an engine and identify the parts discussed in this chapter. Carefully analyze the function of each part as it relates to the others.
3. If the engine used for disassembly is a used one, look for possible defects such as worn bearings, burned valves, broken or worn piston rings, a scored cylinder, or a loose piston pin.
4. Write to manufacturers of small gasoline engines requesting specifications for the models they produce. Write a report on the types of pistons, connecting rods, and crankshafts they use.
5. Prepare a display of the major components of a small gasoline engine. Use actual parts, photos, drawings, and cutaways to show the principal use of each part.

Cutaway view of typical four cycle engine. This particular engine is equipped with overhead valves and an electronic ignition system. (Tecumseh Products Co.)

CHAPTER 3

TWO CYCLE, FOUR CYCLE, AND ROTARY ENGINES

After studying this chapter, you will be able to:

☐ Describe four-stroke cycle engine operation and explain the purpose of each stroke.

☐ Explain the concept of valve timing.

☐ Compare the lubrication system in a four cycle engine to the system in a two cycle engine.

☐ Describe two-stroke cycle engine operation and explain the principles of two cycle operation.

☐ List the advantages and disadvantages of two cycle and four cycle engines.

☐ Identify the components of a Wankel rotary engine.

☐ Summarize rotary engine operation.

A basic design feature that aids in small engine identification is the number of piston strokes required to complete one operating (power) cycle. A four-stroke cycle engine, for example, requires four strokes per cycle; a two-stroke cycle engine requires two.

A *stroke* of the piston is its movement in the cylinder from one end of its travel to the other. Each stroke of the piston, then, is either toward the rotating crankshaft or away from it. Each stroke is identified by the job it performs (intake, exhaust, etc.).

Another type of engine that is used for various applications is the rotary engine. The rotary engine uses a rotor and an eccentric shaft instead of the piston and crankshaft used by the common reciprocating engine.

FOUR-STROKE CYCLE ENGINE

In a *four-stroke cycle engine* (commonly called "four cycle"), four strokes are needed to complete the operating cycle. These strokes are termed: *in-take, compression, power, exhaust*. Two strokes occur during each revolution of the crankshaft. Therefore, a four-stroke cycle requires two revolutions of the crankshaft.

Fig. 3-1 illustrates each of the four strokes taking place in proper sequence.

INTAKE STROKE

Fig. 3-1A shows the piston traveling downward in the cylinder on the *intake stroke*. As piston moves down, the volume of space above it is increased. This creates a partial vacuum that sucks the air-fuel mixture through the intake valve port and into the cylinder.

With the intake valve open during the intake stroke, atmospheric pressure outside the engine forces air through the carburetor. This gives a large boost to the air-fuel induction process. With nature balancing unequal pressures in this manner, it follows that the larger the diameter of the cylinder and the longer the stroke of the piston, the greater the volume of air entering the cylinder on the intake stroke.

Bear in mind that the intake valve, Fig. 3-2, performs several key functions:

• It must open at the correct instant to permit intake of air-fuel mixture.

• It must close at the correct time and seal during compression.

• Its shape must be streamlined, so the flow of gases into combustion chamber will not be obstructed.

The intake valves are not subjected to as high temperatures as the exhaust valve. The incoming air-fuel mixture tends to cool the intake valve during operation.

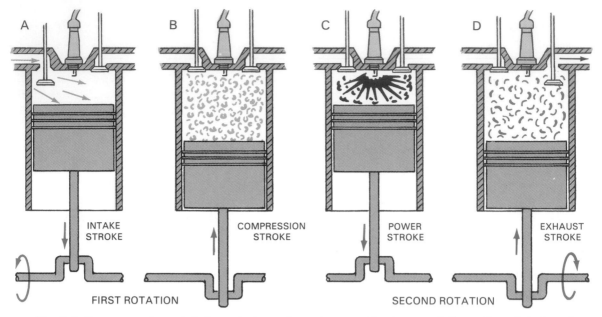

Fig. 3-1. Sequence of events in four-stroke cycle engine, requiring two revolutions of crankshaft and one power stroke out of four.

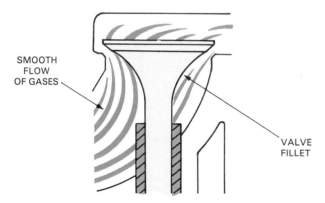

Fig. 3-2. Shape of valve smooths flow of gases around it. Note how flow follows fillet, speeding entry or expulsion. (Cedar Rapids Engineering Co.)

COMPRESSION STROKE

The *compression stroke* is created by the piston moving upward in the cylinder, Fig. 3-1B. Compression is a squeezing action while both valves are closed. On this stroke, the valves are tightly sealed and the piston rings prevent leakage past the piston.

As the piston moves upward, the air-fuel mixture is compressed into a smaller space. This increases the force of combustion for two reasons:

1. When atoms that make up tiny molecules of air and fuel are squeezed closer together, heat

energy is created. Each molecule of fuel is heated very close to its flash point (point at which fuel will ignite spontaneously). When combustion does occur, it is practically instantaneous and complete for the entire air-fuel mixture.

2. The force of combustion is increased because tightly packed molecules are highly activated and are striving to move apart. This energy, combined with expanding energy of combustion, provides tremendous force against the piston. NOTE: It is possible to run an engine on uncompressed mixtures, but power loss produces a very inefficient engine.

POWER STROKE

During the *power stroke*, both valves remain in the closed position, Fig. 3-1C. As the piston compresses the charge and reaches the top of the cylinder, an electrical spark jumps the gap between the electrodes of the spark plug. This ignites the air-fuel mixture, and the force of the explosion (violent burning action) forces the piston downward.

Actually, the full charge does not burn at once. The flame progresses outward from the spark plug, spreading combustion and providing even pressure over the piston face throughout the power stroke.

The entire fuel charge must ignite and expand in an incredibly short period of time. Most engines have the spark timed to ignite the fuel slightly before the piston reaches top dead center (TDC) of the compression stroke. This provides a little more time for the mixture to burn and accumulate its expanding force.

Basically, the amount of power produced by the power stroke depends on the volume of the air-fuel mixture in the cylinder and the compression ratio of the engine (proportionate difference in volume of cylinder and combustion chamber at bottom dead center and at top dead center). If the compression ratio is too high, the fuel may be heated to its flash point and ignite too early.

EXHAUST STROKE

After the piston has completed the power stroke, the burned gases must be removed from the cylinder before introducing a fresh charge. This takes place during the *exhaust stroke*. The exhaust valve opens and the rising piston pushes the exhaust gases from the cylinder. See Fig. 3-1D.

The exhaust valve has to function much like the intake valve. When closed, it must seal. When open, it must allow a streamlined flow of exhaust

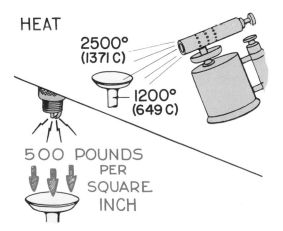

Fig. 3-3. Exhaust valve temperature may range from 1200°F (649°C) to 2500°F (1371°C) due to the hot gases surrounding it. Pressure of combustion may be as high as 500 pounds per square inch. (Briggs and Stratton Corp.)

gases out through the port, Fig. 3-2. The removal of gases from the cylinder is called *scavenging*.

The passageway that carries away exhaust gases is referred to as the exhaust manifold or exhaust port. Like the intake manifold, the exhaust manifold must be designed for smooth flow of gases.

The heat absorbed by the exhaust valve must be controlled or the valve will deteriorate rapidly. Some valve heat is carried away by conduction through the valve stem to the guide. However, the hottest part of the valve, the valve head, transfers heat through the valve seat to the cylinder block, Fig. 3-4.

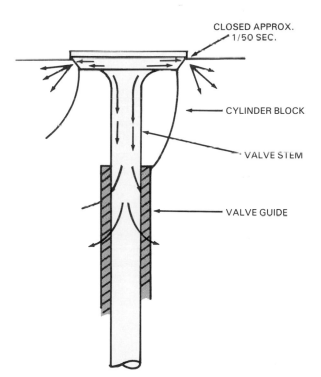

Fig. 3-4. Exhaust valve must cool during an incredibly short period (1/50 sec. at 3600 rpm). Heat is conducted from valve through seat to cylinder block. Some heat conducts down stem and to valve guide.

VALVE TIMING

The degree at which the valves open or close before or after the piston is at top dead center (TDC) or bottom dead center (BDC) varies with different engines. However, if the timing marks on the crankshaft and camshaft gears are aligned, the valve timing will take care of itself. NOTE: Engineers also specify the point at which the spark must occur (see Chapter 6).

Fig. 3-5 shows one complete operating cycle of a four cycle engine. Beginning a point A, the intake valve opens 10° before TDC and stays open through 235°. The exhaust valve closes 30° after TDC. *Valve overlap* occurs when both valves are open at the same time.

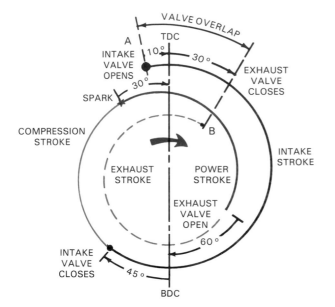

Fig. 3-5. Four-stroke cycle diagram shows exact number of degrees each valve is open or closed and time spark ignition occurs. Note that both valves are open (overlap) through an arc of 40°, permitting exhausting gases to create a partial vacuum in cylinder and help initiate a mixture of fuel in cylinder.

During the compression stroke, the intake valve closes and ignition occurs 30° before TDC, Fig. 3-5. The power stroke continues through 120° past TDC. The exhaust valve opens 60° before BDC and stays open through 270°. During the last 40°, the intake valve is also open and the second cycle has begun.

LUBRICATION

Lubrication of the four cycle engine is provided by placing the correct quantity and grade of engine oil in the crankcase. Several methods are used to feed the oil to the correct locations. The two most common methods are the splash system and the pump system. Some engines employ one or the other; others use a combination of both.

The multiple vee cylinder engine, Fig. 3-6, utilizes a combination splash and pressure lubrication system. The pump picks up the oil from the crankcase and circulates some oil through the filter and directly back to the crankcase. This keeps a clean supply available.

Oil is also pumped through a spray nozzle aimed at the crankshaft, Fig. 3-6. As the shaft rotates, it deflects the oil toward other moving parts. In addition, the splash finger on the bearing cap dips into the crankcase oil and splashes it on various internal surfaces.

Part of the engine oil is pumped through a tube to lubricate the governor assembly above the engine. Oil holes are provided in the connecting rod for lubricating the bearings and piston pin, Fig. 3-6.

Obviously, the oil in a four cycle engine must be drained periodically and replaced with clean oil. Also worth noting, four cycle engines must be operated in an upright position or the oil will flow away from the pump or splash finger, preventing lubrication.

TWO-STROKE CYCLE ENGINE

The *two-stroke cycle engine* (commonly called "two cycle") performs the same cycle of events as the four cycle engine. The main difference is that intake, compression, power, and exhaust functions take place during only two strokes of the piston. The two strokes occur during each revolution of the crankshaft. Therefore, it takes only one revolution of the shaft to complete one two-stroke cycle.

A two cycle engine has several advantages over a four cycle unit. It is much simpler in design than the four cycle engine because the conventional camshaft, valves, and tappets are unnecessary, Fig. 3-7.

Additionally, a two cycle engine is smaller and lighter than a four cycle engine of equivalent horsepower. Unlike the four-stroke cycle engine, the two cycle engine will get adequate lubrication when operated at extreme angles. It receives its

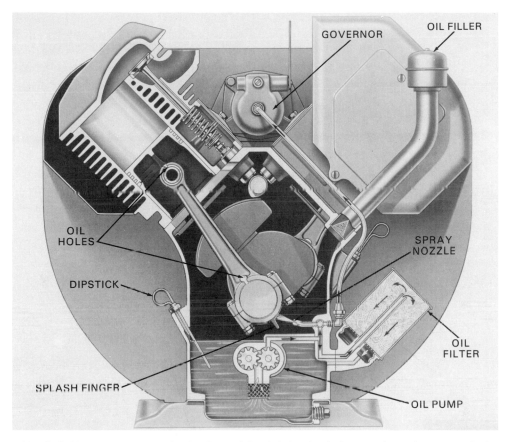

Fig. 3-6. Two common methods of supplying lubrication in four cycle engines are splash system and pressurized system. Engine shown employs both methods. Splash finger churns oil into a mist that makes its way into oil holes and other parts. Gear pump directs oil to remote parts and sprays some on critical parts. (Wisconsin Motors Corp.)

MOVING PARTS–4 CYCLE   MOVING PARTS–2 CYCLE

Fig. 3-7. The number of moving parts in a four cycle engine is greater than in a two cycle engine. Other differences are listed in chart at the end of this chapter. (Lawn-Boy Power Equipment, Gale Products)

lubrication as fuel mixed with oil is passed through the engine.

Installing the correct mixture of fuel and oil is a critical factor in maintaining a two cycle engine in good working condition. The prescribed type and grade of engine oil must be mixed with the fuel in proper proportion before being placed in the fuel tank.

In this way, there is a continuously new, clean supply of oil to all moving parts while the engine is running. The oil eventually burns in the combustion chamber and is exhausted with other gases.

Two cycle engines are popular in lawn mowers, snowmobiles, dune buggies, chain saws, jet boats, and other high-rpm applications.

VARIATIONS IN DESIGN

Two basic types of two cycle engines are in general use. They are the cross-scavenged and loop-scavenged designs, Fig. 3-8.

The *cross-scavenged* engine has a special contour on the piston head, which acts as a baffle to deflect the air-fuel charge upward in the cylinder, Fig. 3-8A. This prevents the charge from going straight out the exhaust port, which is located directly across from the intake port.

Cross-scavenged engines usually employ reed valves or a rotary valve, which is attached to the flywheel, Fig. 3-8B. These valves hold the incoming charge in the crankcase so it can be compressed while the piston moves downward in the cylinder. With this design, the piston acts as a valve in opening and closing intake, exhaust, and transfer ports. The transfer port permits passage of the fuel from the crankcase to the cylinder.

The *loop-scavenged* engine does not have to deflect the incoming gases, so it has a relatively flat or slightly domed piston, as shown in Fig. 3-8C. The fuel transfer ports in loop-scavenged engines are shaped and located so that the incoming air-fuel mixture swirls. This controlled flow

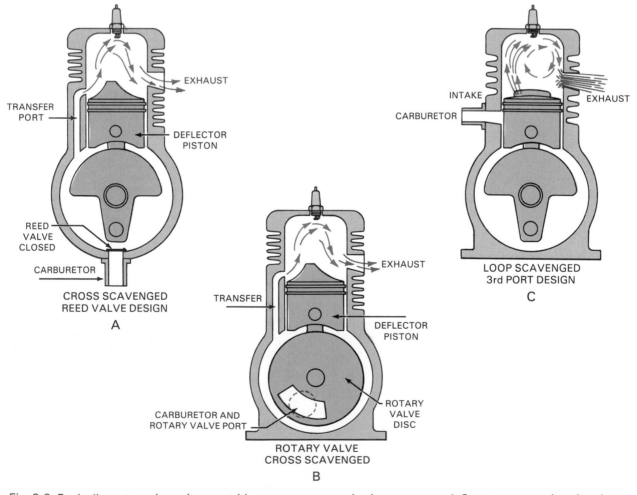

Fig. 3-8. Basically, two cycle engines are either cross scavenged or loop scavenged. Cross-scavenged engines have a contoured baffle on top of piston to direct air-fuel mixture upward into cylinder while exhaust gases are being expelled. Loop-scavenged engines have flat or domed pistons with more than one transfer port. Note three styles of crankcase intake valves. (Kohler Co.)

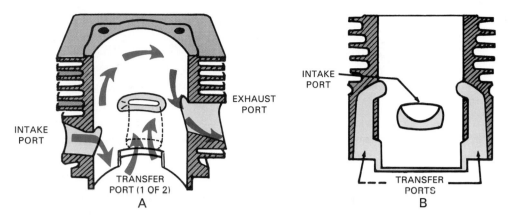

Fig. 3-9. A cutaway cylinder block shows location of intake, exhaust, and transfer ports of a loop-scavenged engine. A—Due to cutaway, only one of two transfer ports in shown. B—Section is revolved 90° to show both ports.

of gas helps force exhaust gases out and permits a new charge of air and fuel to enter.

PRINCIPLES OF OPERATION

The location of the ports in a two cycle engine is essential to correct timing of the intake, transfer, and exhaust functions. The cutaway cylinder in Fig. 3-9A shows the exhaust port at the highest point, the transfer port next, and the intake port at the lowest point. Some engines, particularly loop-scavenged engines, have more than one transfer port. See Fig. 3-9B.

INTAKE INTO CRANKCASE

As the piston moves upward in the cylinder of a two cycle engine, crankcase pressure drops and the intake port is exposed. Because atmospheric pressure is greater than the crankcase pressure, air rushes through the carburetor and into the crankcase to equalize the pressures. See Fig. 3-10A.

While passing through the carburetor, the intake air pulls a charge of fuel and oil along with it. This charge remains in the crankcase to lubricate ball and needle bearings until the piston opens the transfer port on the downstroke.

FUEL TRANSFER

Figs. 3-10C and 3-10D show the piston moving downward, compressing the air-fuel charge in the crankcase. When the piston travels far enough on

the downstroke, the transfer port is opened and the compressed air-fuel charge rushes through the port and into the cylinder. The new charge cools the combustion area and pushes (scavenges) the exhaust gases out of the cylinder.

IGNITION-POWER

As the piston travels upward, Fig. 3-10A, it compresses the air-fuel charge in the cylinder to about one tenth of its original volume. The spark is timed to ignite the air-fuel mixture when the piston reaches TDC. See Fig. 3-10B.

On some small engines, spark occurs almost at TDC during starting, then automatically advances so that it occurs earlier. This is done to get better efficiency from the force of combustion at higher speeds.

Peak combustion pressure is applied against the piston top immediately after TDC. Driving downward with maximum force, the piston transmits straight line motion through the connecting rod to create rotary motion of the crankshaft. See Fig. 3-10C.

EXHAUST

Several things happen during the exhaust phase, Fig. 3-10C. As the piston moves to expose the exhaust port, most of the burned gases are expelled. Complete exhausting of gases from the cylinder and combustion chamber takes place when the transfer ports are opened and the new

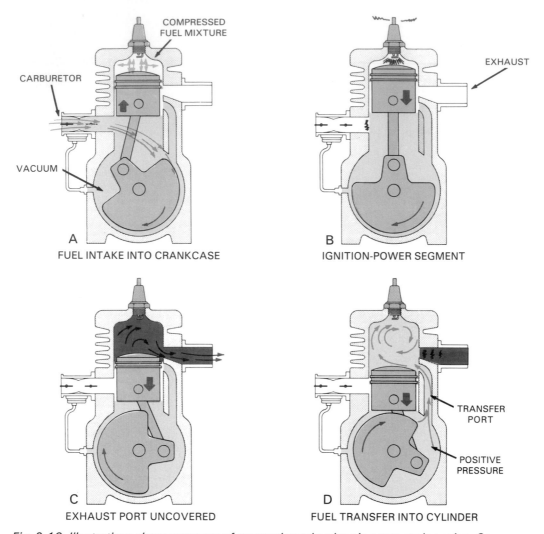

Fig. 3-10. Illustrations show sequence of events that take place in a two cycle engine. Compression and intake occur simultaneously then ignition occurs. Exhaust precedes transfer of fuel during lower portion of power stroke. Piston functions as only valve. (Rupp Industries, Inc.)

air-fuel charge rushes in, Fig. 3-10D. This completes one cycle of operation.

SCAVENGING AND TUNING

When properly designed, the exhaust system scavenges all exhaust gases from the combustion chamber. It allows the new fuel charge to move in more rapidly for cleaner and more complete combustion.

For best efficiency, the fuel charge should be held in the cylinder momentarily while the exhaust port is open. This helps prevent fuel from being drawn out of the cylinder with exhaust gases.

Some well engineered exhaust systems use the energy of sound waves from the exhaust gases for proper tuning. Fig. 3-11 shows a megaphone-like device, which amplifies the sound to speed up scavenging. The sound waves are reflected back into the megaphone to develop back pressure, which prevents the incoming air-fuel mixture from leaving with the exhaust gases. Compare this device with straight pipe operation shown in Fig. 3-12.

ROTARY DISC VALVE ENGINE

Fig. 3-13 illustrates a two cycle engine equipped with a *rotary disc valve*. The intake port is located

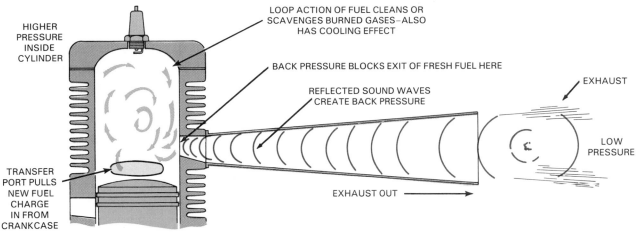

SCAVENGING EFFECT

Fig. 3-11. Pressure pulse exhaust tuning is an effective way of increasing power and efficiency in two cycle engines. Exhaust sound waves reflected back into manifold create a back pressure that stops fuel mixture from leaving cylinder before piston closes port. This system requires precise engineering. (Kohler Co.)

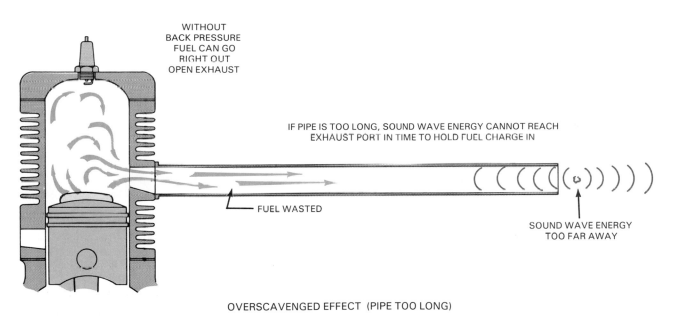

OVERSCAVENGED EFFECT (PIPE TOO LONG)

Fig. 3-12. A straight pipe may sound louder and more powerful than tuned exhaust, but actually is far less efficient. In this illustration, center of sound is too far and lacks amplification to have any beneficial effect on engine.

directly in the crankcase, allowing room for additional transfer ports that promote better fuel transfer and scavenging.

REED VALVE ENGINE

The *reed valve engine*, Fig. 3-14, permits fuel intake directly into the crankcase. The reed is made of thin, flexible spring steel, which is fas-tened at one end, Fig. 3-15. The opposite end covers the intake port. The *reed stop* is thick and inflexible. It prevents the reed from opening too far and becoming permanently bent.

In operation, the reed is opened by atmospheric pressure during the intake stroke. It is closed by the springiness of the metal and the compression in the crankcase on the power stroke. Fig. 3-16A illustrates the air-fuel mixture entering the

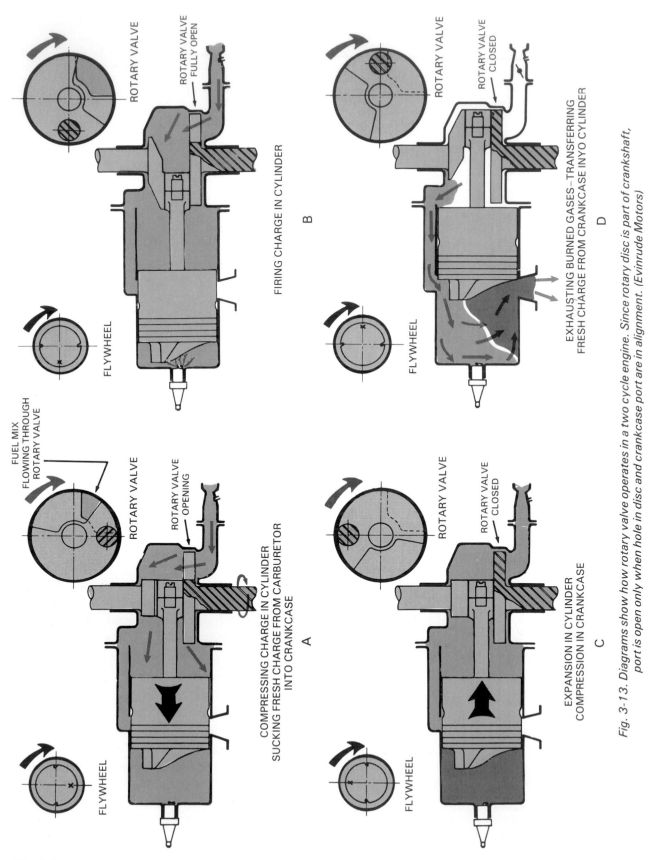

ROTARY VALVE

FUEL MIX
FLOWING THROUGH
ROTARY VALVE

ROTARY VALVE
OPENING

FLYWHEEL

A

COMPRESSING CHARGE IN CYLINDER
SUCKING FRESH CHARGE FROM CARBURETOR
INTO CRANKCASE

ROTARY VALVE

ROTARY VALVE
FULLY OPEN

FLYWHEEL

B

FIRING CHARGE IN CYLINDER

ROTARY VALVE

ROTARY VALVE
CLOSED

FLYWHEEL

C

EXPANSION IN CYLINDER
COMPRESSION IN CRANKCASE

ROTARY VALVE

ROTARY VALVE
CLOSED

FLYWHEEL

D

EXHAUSTING BURNED GASES–TRANSFERRING
FRESH CHARGE FROM CRANKCASE INYO CYLINDER

Fig. 3-13. Diagrams show how rotary valve operates in a two cycle engine. Since rotary disc is part of crankshaft, port is open only when hole in disc and crankcase port are in alignment. (Evinrude Motors)

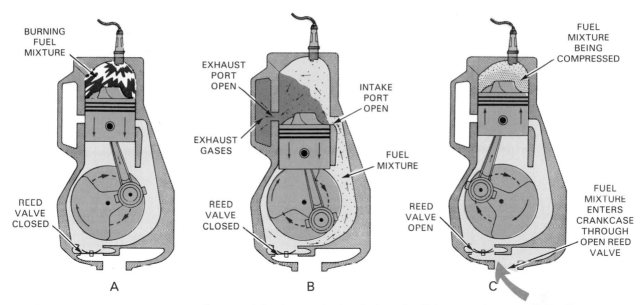

Fig. 3-14. A popular method of crankcase valving is a reed valve designed to fit into crankcase wall. It relies upon difference between atmospheric pressure and crankcase pressure to be opened. At rest position is closed position.

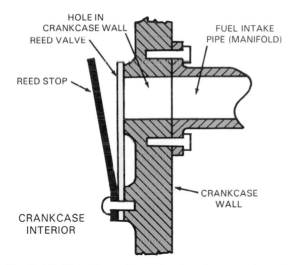

Fig. 3-15. Side view of a reed valve shows spring steel reed covering entry hole. Reed stop controls distance reed may open. This prevents permanent distortion and failure of reed to return snugly against port during time of crankcase compression.

crankcase. Fig. 3-16B shows how the reed valve is closed by crankcase pressure.

There are many reed valve designs. Some typical configurations are illustrated in Fig. 3-17.

FOUR CYCLE VS TWO CYCLE

The advantages and disadvantages of any engine are directly related to the purpose for which

the engine is intended. It cannot be said that one type of engine is better than another without considering every aspect of its application.

The chart in Fig. 3-18. lists the differences between two and four cycle engines.

ROTARY ENGINES

Most small gasoline engines in use today are of the reciprocating piston type. It is unchallenged as a power source wherever light, air-cooled engines are needed.

Another type of engine that you should be familiar with is the *Wankel rotary engine.* As a design, the Wankel has been around for over 30 years. Named for its inventor, Dr. Felix Heinrich Wankel, it was developed out of efforts to create a rotary-type valve for a motorcycle engine.

Wankel did succeed in designing a rotary valve that was used on both German Messerschmidt fighters and Junker bombers during World War II. After the war, Wankel resumed research on the rotary engine and, in 1951, he opened his own laboratory with NSU, a small motorcycle manufacturer.

While Wankel's idea was to be used as a supercharger on the motorcycle, that soon changed. By 1954, NSU and Wankel decided it was possible to

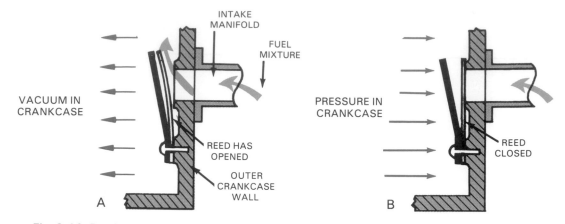

Fig. 3-16. Reed valve action. A–Vacuum in crankcase formed by upward moving piston causes atmospheric pressure to force air-fuel mixture through port opening. B–Downward piston movement compresses fuel mixture in the crankcase to a pressure greater than atmospheric pressure. Springy reed and crankcase pressure act together to close port.

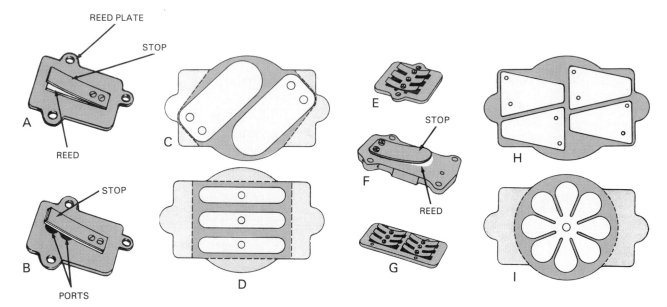

Fig. 3-17. Several forms of reed valves, A–Single reed, closed position. B–Single reed, open position. Note how the reed opening distance is controlled by the stop. C–Twin reed. D–Triple reed. E–Another form of triple reed. F–Single reed. G–Multiple reed. H–Four reed. I–Multiple reed.

build a four-stroke cycle rotary automotive engine. By 1957, the first Wankel engine was being tested. It was small and had a number of imperfections.

One of the main problems was in the seals used on the rotors. Often, just a few minutes of running would wear them out. But these problems were solved and rotary engines are available today in automobiles, snowmobiles, boats, motorcycles, and many other small engine applications.

Wankel engines are known by other names: rotary engines, rotating combustion engines, or, simply, the R.C. engine. There have been other rotary designs, but we will be dealing only with the Wankel-type engine.

FEWER PARTS

Beyond the rotary principle, there are great differences between the rotary and the recipro-

CHARACTERISTICS	FOUR CYCLE ENGINE (equal hp) ONE CYLINDER	TWO CYCLE ENGINE (equal hp) ONE CYLINDER
1. Number of major moving parts	Nine	Three
2. Power strokes	One every two revolutions of crankshaft	One every revolution of crankshaft
3. Running temperature	Cooler running	Hotter running
4. Overall engine size	Larger	Smaller
5. Engine weight	Heavier construction	Lighter in weight
6. Bore size equal hp	Larger	Smaller
7. Fuel and oil	No mixture required	Must be pre-mixed
8. Fuel consumption	Fewer gallons per hour	More gallons per hour
9. Oil consumption	Oil recirculates and stays in engine	Oil is burned with fuel
10. Sound	Generally quiet	Louder in operation
11. Operation	Smoother	More erratic
12. Acceleration	Slower	Very quick
13. General Maintenance	Greater	Less
14. Initial cost	Greater	Less
15. Versatility of operation	Limited slope operation (Receives less lubrication when tilted)	Lubrication not affected at any angle of operation
16. General operating efficiency (hp/ wt. ratio)	Less efficient	More efficient
17. Pull starting	Two crankshaft rotations required to produce one ignition phase	One revolution produces an ignition phase
18. Flywheel	Requires heavier flywheel to carry engine through three non-power strokes	Lighter flywheel

Fig. 3-18. Chart lists the differences between two- and four-stroke cycle engines.

cating engine. The rotary is much lighter and has only a very few moving parts. A six-cylinder automotive piston engine has over 230 basic parts, 166 of them moving. A two-rotor Wankel has about 70; only three move.

This means that less engine fuel is used to overcome friction and drive the engine parts. Also, there are fewer parts to manufacture and replace.

HOW THE ROTARY OPERATES

You are familiar with the four-stroke cycle of the reciprocating piston engine: intake, compression, power, exhaust. The rotary engine has the same familiar four strokes, but it completes all four strokes three times for each revolution of a *rotor*. Also, the shaft that is driven by the rotor

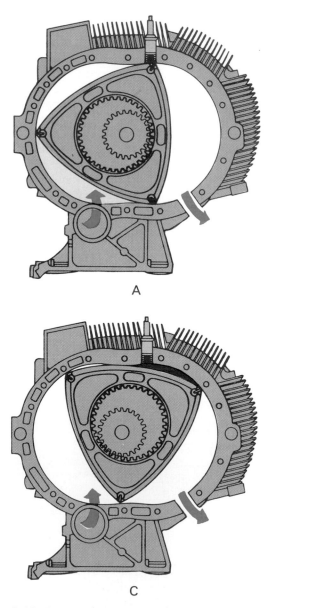

A

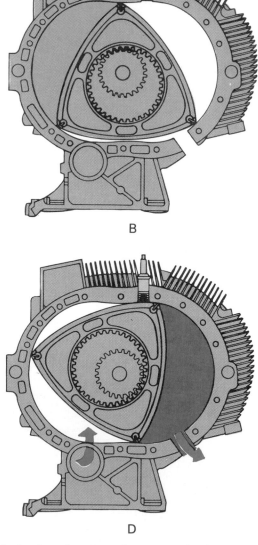

B

C

D

Fig. 3-19. Four cycle sequence of rotary engines. A–Intake stroke begins when rotor tip uncovers intake port. It continues until trailing rotor tip closes intake port, ending one intake cycle and beginning another. B–Compression starts as intake port is closed and reaches highest point in front of spark plug. C–Combustion takes place when charge is most compressed. Ignited air-fuel mixture expands and pushes against turning rotor to keep it rotating. D–Exhaust begins as rotor tip passes exhaust port. Rotor motion provides complete scavenging, pushing spent gases out port before trailing rotor tip closes it. (Outboard Marine Corp.)

revolves three times for each revolution of the rotor.

Fig. 3-19 illustrates the cycle of events as they take place through one revolution of the rotor. Don't forget that each side of the rotor is a separate and independent chamber. Therefore, one rotor does the same work as a three-cylinder, reciprocating engine. Two rotors would increase the engine potential to at least six cylinders, three rotors to nine, and so on.

Beginning with A in Fig. 3-19, the rotor is turning clockwise. The chamber is exposed to the intake port at position A. The increasing volume draws in the air-fuel mixture until, at position B, the trailing apex seal moves past the intake port and compression starts. When the rotor reaches full compression as in C, the intake port is fully closed to that chamber and the volume of the combustion chamber is at its smallest. Then the spark plug ignites the charge to supply force to continue the rotation. When the rotor reaches the point shown in D, the spent gases are pushed out of the exhaust port.

Two spark plugs are used in some rotaries to improve combustion efficiency. The plugs fire one after the other, about 10° apart, igniting the fuel mixture. Remember that while tracing the action of one chamber through one revolution, the other two chambers are following the same sequence 120° and 240° behind. Actually, three intake, three compression, three power and three exhaust actions take place during the one revolution of the rotor.

Notice, though, that the eccentric portion of the shaft rotates the shaft three times. This means there are three power pulses per rotor revolution. If there were two rotors overlapping, there would be six power pulses per shaft rotation, and greater power produced.

THE DRIVING MECHANISM

Fig. 3-20 illustrates the way that the eccentric shaft is driven by the rotor. First, the stationary gear and internal rotor gear do not transmit torque. The purpose of the gears is to control the motion and position of the rotor while maintaining proper ignition timing since the ignition distributors are driven from the eccentric shaft. Torque is transmitted from the rotor by the eccentric as shown in B. The *eccentric shaft* replaces the familiar crankshaft used in the reciprocating engine.

SEALS

Instead of piston rings, the rotor requires *side seals* and *apex seals* to prevent leakage of compression and combustion gases, Fig. 3-20. One of

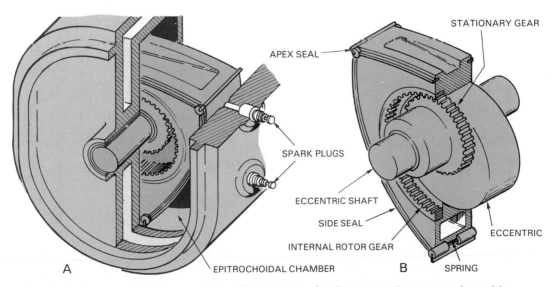

Fig. 3-20. As rotor turns, eccentric is driven by rotor. Stationary gear keeps rotor in position and does not transmit torque.

the major problems encountered with early rotary engines was extreme wear of the apex seals which slide over the inner surface of the epitrochoidal chamber.

Although various materials are being tried, current rotary engine seals are not as durable as conventional piston rings. Hard carbon and aluminum impregnated carbon have provided the best results so far. New sealing materials have been developed that can provide more than 100,000 miles of service in automobiles.

Side seals, Fig. 3-20B, are just as important as apex seals but do not provide any particular technical difficulty. Fig. 3-21 shows spring loading and relative position of the apex and side seals.

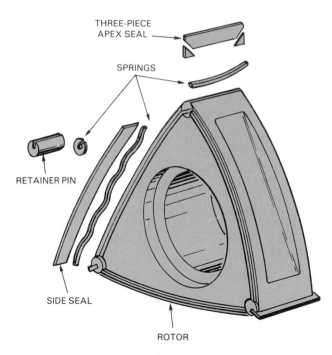

THREE-PIECE APEX SEAL

SPRINGS

RETAINER PIN

SIDE SEAL

ROTOR

Fig. 3-21. Seals are an important part of rotor. Springs are used to keep seals in sliding contact with chamber walls.

WEARABILITY

Due to the sliding friction created by the seals, it is necessary to use some material that will resist wear to both the seals and the inner walls they slide on. The inner walls of the epitrochoid chamber are faced with tungsten carbide, hard chrome, or nickel chromium carbide.

COOLING

The rotary engine can be either air cooled or water cooled. In applications requiring small rotaries, air cooling is used. Heat is concentrated in the combustion area of the chamber around the spark plug and in the rotor.

Cooling of the rotor is difficult. It is completely enclosed in the chamber and does not get any cooling from the crankcase as in the reciprocating engine. Heat is controlled by circulating cool oil from the sump.

LUBRICATION

Lubricating oils in rotaries are not contaminated by blow-by. Thus, periodic oil changes are not needed. To lubricate the rotor seals and housing, oil should be added as needed to replace that which is lost during normal operation.

Bearings that surround the output shaft are lubricated by oil from the sump, which is under pressure from a gear-type oil pump. In some cases, oil has been mixed with the fuel to provide lubrication of seals, bearings and chambers. More recent designs, however, have an independent feed system, which introduces measured amounts of oil at the intake ports.

SERVICE

Rotary engines have no valves to burn or stick. There are no rings, but there are seals that must be replaced occasionally. These seals are generally as durable as rings. Carburetors and ignition systems are very accessible and easy to service. Ignition units and carburetors are of conventional design.

SUMMARY

The stroke of a piston is its movement in the cylinder from one end of its travel to another. Four-stroke cycle engines need four strokes to complete the operating cycle: intake, compression, power, and exhaust. Lubrication of four cycle engines is generally provided by a splash system or a pump system.

In a two-stroke cycle engine, the intake, compression, power, and exhaust functions take place during two strokes of the piston. Two cycle engines have many advantages over four cycle units. They do not have conventional valves, tappets, or a camshaft. Two cycle engines are smaller and lighter than four cycle engines of equivalent horsepower.

The two cycle engine receives its lubrication as a fuel-oil mixture is passed through the engine. Therefore, it will receive adequate lubrication when operated at extreme angles.

The rotary engine has several advantages over the reciprocating engine. The rotary is much lighter and has very few moving parts. Lubricating oils in rotary engines are not contaminated by blow-by. Therefore, periodic oil changes are not needed. Rotary engines have no valves to burn or stick. Although there are no rings in the rotary engine, there are seals that require occasional replacement.

KNOW THESE TERMS

Stroke, Four-stroke cycle engine, Intake stroke, Power stroke, Compression stroke, Exhaust stroke, Scavenging, Valve timing, Valve overlap, Two-stroke cycle engine, Cross scavenged, Loop scavenged, Rotary disc valve, Reed valve, Reed stop, Wankel rotary engine, Rotor, Side seals, Apex seals, Eccentric shaft.

REVIEW QUESTIONS–CHAPTER 3

1. Name the four strokes of a four cycle engine in proper order.
2. Name three important intake valve functions.
3. Explain why a four cycle engine runs cooler than a two cycle engine.
4. Why is there a difference in temperature between the intake and exhaust valve?
5. The exhaust valve is cooled mainly by:
   a. Radiation.
   b. Conduction.
   c. Convection.
   d. Air-fuel circulation.
6. How does compression increase the engine power?

7. The compression ratio must be limited in gasoline spark ignition engines because:
   a. There is no power advantage after compressing the fuel to a certain point.
   b. The engine becomes too difficult to start.
   c. Mechanically it is not possible to increase the compression ratio.
   d. The heat of compression will ignite the air-fuel mixture too soon.
8. What are the two methods employed for lubricating four cycle engines?
9. What are the two types of scavenging systems used in two cycle engines?
10. Why can two cycle engines be run in any position?
11. Name three valving systems employed in two cycle engines.
12. The baffle on the contoured piston is for:
    a. Creating turbulent flow of gases.
    b. Slowing the air-fuel mixture entering the combustion chamber.
    c. Directing the flow of air-fuel mixture upward in the cylinder
    d. Directing oil evenly to the cylinder walls.
13. The _____ _____ type of two cycle engine requires a contoured piston.
14. In a properly tuned exhaust system, _____ _____ prevent the air-fuel mixture from leaving with the exhaust.
15. What advantage is there in having the intake port lead directly into the crankcase?
16. The time during the four-stroke cycle when both valves are open is called _____ _____.
17. A four cycle engine accelerates slower than a two cycle engine because:
    a. There is only one power stroke in four.
    b. The flywheel is heavier to carry the engine through three non-power strokes.
    c. There are more moving parts to be driven by the engine.
    d. All of the above are true.
18. A two-rotor rotary engine would produce power pulses equivalent to a:
    a. Single-cylinder reciprocating engine.
    b. Two-cylinder reciprocating engine.
    c. Four-cylinder reciprocating engine.
    d. Six-cylinder reciprocating engine.

19. When the rotor revolves once, the eccentric shaft revolves:
    a. Once.
    b. Three times.
    c. 120.
    d. Two times.
20. Why do some rotary engines use two spark plugs?
21. Apex seals in a rotary engine are made of hard carbon and carbon impregnated with what other material?

SUGGESTED ACTIVITIES

1. Look up additional information about the development of internal combustion engines. Some names to look up are: Christian Huygens, Philip Lebon, Samuel Brown, William Barnett, Pierre Lenoir, Beau DeRochas, Dr. N. A. Otto, Atkinson, Gottlieb Daimler, Priestman and Hall, Herbert Akroyd Stuart, Rudolph Diesel.
2. Begin a collection of engine repair and service manuals by writing to various engine manufacturers.
3. Using a worn out engine, cut away portions that will make the working parts visible while still enabling them to move. Make a report on the operation and timing of each part.
4. After further study, replace the spark plug of the cutaway engine with a small light bulb switched on and off by the breaker points to simulate ignition.
5. Make a bulletin board display that illustrates the main principles of four and two cycle engines.

This leaf blower is equipped with a 21.2 cc, 2 cycle engine. Two cycle engines are small, lightweight, and extremely versatile. (Deere & Co.)

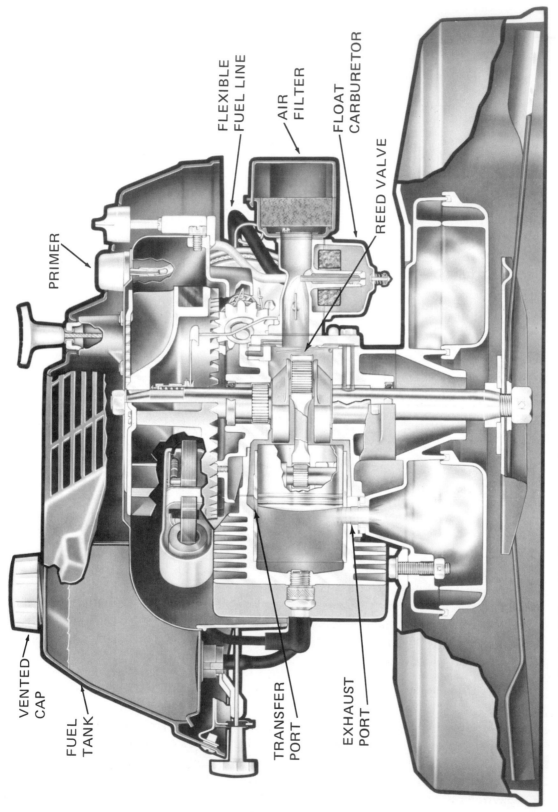

PRIMER

VENTED CAP

FUEL TANK

TRANSFER PORT

EXHAUST PORT

FLEXIBLE FUEL LINE

AIR FILTER

FLOAT CARBURETOR

REED VALVE

Fig. 4-1. Callouts identify components of a typical fuel system on a two cycle engine used in a power mower application.

CHAPTER 4

FUEL SYSTEMS

After studying this chapter, you will be able to:
- ☐ Name various types of fuel that can be used in a small engine and list practical applications for each.
- ☐ Explain the importance of proper fuel-oil mixture in a two cycle engine.
- ☐ Describe the purpose of fuel filters.
- ☐ Explain fuel pump operation.
- ☐ Describe the operation of a pressurized fuel system.

Small gas engines can be designed to operate efficiently on gasoline, liquefied petroleum gas (LP-Gas), natural gas, kerosene, or diesel fuel. See Fig. 4-1. Gasoline is the most popular of all small engine fuels. *In addition to its power potential, gasoline is readily available and easily transported. Tank refilling is also simple.* See Fig. 4-2.

GASOLINE

Most small engine manufacturers specify the use of regular grade, unleaded gasoline with an octane rating around 90. Occasionally, premium fuels are recommended for use in hot climates. This practice may prevent detonation or "dieseling" (after-run). However, a heavier buildup of solid materials in the combustion chamber can be expected from premium fuels because they contain more additives than regular grade fuels.

Gasoline should be clean, free from moisture, and reasonably "fresh." After prolonged storage, especially in small quantities, gasoline tends to become "stale." This is caused by oxidation that forms a sticky, gum-like material. This gum can clog small passageways in the carburetor and cause poor engine performance or hard starting.

Fig. 4-2. Refueling a four cycle engine requires use of an approved fuel can and a fresh fill of regular gasoline.

LP-GAS AND NATURAL GAS

LP-Gas may be propane, butane, or a mixture of both. Properly designed fuel systems will allow the use of LP-Gas with no appreciable loss of horsepower, as compared to a similar engine burning gasoline. LP-Gas burns cleanly and leaves few combustion chamber deposits. Because they emit fewer noxious fumes, these engines are often used in warehouses, factories, etc. LP-Gas also has a high anti-knock rating.

Natural gas generally causes a horsepower loss of around 20 percent when compared with gasoline. Both LP-Gas and natural gas require a different fuel system setup than the conventional

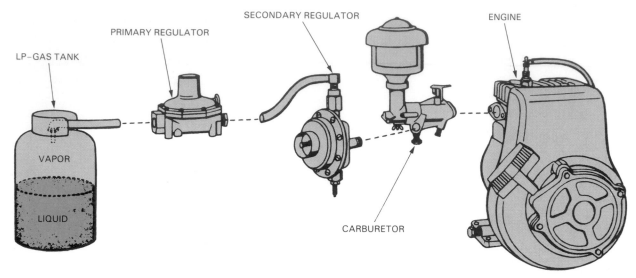

Fig. 4-3. Typical LP-Gas fuel system using vapor withdrawal. (Clinton Engine Corp.)

type used for supplying gasoline. Fig. 4-3 shows the components of one type of LP-Gas system.

COMBUSTION OF LP-GAS

LP-Gas burns slower than gasoline because it has higher ignition temperatures. For this reason, the timing is often advanced on LP-Gas engines.

Due to the higher ignition temperatures, greater voltage at the spark plugs may be needed for LP-Gas combustion. "Colder" plugs or smaller spark plug gaps may solve this problem. Check the engine manual for recommendations.

Less heat is required at the intake manifold to vaporize LP-Gas than gasoline. LP-Gas vaporizes at much lower temperatures than gasoline. It vaporizes at room temperature. This results in less wasted heat and more heat being converted to engine power.

ADVANTAGES OF LP-GAS

- Cheaper, especially when close to the source (refinery).
- Less oil consumption due to engine wear.
- Reduced maintenance costs–longer engine life between overhauls.
- Smoother power from the slow, even burning of LP-Gas.
- Fewer noxious or poisonous exhaust gases, such as deadly carbon monoxide gas.

DISADVANTAGES OF LP-GAS

- Initial equipment costs are high. Bulk fuel storage and carburetion equipment are costly.
- Fewer accessible fuel points (gas stations).
- Harder to start LP-Gas engines in cold weather–0°F (-18°C) or below.

KEROSENE AND DIESEL FUELS

Some non-diesel type small gas engines can be converted to operate successfully on *kerosene* or fuel oil through the installation of a low compression cylinder head and a special carburetor. These engines are started and operated on gasoline until fully warm, then switched over to the kerosene or fuel oil. These fuel installations are generally limited to heavy-duty, industrial engines.

A true diesel engine uses *diesel fuel* injected into the cylinder where it is ignited by the heat of compression. It is not unusual to have compression ratios as high as 20 to 1. Currently, however, diesel application in the small engine field is somewhat limited. It is only practical on applications where continuous use for long periods of time are common.

TWO CYCLE FUEL MIXTURES

Most two cycle engines receive lubrication only from the oil mixed with the gasoline. Because of this, it is important that the correct quantity and proper quality of oil is thoroughly mixed with a specific amount of gasoline. *Always follow the manufacturer's specifications as to the type and quantity of oil to use.*

Too little oil can cause the engine to overheat. Overheating, in turn, causes expansion of parts and possible scoring of machined surfaces. Eventually, the pistons may seize (bind, then stick) in the cylinders. Excessive oil, on the other hand, will cause incomplete combustion and rapid buildup of carbon, fouling the spark plugs and adding weight to the pistons.

TANKS, LINES, AND FITTINGS

Small engine fuel tanks are made of metal or plastic. Some are mounted away from the engine, Fig. 4-4. Others are contoured to fit snugly around the engine, Fig. 4-5.

The tank filler cap is vented. If the vent becomes clogged, the engine will create enough vacuum in the tank to cause fuel starvation. Most filler caps have baffles and filters. See Fig. 4-6.

The purpose of a fuel tank filler cap with a screw vent is to prevent fuel evaporation when the vent is closed. The vent should be opened before starting the engine. A variety of cap styles are shown in Fig. 4-7.

Fuel tanks used in all terrain vehicles (ATVs) and snowmobiles often have the *fuel pick-up line* inserted from the top of the tank. The pick-up line usually is very flexible and weighted at the bottom, so the line will always be where the fuel is deepest in the tank when the vehicle is at a steep angle.

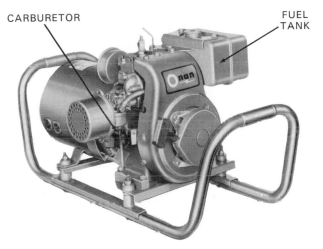

Fig. 4-4. Portable engine-driven generator with the fuel tank mounted on top side opposite carburetor. (Onan Corp.)

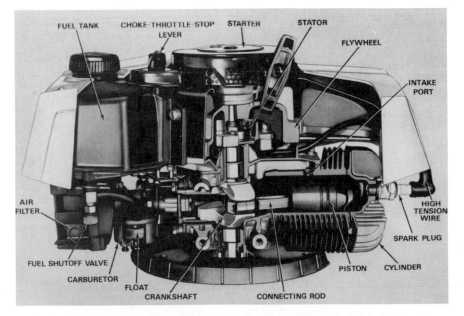

Fig. 4-5. Cutaway view of vertical shaft engine shows a plastic fuel tank contoured to fit snugly around engine. (Jacobsen Mfg. Co.)

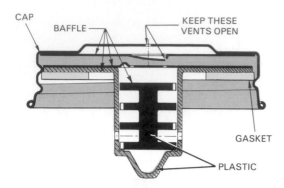

Fig. 4-6. Vented fuel filler caps are baffled to prevent dirt and dust from entering fuel tank. (Clinton Engine Corp.)

Fig. 4-7. Fuel filler caps: A – Three-piece plastic cap showing maze-type baffle with fiber gaskets. B – Plastic cap with plastic fabric as a filter and a perforated fiber disc. C – Cap with a threaded screw vent, which will seal tank. D – Standard cap with a single vent hole.

FUEL FILTERS

Some small engines have a fuel line fitting in the bottom of the tank. A filter screen is placed in the tank fitting or at the end of the pick-up line, Fig. 4-8. A top mounted pick-up line with the filter element at the bottom end is shown in Fig. 4-9.

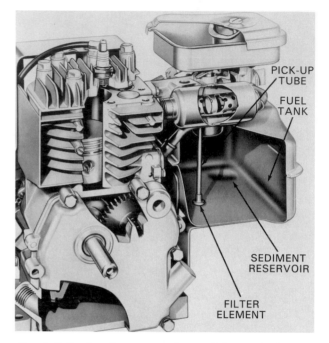

Fig. 4-9. Fuel tank cutaway shows pick-up tube in tank with ball check valve and filter element attached. (Briggs and Stratton Corp.)

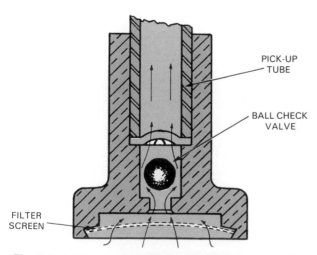

Fig. 4-8. A filter and ball check valve attached to end of a fuel pick-up tube. Ball check valve prevents fuel from draining back into tank when engine is running. (Clinton Engine Corp.)

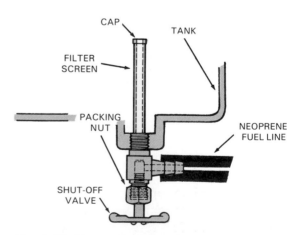

Fig. 4-10. Shutoff valve in tank is necessary to stop loss of fuel whenever a part of fuel system is undergoing work. Filter screen is permanent part of valve. Care during installation is necessary. (Lawn-Boy Power Equipment, Gale Products)

Other engines have a bottom mounted fuel fitting with a shutoff valve threaded into the tank, Fig. 4-10.

Older small engines have a more elaborate filter incorporated in a glass sediment bowl, Fig. 4-11. The gasket, screen, and bowl can be removed for inspection and cleaning. A similar filter is shown in Fig. 4-12. It is mounted directly on the fuel tank and has a shut-off valve.

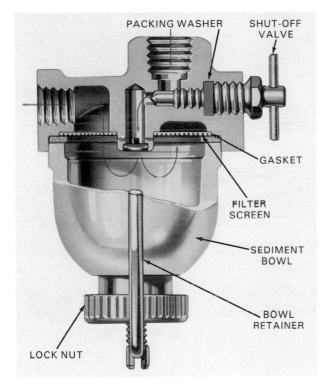

Fig. 4-11. Some small engines have remote fuel strainers located somewhere along fuel line. Glass bowl permits visual inspection without dismantling. (Wisconsin Motors Corp.)

FUEL PUMPS

Fuel pumps are used on engines having the fuel tank mounted in such a way that a gravity fuel supply system will not operate. In these applications, the tank and fuel level is lower than the carburetor, or the fuel level may be above the carburetor at times and below the rest of the time. For example, an all terrain vehicle (ATV) may have the fuel tank mounted away from the engine and the angle of the vehicle may change constantly.

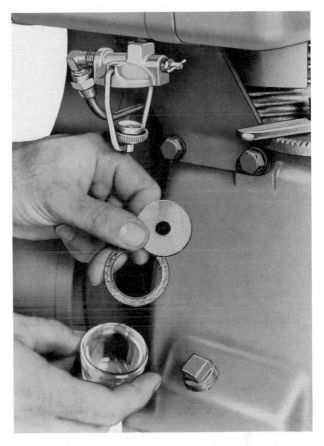

Fig. 4-12. When moisture or dirt is found in sediment bowl, fuel strainer can easily and quickly be taken apart for cleaning. Fuel shutoff valve is closed before removing bowl.

Fuel pumps provide constant, pressurized fuel flow to the carburetor under changing conditions. They help ensure that the engine can always provide quick acceleration and constant full power.

MECHANICAL FUEL PUMPS

The mechanical fuel pump used on small engines is basically the same as the type used on automobile engines. It may include a filter as part of the pump design. Fig. 4-13 is a cutaway view of a combination fuel pump and filter system. Trace the arrows to follow the flow of fuel.

FUEL PUMP OPERATION

The typical mechanical fuel pump shown in Fig. 4-13 operates by means of a diaphragm and

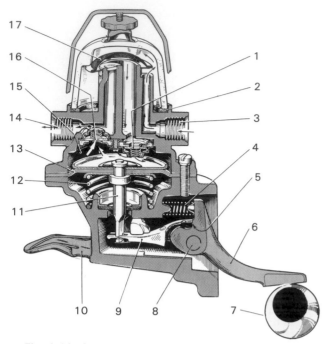

Fig. 4-13. Cutaway of a mechanical fuel pump with a combined fuel strainer on top. Pumps are activated by camshaft (7) as shown. Diaphragm (13) pulsates and forces fuel through check valves (1) and (16).

atmospheric pressure on the surface of the fuel in the tank. As the engine camshaft revolves, an eccentric (7) actuates the fuel pump rocker arm (6) pivoted at 8. This pulls the pull rod (11) and diaphragm (13) down against spring pressure (12), creating a depression in the pump chamber (15). Fuel drawn from the tank enters the glass bowl from the pump intake (3). After passing through the filter screen (17) and the inlet valve (1), the fuel enters the pump chamber (15).

On the return stroke, pressure of the spring (12) pushes the diaphragm (13) upward, forcing fuel from the chamber (15) through the outlet valve (16) and outlet (14) to the carburetor. When the carburetor bowl is full, the carburetor float will seat the needle valve, preventing any flow from the pump chamber (15). This will hold the diaphragm (13) down against the spring pressure (12). It will remain in this position until the carburetor requires additional fuel and the needle valve opens. The rocker arm (6) operates the connecting link (9) by making contact at 5. This construction allows idling movement of the rocker arm without moving the fuel pump diaphragm. The spring (4) keeps the rocker arm in constant contact with the eccentric (7) to eliminate noise.

FUEL PUMP HAND PRIMER

The *hand primer* shown as (10) in Fig. 4-13 is used when the carburetor float bowl or pump bowl has become empty. By pulling the hand primer upward, the float bowl will fill and ensure easy starting without prolonged use of the starter.

Because of the special construction of the pump, it is impossible to overprime the carburetor. After several strokes of the hand primer, its handle will become free acting. This indicates that the float bowl is full.

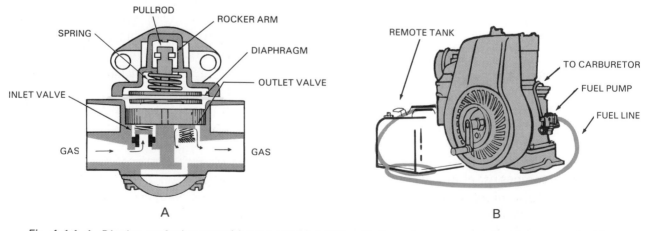

Fig. 4-14. A—Diaphragm fuel pump without a combined filter. B—Pump is cam operated and can pump fuel from portable fuel tanks if necessary. (Clinton Engine Corp.)

FUEL PUMP WITHOUT A FILTER SYSTEM

The fuel pump in Fig. 4-14A is a mechanical pump without a filter chamber. It is mounted on the crankcase, Fig. 14-14B. A separate filter may be installed somewhere along the fuel line. In some applications, the system may rely solely on the pick-up line filter, a filter in the carburetor, or a combination of both. Fig. 4-15 shows a combination fuel pump and filter system. Note shut off needle valve and primer lever.

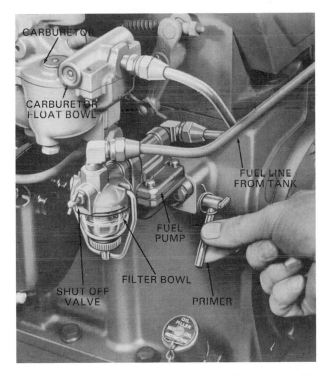

Fig. 4-15. Combination fuel pump and filter mounted in tandem. Fuel pump has manual primer used to initially get fuel to carburetor. (Wisconsin Motors Corp.)

IMPULSE DIAPHRAGM FUEL PUMPS

One type of diaphragm fuel pump sometimes used on small gas engines is activated by the pulsing vacuum in the intake manifold or crankcase. Four cycle engines use the intake manifold vacuum; two cycle engines use crankcase vacuum.

A typical impulse diaphragm pump is illustrated in Fig. 4-16. When vacuum draws the diaphragm upward against spring tension, the inlet check valve opens to allow fuel to flow in. When

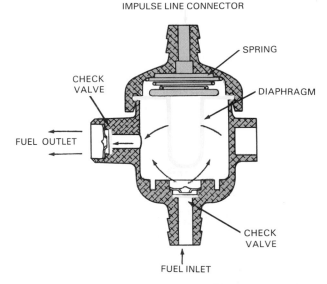

Fig. 4-16. This fuel pump is operated by vacuum pulses transferred from engine crankcase. It can be mounted in any convenient location on engine. (Clinton Engine Corp.)

vacuum is relieved, the spring pushes the diaphragm downward to force fuel through the outlet check valve. This process is repeated as long as the engine is running.

It is common practice today to design the impulse diaphragm fuel pump directly into the carburetor. This design principle will be explained under the section on carburetors.

PRESSURIZED FUEL SYSTEM

A *pressurized fuel system* is used when the fuel tank is located a considerable distance below the carburetor, Fig. 4-17. Outboard engines, for example, often operate from portable tanks resting in bottom of the boat.

The pressurized fuel system shown in Fig. 4-17 operates as follows:
1. The carburetor (1) is connected to the fuel tank (15) through the fuel filter by fuel lines (2 and 5).
2. Fuel flow from the tank to the carburetor is induced by pressure transmitted from the crankcase to the air space above the fuel line level via the air line (10), which runs from pressurized valve (11) to the twist connector (4) and into the top of the fuel tank (15).

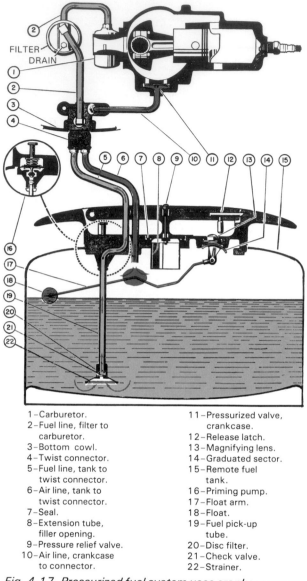

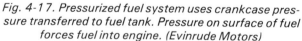

1–Carburetor.
2–Fuel line, filter to carburetor.
3–Bottom cowl.
4–Twist connector.
5–Fuel line, tank to twist connector.
6–Air line, tank to twist connector.
7–Seal.
8–Extension tube, filler opening.
9–Pressure relief valve.
10–Air line, crankcase to connector.
11–Pressurized valve, crankcase.
12–Release latch.
13–Magnifying lens.
14–Graduated sector.
15–Remote fuel tank.
16–Priming pump.
17–Float arm.
18–Float.
19–Fuel pick-up tube.
20–Disc filter.
21–Check valve.
22–Strainer.

Fig. 4-17. Pressurized fuel system uses crankcase pressure transferred to fuel tank. Pressure on surface of fuel forces fuel into engine. (Evinrude Motors)

3. For starting, initial flow to the carburetor is induced by the hand-operated priming pump (16).
4. A disc filter (20) is incorporated in the bottom of the fuel pick-up tube (19).
5. The fuel level is indicated by a graduated sector (14), actuated by a float (18) attached to arm (17).
6. The pressure relief valve (9) in the center of the carrying handle permits relieving pressure when necessary.

7. The check valve in the twist connector (4) permits disconnecting the air-fuel line without loss of tank pressure.
8. The tank pressure forces fuel up through the pick-up line (19), through the filter, and into the carburetor (1).
9. The check valve (21) is essential to the operation of the priming pump (16).

VAPOR RETURN FUEL SYSTEMS

If the temperature of the air around or inside a carburetor becomes high enough to vaporize the gasoline, pockets of vapor will stop all flow of fuel. When this occurs, the engine will become *vapor locked*. It will not run until the temperature drops low enough for the vapor to condense (return to a liquid).

One of the best ways to prevent vapor lock is to use a carburetor with a vapor return line. In these systems, any vapor that forms is directed back into the fuel tank where the pressure is vented to the atmosphere.

A diagram of a typical vapor return fuel system is shown in Fig. 4-18. The carburetor in this system has a built-in diaphragm fuel pump. The impulse tube operates the pump.

SUMMARY

Small gas engines can be designed to operate on gasoline, LP-Gas, natural gas, kerosene, or diesel fuel.

Most manufacturers specify the use of unleaded gasoline with an octane rating around 90. Gas should be clean, free from moisture, and reasonably fresh.

Two cycle engines receive lubrication from oil that is mixed with fuel. Always follow the manufacturer's specifications for the type and quantity of oil to use.

Small engine fuel tanks are made of metal or plastic. Various types of fuel filters are used in small engines.

Fuel pumps are used on engines that do not have a gravity fed fuel supply system. Fuel pumps provide constant, pressurized fuel flow to the carburetor under changing conditions. Mechanical fuel pumps are usually driven by the camshaft.

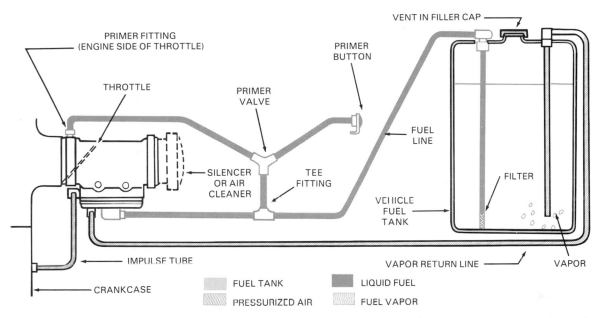

Fig. 4-18. Vapor return fuel system is one of the best methods for preventing a vapor lock. Vapors formed by heat are directed back to fuel tank where they are cooled and condensed to liquid form. (Kohler Co.)

Diaphragm fuel pumps are activated by the pulsing vacuum in the intake manifold or the crankcase. Pressurized fuel systems are used when fuel tanks are located a considerable distance below the carburetor.

KNOW THESE TERMS

Gasoline, LP-gas, Kerosene, Diesel fuel, Two cycle mixture, Filler cap vent, Fuel pick-up line, Fuel filter, Fuel pump, Hand primer, Impulse diaphragm fuel pump, Pressurized fuel system, Vapor return fuel system, Vapor lock.

REVIEW QUESTIONS–CHAPTER 4

1. In addition to the power available from gasoline, give three other reasons for its wide acceptance for engine use.
2. Most manufacturers specify regular grade, unleaded gasoline for small engines. True or False?
3. Premium fuels are sometimes recommended for use in hot climates. True or False?
4. A greater build-up of solid materials in the combustion chamber could be expected from using regular grade fuel. Yes or No?
5. Premium fuels contain more additives than regular grade fuels. True or False?
6. LP-Gas is either _____ or _____ or a mixture of both.
7. Natural gas used as a small engine fuel is generally accompanied by a horsepower loss of _____ percent.
8. If excessive oil is mixed with the fuel for a two cycle engine:
   a. Overheating may result.
   b. Spark plugs may become overheated.
   c. Incomplete combustion may occur.
   d. Seizing will result.
9. Filler caps with screw vents are for the purpose of:
   a. Preventing fuel evaporation when closed.
   b. Preventing fuel starvation when open.
   c. Preventing contamination in the tank.
   d. All of the above are correct.
10. The two types of fuel pumps discussed in this chapter are:
    a. Atmospheric pressure and gravity vacuum.
    b. Impulse diaphragm and mechanical.
    c. Gravity vacuum and mechanical.
    d. Gravity vacuum and impulse diaphragm.
11. When a carburetor has been removed and replaced, the engine will be slow starting

because of lack of fuel. This problem can be overcome if the engine has a fuel pump with a _____.

12. One satisfactory fuel system that prevents vapor lock is the _____ _____ system.

SUGGESTED ACTIVITIES

1. Collect a variety of tank filler caps. Either cut them in half or disassemble them. Make a display board showing the baffle and filter system.

2. Make a display board of cutaway drawings of fuel tanks with gravity feed fuel lines and top mounted pick-up lines.

3. Obtain and cut away some old fuel pumps so that internal parts can be seen and worked. Note the function and location of each internal component.

4. Cut away parts of an old fuel filter so that the fuel circuit can be traced.

A 7 1/2 hp outboard engine provides a pleasant mixture of fuel ''miser'' economy, light weight and twin-cylinder ''muscle.'' (Evinrude Motors)

CHAPTER **5**

CARBURETION

After studying this chapter, you will be able to:
☐ List and explain the principles of carburetion.
☐ Identify the three basic types of carburetors.
☐ Explain float-type carburetor operation.
☐ Explain the operation of diaphragm-type carburetors.
☐ Define manual throttle controls.
☐ List the basic functions of a governor.
☐ Adjust and maintain common governors.
☐ Describe the purpose of an air cleaner.

A carburetor's primary purpose is to produce a mixture of fuel and air to operate the engine. This function, in itself, is not difficult. It can be done with a simple mixing valve.

The mixing valve, however, is limited in efficiency. It cannot, for example, provide economical fuel consumption and smooth engine operation over a wide range of speeds. Meeting these performance goals requires a much more complex mechanism. This is the main reason why there are so many styles and designs of carburetors.

PRINCIPLES OF CARBURETION

Gasoline engines cannot run on "liquid" gasoline. The carburetor must vaporize the fuel and mix it with air in the proper proportion for varying conditions:
• Cold or hot starting.
• Idling.
• Part throttle.
• Acceleration.
• High speed operation.
Basically, air enters the top of the carburetor and is mixed with liquid fuel, which is fed through carburetor passages and sprayed into the air-

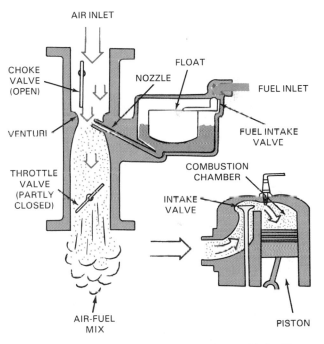

Fig. 5-1. Air entering carburetor mixes with fuel in proper proportion, and mixture flows into combustion chamber. (Deere & Co.)

stream. The *air-fuel mixture* that results is forced into the intake manifold by atmospheric pressure and burned in the combustion chamber of the engine.

Fig. 5-1 shows how a typical carburetor operates. In this particular small gasoline engine application, the engine is at part throttle operation. Note that the choke valve is open, while the throttle valve is partly closed.

AIR-FUEL MIXTURE

The amount of air needed for combustion is far greater than the amount of fuel required. The usual weight ratio is 15 parts of air to 1 part of fuel, since one pound of air would take up a much greater space than one pound of fuel. Therefore, by volume, one cubic foot of gasoline would have to be mixed with 9000 cubic feet of air to establish a 15 to 1 weight ratio.

Small gasoline engines use varying air-fuel ratios, depending on engine speed and load. The chart in Fig. 5-2 shows how the mixture changes for various operating conditions.

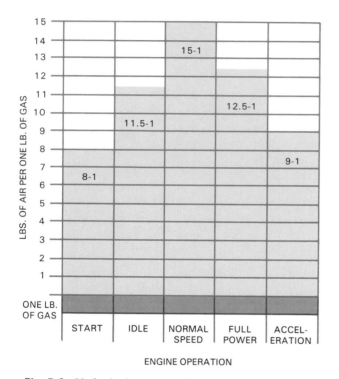

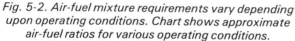

Fig. 5-2. Air-fuel mixture requirements vary depending upon operating conditions. Chart shows approximate air-fuel ratios for various operating conditions.

CARBURETOR PRESSURE DIFFERENCES

A carburetor is a device that is operated by pressure differences. When discussing pressure differences, several terms are commonly used. They are *vacuum, atmospheric pressure,* and *venturi principle.*

Vacuum

An absolute vacuum is any area completely free of air or atmospheric pressure. This condition is difficult to obtain and is never reached in a small gasoline engine. Therefore, any pressure less than atmospheric pressure generally is referred to as a *vacuum.*

Atmospheric pressure

The pressure produced by the weight of air molecules above the earth is called *atmospheric pressure.* The amount of atmospheric pressure varies with altitude. A person standing on a beach at sea level, for example, would be under a higher vertical column of air than a person standing on a mountaintop. Therefore, the total weight of air molecules would be greater at sea level, Fig. 5-3.

MEAN ATMOSPHERIC PRESSURE at 68° F (20 C)	
Altitude (ft.)	Atmospheric Pressure (in. Hg.)
9000	20.92
8000	21.92
7000	22.92
6000	23.92
5000	24.92
4000	25.92
3000	26.92
2000	27.92
1000	28.92
Sea Level	29.92

Fig. 5-3. Weight of air exerted on a given object is determined by height and density of a column of air above object. Air is less dense at higher altitudes. For every 1000 ft. above sea level, mercury column pressure is reduced by 1.0 in.

Furthermore, any time air molecules are removed from a particular space, a vacuum is created. If conditions permit, this space immediately fills with air under atmospheric pressure.

The effect of atmospheric pressure can be related to small gasoline engines. The downward movement of the piston creates a partial vacuum in the cylinder. As soon as the intake valve opens or the intake port is uncovered, atmospheric pressure forces air through the carburetor and manifold to fill that vacuum.

Venturi principle

The carburetor creates a partial vacuum of its own by means of a venturi for the purpose of drawing fuel into the airstream. A *venturi* is a restriction in a passage, which causes air to move faster (increase velocity). The gage shown at the top in Fig. 5-4 indicates no change in velocity. Therefore there is no change in pressure. The area in which the air is moving faster (middle gage) develops a lower pressure.

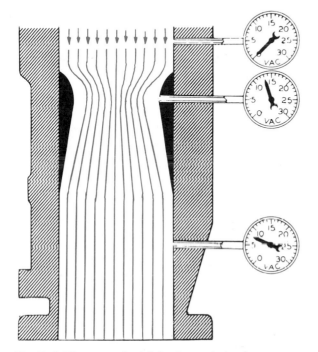

Fig. 5-4. The venturi priciple. A restriction in a passage will cause incoming air to increase its velocity, while pressure will be reduced. Reduction in pressure draws fuel into airstream.

Fig. 5-5 shows a simple carburetor with fuel being drawn from the float bowl through the main discharge nozzle. This nozzle is located so that its outer end is in the low pressure area of the venturi section. Fuel coming from the discharge nozzle is still in relatively large liquid droplets that do not burn well. To further atomize the fuel, an air bleed passage is built into the air horn, Fig. 5-6. A small portion of the air rushing through the carburetor is forced through the air bleed passage to the main discharge tube. This air mixes with the stream of fuel, breaking it into small particles before it reaches the venturi. The small particles of fuel are broken into finer particles by the air rushing through the venturi.

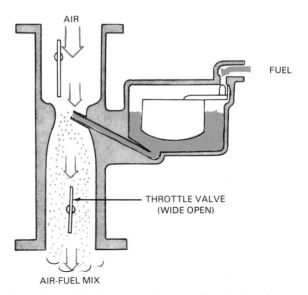

Fig. 5-5. Air flowing through venturi has reduced pressure around nozzle. Fuel is drawn up nozzle by vacuum and mixes in airstream. (Deere & Co.)

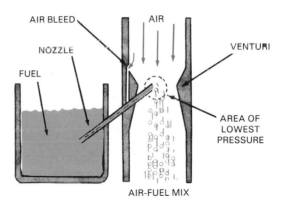

Fig. 5-6. To atomize fuel into finer particles, an air bleed is used. Higher pressure of air horn forces some air to enter at a port midway in nozzle, so fuel is partly atomized before leaving nozzle.

When the fuel moves into the intake manifold (which is under partial vacuum), the boiling point of the gasoline is lowered. This causes many of the atomized particles to "boil" or "flash" into a vapor. See Fig. 5-7A. As the partially vaporized fuel moves through the manifold, it is warmed by the heat of the manifold walls. This causes further vaporization, Fig. 5-7B. When the mixture enters the combustion chamber, the swirling motion and the sudden increase in temperature due to the compression stroke complete the vaporization of the fuel.

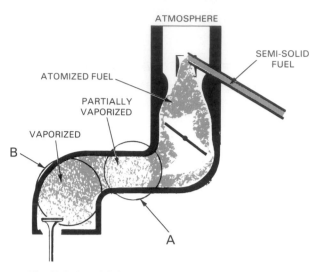

Fig. 5-7. In addition to air bleed and venturi, fuel is further vaporized: A–By vacuum in manifold. B–Engine heat.

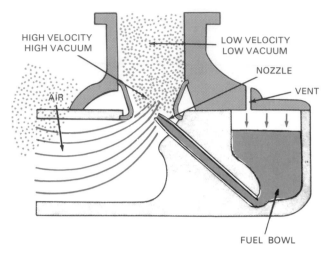

Fig. 5-9. Air flowing through updraft carburetor moves vertically upward into venturi. Passages must be comparatively smaller than those in the downdraft carburetor to increase air velocity so it will carry fuel upward.

TYPES OF CARBURETORS

The three basic types of carburetors are named according to the direction that air flows from their outlets to the engine manifold:
- Natural draft or side draft, Fig. 5-8.
- Updraft, Fig. 5-9.
- Downdraft, Fig. 5-10.

The *natural draft carburetor* is used when there is little space on top of the engine. The air flows horizontally into the manifold.

Updraft carburetors are placed low on the

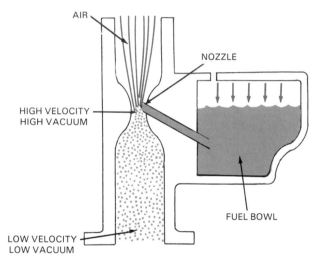

Fig. 5-10. Downdraft carburetor has downward flow of air through venturi. Since it can operate with lower velocities, it has larger passages.

engine and use a gravity-fed fuel supply. However, the air-fuel mixture must be forced upward into the engine. The air velocity must be high, so small passages must be used in the carburetor and manifold.

Downdraft carburetors operate with lower air velocities and larger passages. This is because gravity assists the air-fuel mixture flow to the cylinder. The downdraft carburetor can provide large volumes of fuel when needed for high speed and high power output.

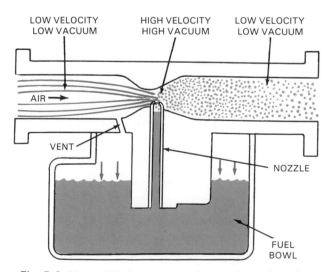

Fig. 5-8. Natural draft carburetor has horizontal air flow through it. (Deere & Co.)

FLOAT-TYPE CARBURETOR

The carburetor *float* is a small sealed vessel made of brass or plastic. Some floats are made of solid flotation materials that eliminate the possibility of leakage. See Fig. 5-11.

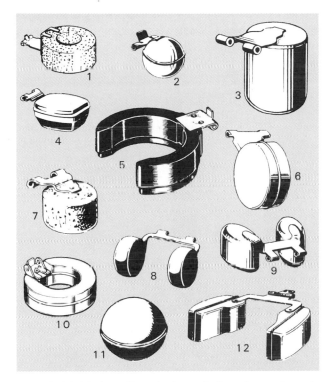

Fig. 5-11. Various float designs. 1–Doughnut-shaped cork. 2–Ball-shaped metal. 3–Cylindrical metal. 4–Rectangular metal. 5–Horseshoe-shaped plastic. 6–Cylindrical metal. 7–Round cork. 8 and 9–Twin-type metal. 10–Doughnut-shaped metal. 11–Ball-shaped metal. 12–Twin-type plastic.

The purpose of the carburetor float is to maintain a constant level of fuel in the float bowl. The float rises and falls with the fuel level. As fuel is used from the float bowl, the float lowers and unseats a needle valve, which lets fuel enter the bowl. This, in turn, raises the float, seating the needle and shutting off fuel supply to the bowl.

The closed position of the needle valve is illustrated in Fig. 5-12. The needle valve illustrated in Fig. 5-13 shows valve action in greater detail. The neoprene needle point is soft and seats well in the valve. Also, it is not as likely to wear out as a brass needle point.

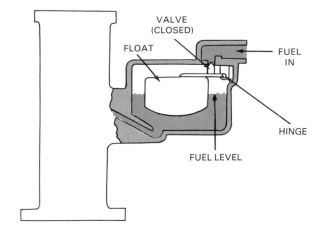

Fig. 5-12. Float in float bowl maintains a constant fuel level. When fuel level rises, float closes needle valve, stopping incoming fuel. When fuel level lowers, float unseats needle and lets more fuel in. (Deere & Co.)

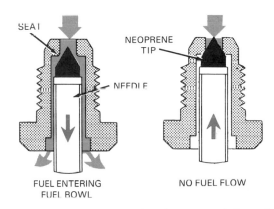

Fig. 5-13. Needle in float bowl opens and closes fuel passage into chamber. Needle is operated by hinged arm of float. (Evinrude Motors)

FLOAT BOWL VENTILATION

Most carburetors are sealed and balanced to maintain equal air pressure. The air pressure above the fuel in the bowl and the air pressure entering the carburetor are equalized by a vent in the float bowl. Refer to Fig. 5-9. This vent assures a continuous, free flow of fuel.

CHOKE SYSTEM

The carburetor *choke* is a round disc mounted on a shaft located at the intake end of the carburetor, Fig. 5-14. When closed, the choke provides a rich air-fuel mixture, which is necessary when starting a cold engine. It allows less air to enter

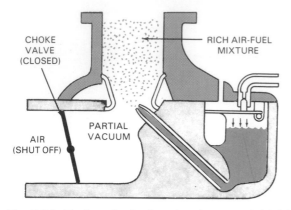

Fig. 5-14. Choke valve is closed and vacuum is high in carburetor. Fuel mixture entering intake manifold is extremely rich.

the carburetor. The manifold vacuum draws harder on the fuel nozzle. Therefore, more fuel and less air enters the combustion chamber.

THROTTLE SYSTEM

Like the choke, the *throttle* is a round disc mounted on a shaft. This valve, however, is located beyond the main fuel nozzle, Fig. 5-15.

The main purpose of the throttle valve is to regulate the amount of air-fuel mixture entering the cylinders. It also permits the operator to vary engine speed to suit conditions or to maintain a uniform speed when the load varies.

On many engines, a linkage connects the throttle valve to a governor. The governor, in turn, is connected to a speed control lever. When the

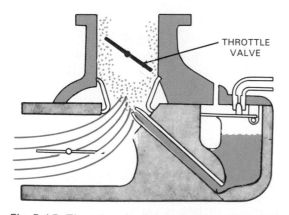

Fig. 5-15. Throttle valve is located beyond main fuel nozzle. Throttle regulates amount of air-fuel mixture entering engine.

speed control lever is set for a given speed, the governor will maintain that speed until the engine reaches its limit of power.

When the load on the engine increases, the governor automatically opens the throttle valve. This permits more air-fuel mixture to enter the engine, providing increased power to maintain a uniform speed. When the load decreases, the governor closes the throttle to reduce engine power. More details on governors is presented later in this chapter.

LOAD ADJUSTMENT

The amount of fuel entering the main discharge nozzle is sometimes regulated by a load adjusting needle, Fig. 5-16. Many carburetors have a fixed jet or orifice, which is preset to allow proper fuel flow for maximum power and economy. Carburetors equipped with a fixed jet are nonadjustable.

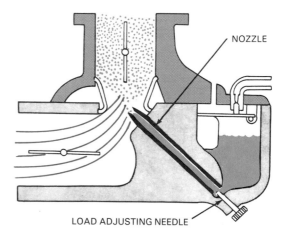

Fig. 5-16. A load adjusting needle, located as shown, regulates amount of fuel entering main nozzle.

ACCELERATION SYSTEM

When the throttle valve is opened quickly for acceleration, a large amount of air is allowed to enter. Unless some method is used to provide additional fuel to maintain a satisfactory air-fuel ratio, the engine will slow down and possibly stop. On larger engines and multi-cylinder engines, a mechanical plunger-type pump is connected to the throttle linkage. When the throttle valve is opened on acceleration, the pump automatically depresses and forces fuel into the carburetor.

ACCELERATION WELL

An *acceleration well* is a reservoir of fuel. During idling (when load nozzle is inactive), fuel rises inside the nozzle. The fuel flows through holes in the side of the nozzle and into the acceleration well, Fig. 5-17.

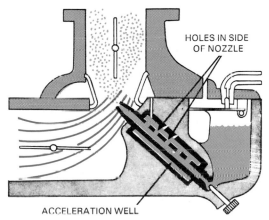

Fig. 5-17. Acceleration well stores fuel for use during rapid acceleration. When fuel has been used from acceleration well, nozzle holes act as air bleeds. (Deere & Co.)

When the throttle valve is opened quickly, the stored fuel rushes through the holes in the nozzle without being metered by the adjusting needle. This fuel combines with the fuel in the nozzle, and the double charge enters the airstream. This provides a much richer air-fuel mixture when there is a sudden need for more power. As the fuel supply decreases in the accelerating well and the holes are uncovered, they become air bleeds for the main nozzle.

ECONOMIZER CIRCUIT

During part throttle operation, the full capacity of the main nozzle is not required. To reduce capacity, some carburetors are equipped with *economizer circuits. The economizer circuit is designed to retard fuel flow to the engine at part throttle.*

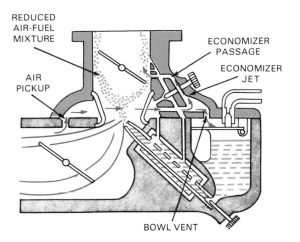

Fig. 5-18. Economizer system creates a reduced pressure in float bowl during part throttle operation, which retards amount of fuel discharged from main nozzle.

The basic "economizing" process is the same for all carburetors. Fig. 5-18 shows an updraft carburetor with the bowl vent passage extended to a point near the throttle valve. When the throttle valve is partially open, the economizer passage is on the engine side of the plate. This permits the engine to draw air through the passage, reducing air pressure in the bowl and cutting down on fuel flow from the nozzle.

IDLING CIRCUIT

During idling operation, the throttle valve is closed. In this condition, the idling system of any type of carburetor supplies just enough air-fuel mixture to keep the engine running. However, actual idling system operation varies in updraft, downdraft, and natural draft carburetors.

The *updraft carburetor* in Fig. 5-19 is in the idling mode of operation. The choke is partially closed, directing airflow through the pickup. Since the throttle valve is closed, the air moves through a passage outside of the venturi to the idle orifice. At this point, the idle adjusting needle regulates the amount of air mixing with the fuel in the idle orifice. Less air provides a richer mixture, more air produces a leaner mixture.

At slow idle, the throttle valve is closed. Only the primary orifice is exposed to allow fuel to enter into the manifold. At fast idle, the throttle valve opens slightly to expose both primary and

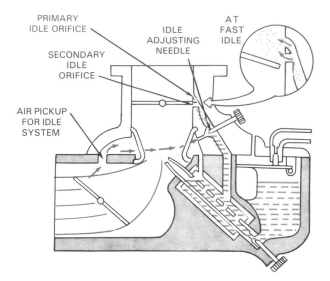

Fig. 5-19. During idling, some incoming air is directed through a passage around venturi. This air mixes with fuel and is drawn out primary and secondary idle orifices. Throttle valve is closed for idle and slightly opened for fast idle.

secondary orifices. Remember, the speed and power of the engine is directly related to the amount of air-fuel mixture allowed to enter the cylinder. Note that at idling speed, the main discharge nozzle is inoperative due to lack of airflow through the venturi.

The *downdraft carburetor* in Fig. 5-20 is in the idling mode. The air bleed is located above the

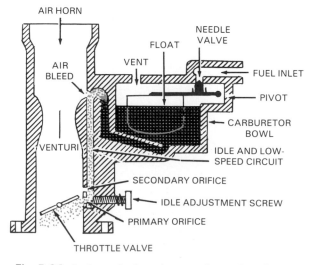

Fig. 5-20. In downdraft carburetor, incoming air enters in air bleed above venturi and travels with fuel to idle orifice. Carburetor is in idling state, since throttle valve has uncovered primary orifice only. (Deere & Co.)

venturi and serves both the idling ports and main discharge nozzle. NOTE: Main discharge nozzle is not shown in Fig. 5-20 for purpose of clarity. It would be located as shown in Fig. 5-10. The idle adjustment screw in this carburetor regulates flow of air-fuel mixture.

The *natural draft carburetor* in Fig. 5-21 is in the idling mode. The throttle valve is closed, and the engine is running from the primary idle discharge hole. The choke valve is wide open. The engine is idling.

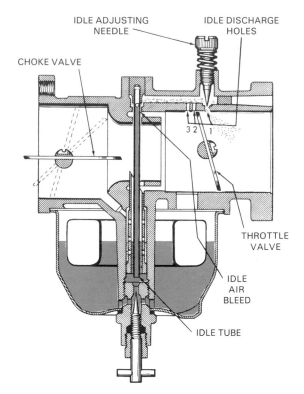

Fig. 5-21. Idling. Throttle valve is closed, and engine is operating from primary idle orifice. (Zenith Div., Bendix Corp.)

PART THROTTLE, FULL THROTTLE SEQUENCE

Beyond idling speed, the carburetor has other circuits for part throttle and full throttle operation. In Fig. 5-22, the throttle valve in this natural draft carburetor is partly open. Primary and secondary discharge holes are open, allowing more air-fuel mixture to enter. The engine is running at part throttle.

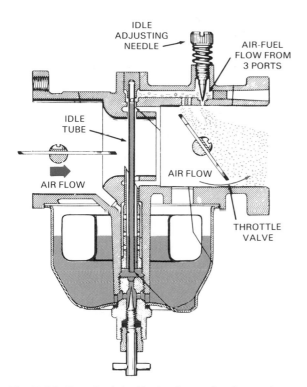

Fig. 5-22. Part throttle. Engine is running from primary and secondary orifices

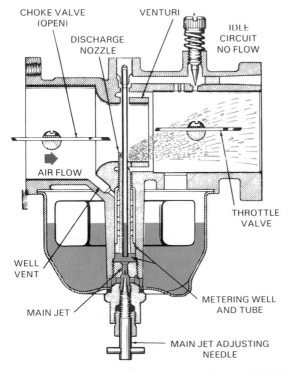

Fig. 5-23. Full throttle. Idle orifices have stopped feeding fuel due to reduced vacuum in that part of carburetor. A full flow of fuel is being drawn from main nozzle.

Refer to Fig. 5-23 for full throttle mode of operation. The throttle is wide open, and the maximum amount of air is flowing through the venturi. The main discharge nozzle is operating because of high vacuum in the nozzle area. The maximum air-fuel mixture is entering the cylinders, and the engine is developing full speed and power.

Fig. 5-24 is an exploded view of the natural draft carburetor shown in Figs. 5-21 through 5-23.

THE PRIMER

Some carburetors are equipped with primers. The *primer* is a hand-operated plunger, which,

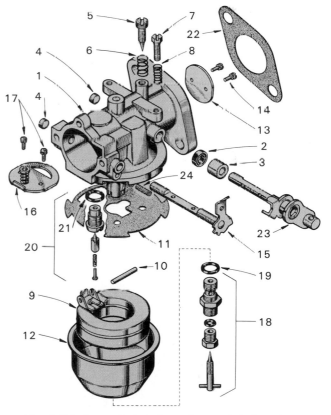

Fig. 5-24. An exploded view of natural draft carburetor: 1–Throttle body. 2–Seal. 3–Retainer. 4–Cup rings. 5–Idle adjustment needle. 6–Spring. 7–Throttle stop screw. 8–Spring. 9–Float and hinge assembly. 10–Float pin. 11–Gasket. 12–Fuel bowl. 13–Throttle valve. 14–Screw. 15–Lever and shaft assemble for choke. 16–Choke valve. 17–Screw. 18–Main jet and adjustment assembly. 19–Washer. 20–Fuel valve and seat assembly. 21–Gasket. 22–Flange gasket. 23–Throttle shaft and lever assembly. 24–Spring. (Zenith Div., Bendix Corp.)

when depressed, forces additional fuel through the main nozzle prior to starting a cold engine.

In operation, the primer pumps air pressure into the float bowl, forcing fuel up the nozzle. A primer mounted on a float-type carburetor is shown in Fig. 5-25.

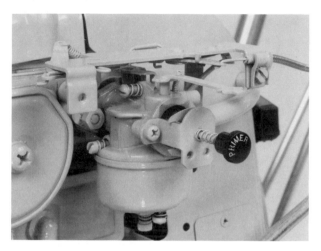

Fig. 5-25. A primer plunger mounted on the carburetor. (Deere & Co.)

DIAPHRAGM-TYPE CARBURETORS

The *diaphragm carburetor* does not have a float system. Instead, the difference between atmospheric pressure and the vacuum created in the engine pulsates a flexible diaphragm. A diaphragm, control needle, and needle seat are shown in Fig. 5-26. The diaphragm draws fuel

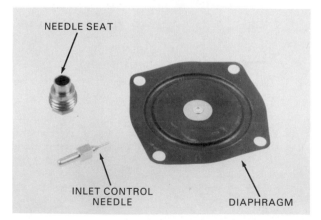

Fig. 5-26. A diaphragm, control needle, and needle seat used in a diaphragm carburetor system. (Deere & Co.)

into a chamber of the carburetor from which it is readily drawn into the venturi.

The carburetor shown in Fig. 5-27A is a diaphragm-type, natural draft carburetor. Views B and C illustrate the operating system of the carburetor.

In Fig. 5-27B, vacuum created in the manifold draws fuel from the upper chamber through the check valve into the venturi. Then, reduced pressure in the upper chamber allows atmospheric pressure to lift the diaphragm, compressing the inlet tension spring. Finally, movement of the diaphragm opens the fuel valve, permitting fuel to flow into the upper chamber. Remember, this action takes place on the intake stroke of the piston.

In Fig. 5-27C, manifold pressure increases to equal atmospheric pressure when the piston rises

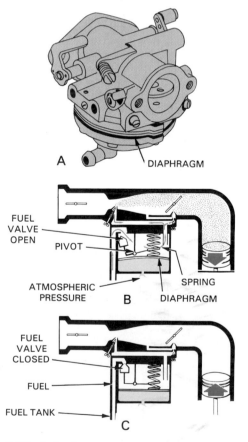

Fig. 5-27. A diaphragm-type, natural draft carburetor. Diaphragm is lifted by manifold vacuum while fuel is being drawn from jets. When vacuum is reduced, diaphragm returns to normal, drawing new fuel into upper fuel chamber.

on the compression stroke. Since there is no difference in pressure between the upper chamber and the lower chamber, the inlet tension spring closes the fuel valve and returns the diaphragm to a neutral position. The check valve closes immediately when the pressure is equalized. Therefore, the retracting diaphragm draws fuel into the upper chamber before the fuel valve closes completely.

The pulsation of the diaphragm takes place on every intake and compression stroke, regardless of the number of engine cylinders. On four cycle engines, fuel is drawn into the cylinder on the downstroke of the piston. On two cycle engines, the fuel is drawn into the crankcase during the upstroke of the piston.

In some applications, the diaphragm spring is adjustable (adjustment screw not shown) to balance the force of the inlet tension spring.

DIAPHRAGM CARBURETOR OPERATION

A study of the various circuits of a typical diaphragm-type carburetor will help clarify operating principles. The carburetor shown in Fig. 5-28 is in starting condition with the choke valve closed. Follow the arrows that indicate direction of fuel flow.

Fuel is drawn from the idle discharge ports and main nozzle because manifold vacuum is high. The carburetor diaphragm is drawn upward during the intake stroke of the engine piston, unseating the fuel inlet needle to allow fuel to flow. Note that the natural draft carburetor in Fig. 5-30 also incorporates a fuel pump diaphragm in its body.

Fig. 5-29 illustrates idling operation with the choke valve open and the throttle valve closed. Vacuum is in effect on the engine side of the throttle valve. Since only one idle discharge port is exposed, a small quantity of fuel is being used, and the engine runs slowly.

Fig. 5-30 shows the throttle partially open for intermediate speed. Airflow through the primary venturi is still not great enough to draw fuel up the main nozzle. Three idle discharge ports are feeding fuel for medium speed. These extra idle discharge ports are termed off-idle ports. They

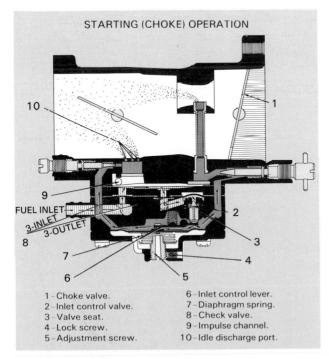

STARTING (CHOKE) OPERATION

FUEL INLET
3-INLET
3-OUTLET

1 – Choke valve.	6 – Inlet control lever.
2 – Inlet control valve.	7 – Diaphragm spring.
3 – Valve seat.	8 – Check valve.
4 – Lock screw.	9 – Impulse channel.
5 – Adjustment screw.	10 – Idle discharge port.

Fig. 5-28. With choke plate closed, a very strong vacuum is formed in air horn. A large quantity of gasoline is "sucked" out of idle jets and main nozzle. A rich mixture results, which can support cold engine operation.

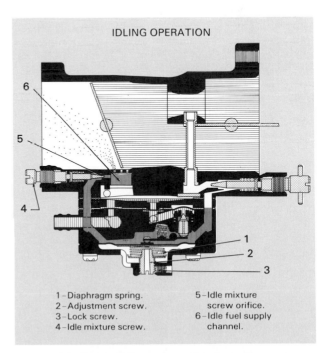

IDLING OPERATION

1 – Diaphragm spring.	5 – Idle mixture
2 – Adjustment screw.	screw orifice.
3 – Lock screw.	6 – Idle fuel supply
4 – Idle mixture screw.	channel.

Fig. 5-29. During idling or very slow speed operation, only the primary orifice (passage) is feeding fuel to engine. Air velocity is not high enough to draw fuel out of the high speed circuit. Remember, this circuit controls fuel mixture when an engine is idling.

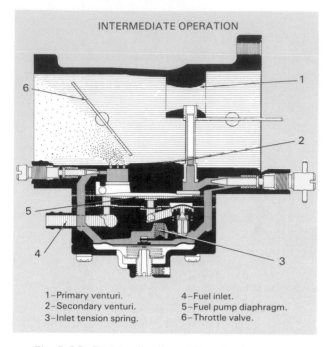

INTERMEDIATE OPERATION

1–Primary venturi.
2–Secondary venturi.
3–Inlet tension spring.
4–Fuel inlet.
5–Fuel pump diaphragm.
6–Throttle valve.

Fig. 5-30. Fuel feeding from idling discharge ports provides intermediate speed operation. (Rupp Industries, Inc.)

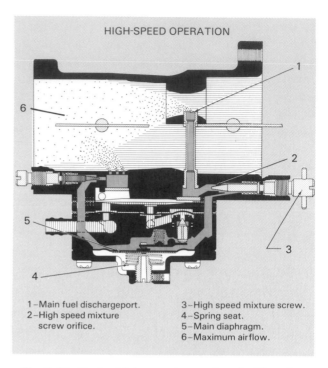

HIGH-SPEED OPERATION

1–Main fuel dischargeport.
2–High speed mixture screw orifice.
3–High speed mixture screw.
4–Spring seat.
5–Main diaphragm.
6–Maximum airflow.

Fig. 5-31. During high-speed operation, fuel flow from main nozzle and idle jets combines with maximum airflow.

must supply more fuel than the single idle port, yet not as much as the main discharge port. The intermediate circuit must provide fuel for transition from idle to high-speed operation.

Fig. 5-31 illustrates high-speed operation with maximum air and fuel flowing through the carburetor. All idle ports and the main nozzle are feeding fuel. NOTE: Choke and throttle valves are open.

MANUAL THROTTLE CONTROLS

A basic *manual throttle control* consists of either mechanical linkage or a flexible cable. One end of the control is attached to the throttle shaft lever. The other end is connected to a lever, slide, or dial that is operated manually to open and close the throttle valve.

The manual throttle can be used as the sole control for positioning the throttle valve. Typical applications of this type are chain saws, motorcycles, snowmobiles, and outboard engines. In some installations, the manual control is used in conjunction with a governor. This setup permits governed speed to be changed when desired.

Fig. 5-32 pictures a manual remote control using a flexible cable to transmit motion from the speed control lever to the throttle valve. In this case: A–Motion changes tension on the governor spring. B–Tension on governor spring changes the throttle valve position.

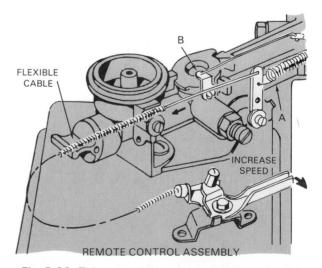

FLEXIBLE CABLE

INCREASE SPEED

REMOTE CONTROL ASSEMBLY

Fig. 5-32. This manual throttle control uses a flexible cable to transmit motion from hand lever to governor spring lever. (Briggs and Stratton Corp.)

Fig. 5-33 shows a throttle control that varies governor spring tension, positions the throttle valve, and actuates the choke for starting. In Fig. 5-33A, the control knob is turned to START, which rotates the choke valve shaft to the choked position. When the engine is started, the control knob is turned to RUN.

To stop the engine, the knob is turned to STOP, as shown in Fig. 5-33B. The stop switch grounds the ignition system, cutting off the flow of electricity to the engine. NOTE. This switch is illustrated in detail in the chapter on ignition systems.

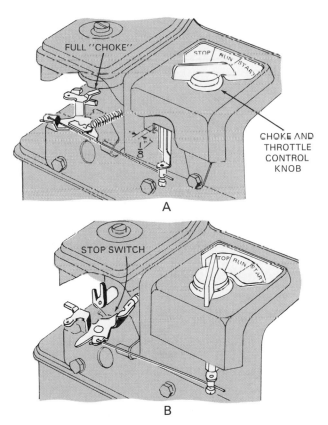

Fig. 5-33. Combined manual throttle and choke control. A–Control knob turned to START activates choke. B–Knob turned to STOP closes stop switch.

GOVERNOR THROTTLE CONTROLS

In many small gasoline engine applications (lawn mowers, generators, and garden tractors), the load on the engine can change instantly. The change in load would require constant throttle changes on the part of the operator. Instead, *governors* are used to provide a smooth, constant speed, regardless of engine loading.

WHAT AN ENGINE GOVERNOR DOES

Governors can be designed to serve three basic functions:
- Maintain a speed selected by operator which is within range of governor.
- Prevent overspeeding that may cause engine damage.
- Limit both high and low speeds.

In Fig. 5-34, observe how tractor speed varies without a governor, but stays constant with a governor.

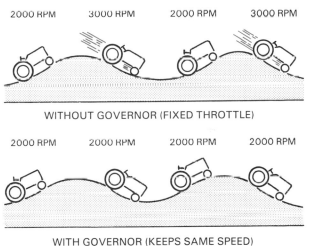

Fig. 5-34. Notice what the governor does for the engine. (Deere & Co.)

Small engine governors are generally used to maintain a fixed speed not readily adjustable by the operator or to maintain a speed selected by means of a throttle control lever. In either case, the governor protects against overspeeding. If the load is removed, the governor immediately closes the throttle. If the engine load is increased, the throttle will be opened to prevent engine speed from being reduced.

For example, a lawn mower normally has a governor. When mowing through a large clump of grass, engine load increases suddenly. This tends to reduce engine speed. The governor reacts by opening the carburetor throttle valve. Engine power output increases to maintain cutting blade speed. When the mower is pushed over a sidewalk (no grass or engine load), engine speed tends to go up. The governor reacts by closing the carburetor throttle valve. This limits maximum cutting blade speed. As a result, mower engine and cutting blade speeds stay relatively constant.

TYPES OF GOVERNORS

There are several types of engine governors: *air vane* (also called *pneumatic*), *centrifugal* (also called *mechanical*), Fig. 5-35, and *vacuum*. Most modern governors are air vane types, which adjust fuel intake according to engine demands, Fig. 5-36. However, centrifugal governors are also fairly common. Vacuum governors are usually found on farm and industrial engines. Basically, all accomplish the same purpose–to protect the engine from overspeeding and to maintain a constant speed, independent of load. However, different speed sensing devices are used.

AIR VANE GOVERNOR

The air vane governor is operated by the stream of air created by the flywheel cooling fins. The force developed by the airstream is in direct proportion to the speed of the engine.

A lightweight, thin strip of metal called an air vane is placed in the direct path of the airstream. It is pivoted on a pin or shaft set near one end. The vane is connected, with linkage, to the throttle shaft lever. When the engine is running, the air-

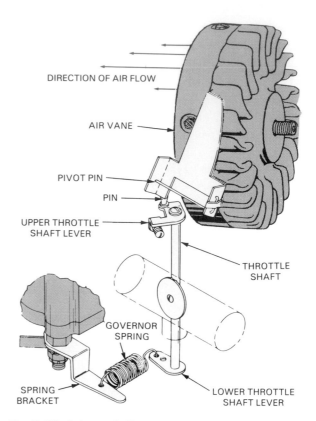

Fig. 5-36. Schematic illustrates operation of an air vane governor. Vane tries to close throttle valve, while governor spring tries to open it. Balance between these two forces determines throttle position.
(Tecumseh Products Co.)

stream pivots the vane and attempts to close the throttle valve, Fig. 5-36.

The governor spring is attached to the throttle lever or to linkage from the vane. This spring is designed to pull the throttle valve to wide open position.

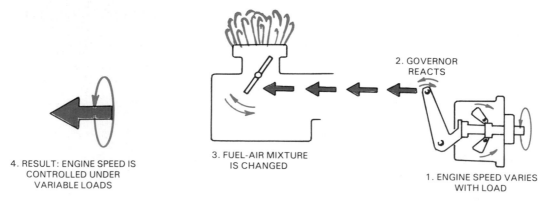

Fig. 5-35. Centrifugal governor controls engine speed by varying fuel mixture. (Deere & Co.)

Note in Fig. 5-36 how the airflow pivots the vane, causing it to exert rotary pressure on the upper throttle shaft lever while the governor spring tries to pull the throttle valve open. The ratio of pressure developed by the vane, as opposed to the tension of the governor spring, determines throttle valve position. When the engine is stopped, the airstream ceases and the throttle valve is pulled to wide open position by the governor spring.

The governor spring bracket can be moved to vary the amount of tension exerted by the spring. This will alter vane spring pressure balance and establish a new throttle setting, Fig. 5-36.

Governor spring tension is carefully calibrated by the manufacturer. If the spring is stretched or altered in any way, it should be replaced with a new spring designed for that make and model engine. If the linkage is bent, worn, or damaged, it should be straightened, repaired, or replaced. Look at Fig. 5-37.

If it is necessary to replace any component on an air vane governor, the top no-load rpm should be checked with an accurate tachometer. The top speed must NOT exceed the maximum recommended rpm for the implement being driven.

In the case of lawn mowers, blade tip speed should not exceed 19,000 feet per minute in a no-load condition. If necessary, change the governor spring or adjust the top speed limit device so the engine stops accelerating at the recommended rpm, which is based on blade length. See Fig. 5-38.

BLADE LENGTH IN INCHES (MILLIMETRES)	MAXIMUM ROTATIONAL RPM
18 (460)	4032
19 (485)	3820
20 (510)	3629
21 (535)	3456
22 (560)	3299
23 (585)	3155
24 (610)	3024
25 (635)	2903
26 (660)	2791

Fig. 5-38. Chart lists various lengths of lawn mower blades and maximum rotational speeds, which produce blade tip speeds of 19,000 feet per minute. It is recommended that top speeds be set 200 rpm less than shown.

Fig. 5-37. Parts of a typical governor system for actual engine. Note location of idle speed screw and needle valve screw.

Since blade tip speed is a function of blade length and engine rpm, longer blades require lower engine speeds. It is suggested that top governed engine speed be adjusted at least 200 rpm lower than the speeds shown in Fig. 5-38 to account for tachometer inaccuracy.

FIXED SPEED

If the engine is designed to run at only one specific rpm setting, the tension of the governor spring is carefully adjusted until the speed is correct. Then it is left at this setting, Fig. 5-39. Fixed speed engines of this type have a limited range of governor spring adjustment. When the engine is started, the force of the airstream on the vane closes the throttle until that force is equal to spring tension.

VARIABLE SPEED

It is often desirable to have an engine operate at many different speeds that may be quickly and easily set by the operator. In this case, a variable speed air vane governor is used, Fig. 5-40. Engine rpm is changed by pivoting the governor spring bracket. Remember, the throttle control adjusts governor spring tension. The spring is not connected directly to the throttle lever.

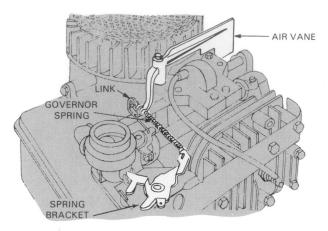

Fig. 5-40. Variable speed air vane governor. To change rpm, operator alters governor spring tension by rotating spring bracket.

CENTRIFUGAL OR MECHANICAL GOVERNOR

Like the air vane governor, the centrifugal (mechanical) governor also controls engine speed. The centrifugal governor, however, utilizes pivoted flyweights that are attached to a revolving shaft or gear driven by the engine. With this setup, governor rpm is always directly proportional to engine rpm.

Figs. 5-41A and 5-41B show how centrifugal governors operate. When the engine is stopped,

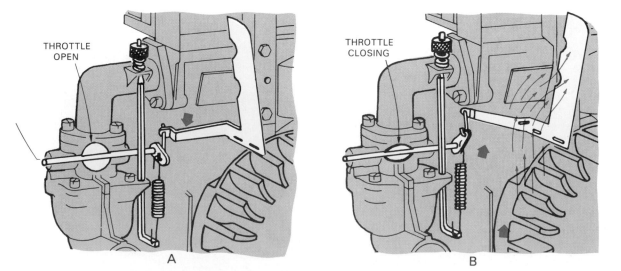

Fig. 5-39. Air vane governor. A—Engine stopped. Spring holds throttle open. B—Engine running. Air pressure pivots vane of fixed speed governor and shuts throttle valve until spring pressure and vane pressure are balanced. Knurled nut alters spring tension and adjusts speed. (Briggs and Stratton Corp.)

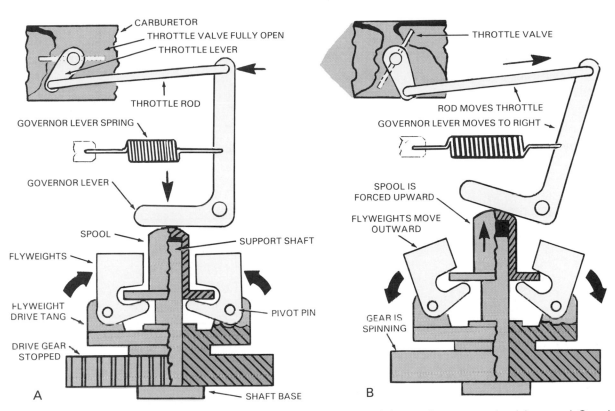

Fig. 5-41. A centrifugal-type governor. Centrifugal force causes flyweights to pivot outward, raising spool. Spool rotates governor lever which closes throttle valve. Balance between centrifugal force and governor spring tension determines throttle valve setting.

the heavy ends of the flyweights are held close to the shaft by the governor spring. The throttle valve is held fully open as illustrated in Fig. 5-41A.

When the engine is started, the governor is rotated. As its speed increases, centrifugal force increases and causes the flyweights to pivot outward. This forces the spool upward, raising the governor lever until spring tension equals the centrifugal force on the weight. This action partially closes the throttle valve shown in Fig. 5-41B.

If the engine is subjected to a sudden load that reduces rpm, the reduction in speed lessens centrifugal force on the flyweights. The weights move inward, lowering the spool and governor lever. This series of actions opens the throttle valve and lost rpm is regained.

CHANGING GOVERNOR SPEED SETTING

The centrifugal governor speed setting can be changed by turning a knurled adjusting nut on the end of the tension rod, Fig. 5-42. This system

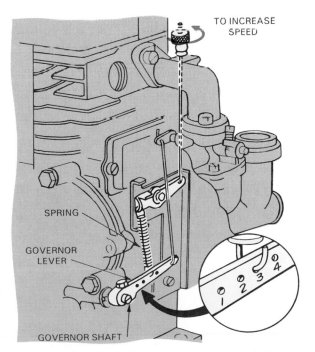

Fig. 5-42. Single speed setting, using knurled nut to provide a limited governor speed range. (Briggs and Stratton Corp.)

is used when the engine is expected to run at a constant speed setting for long periods of time.

A remote, hand-controlled cable that alters spring tension also can be used to set governor speed, Fig. 5-43.

Movement of the control handle increases or decreases spring tension, which speeds up or slows down the engine. The operator can quickly select any speed within the range of the governor.

Another type of governor speed adjusting arrangement is shown in Fig. 5-44. Movement of the governor adjusting lever changes spring tension and engine rpm.

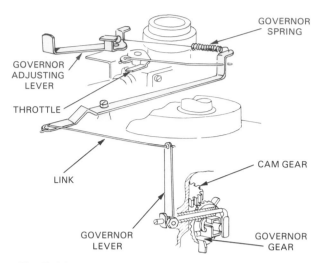

Fig. 5-44. Method of setting governed speed by using an adjusting lever to control governor spring tension.

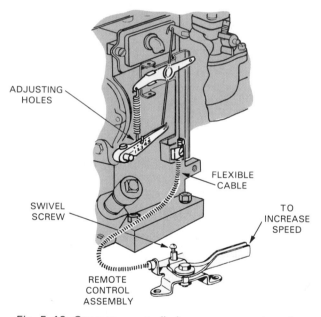

Fig. 5-43. Operator-controlled governor speed setting device allows quick, wide, governed speed changes. (Briggs and Stratton Corp.)

HUNTING OF CENTRIFUGAL GOVERNORS

Frequently, when an engine is first started or is working under load, its speed becomes erratic or oscillates. The engine speeds up rapidly; the governor responds and engine speed drops quickly. The governor stops functioning and engine speed again increases. The governor responds and this action is repeated over and over. This condition is known as *hunting*.

Hunting is usually a result of improper carburetor adjustment. Leaning or richening the fuel mixture can often correct the problem. Also, the governor may cause hunting if it is too stiff or binds at some point. It must work freely.

VACUUM GOVERNORS

Farm and industrial engines are often equipped with a vacuum governor for regulating maximum engine speed. The vacuum governor, Fig. 5-45, is located between the carburetor and the intake manifold. It senses changes in intake manifold pressure (vacuum). There is no other mechanical connection between the governor and other parts of the engine.

As engine speed and suction (vacuum) increase, the governor unit closes the throttle butterfly valve. This causes a decrease in fuel flow and engine speed.

When engine speed and vacuum decrease, the spring opens the throttle valve. This action causes the fuel flow and engine speed to increase. An adjustment of spring tension is used to set the desired speed range.

GOVERNOR FEATURES

The operating principles of the governor mechanism are quite simple and reliable. Governors provide accuracy and efficiency of operation combined with convenience and comfort for the operator. Two of the most important operat-

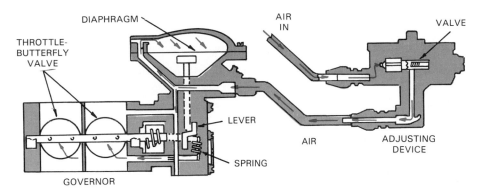

Fig. 5-45. A vacuum governor must maintain a preset maximum engine speed, independent of engine load.

ing features of engine speed and power output governors are stability and sensitivity. *Stability* is the ability to maintain a desired engine speed without fluctuating. Instability results in hunting or oscillating due to over-correction. Excessive stability results in a dead-beat governor (one that does not correct sufficiently for load changes). *Sensitivity* is the percent of speed change required to produce a corrective movement of the fuel control mechanism. High governor sensitivity will help keep the engine operating at a constant speed.

AIR CLEANERS, AIR FILTERS

An engine breathes a tremendous quantity of air during its normal service life. If the incoming air is not thoroughly cleaned by passing it through a filtering device, dirt and grit entering the cylinder would cause rapid wear and scoring of machined parts throughout the engine. Engine life, under severe dust conditions, actually could be reduced to minutes.

Three types of air cleaners widely used in small gasoline engines are the *oil bath, oil-wetted*, and *dry types*.

OIL BATH AIR CLEANER

Oil bath air cleaners usually consist of two housings, Fig. 5-46. The upper housing has a built-in filtering element. The lower housing is constructed with an oil reservoir at the bottom.

Oil bath air cleaners, Fig. 5-47, draw air down between the inner and outer housings. The air is

Fig. 5-46. Oil bath air cleaner. Oil sump traps heavier particles of dirt as airflow changes direction. Many of remaining contaminants are trapped as air passes through filter element. (Kohler Co.)

moving at high speed and must suddenly reverse its direction when it strikes the bottom, which is covered with a pool of engine oil. When the air reverses direction, airborne dust and grit (heavier than air) is thrown into the oil, where it sinks to the bottom.

A small amount of oil is drawn into the filter element by the airstream. Dirt small enough to make the high speed change of direction is trapped by the oil-moistened filter. When the

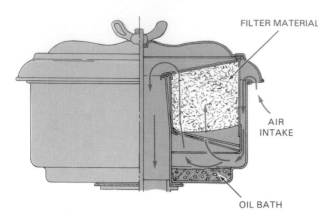

FILTER MATERIAL

AIR
INTAKE

OIL BATH

Fig. 5-47. Cutaway of an oil bath air cleaner. Note oil level. If it is too low, poor filtering results. Too high, engine draws excess oil through carburetor, causing carbon deposits in engine. (Wisconsin Motors Corp.)

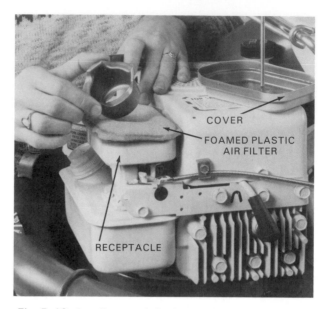

COVER

FOAMED PLASTIC
AIR FILTER

RECEPTACLE

Fig. 5-48. An oil-wetted air cleaner. Polyurethane foam is dampened with oil and contained in a vented case attached to carburetor. Filter can be cleaned and re-oiled.

engine is stopped, the oil in the filter element drains back into the oil sump, carrying much of the dirt with it.

The construction of a typical oil bath air cleaner is shown in Fig. 5-47. Note how incoming air changes direction above the oil sump, then passes up through the filter pack and down into the carburetor. The filter material must be rinsed and dried, and the oil in the sump must be changed periodically.

OIL-WETTED AIR CLEANER

The oil-wetted air cleaner utilizes a filtering element (crushed aluminum, polyurethane foam, etc.) dampened with engine oil. Fig. 5-48 shows a polyurethane foam oil-wetted element, reservoir, and cover. In operation, the air is drawn directly through the oil-wetted element where the damp material effectively filters out contaminants. This type of element can be reused by rinsing in cleaning solvent, drying, and re-oiling.

DRY-TYPE AIR CLEANER

Dry-type air cleaners pass the airstream through treated paper, felt, fiber, or flocked screen. Some filter elements (flocked screen) can be cleaned, but most are designed to be thrown away when they become dirty.

A typical treated paper air cleaner element is shown in Fig. 5-49. You can clean this filter by

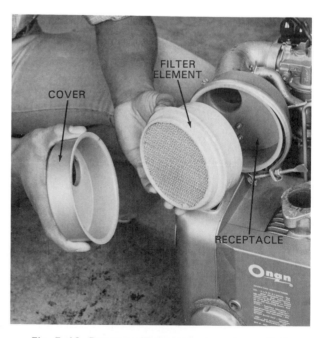

FILTER
ELEMENT

COVER

RECEPTACLE

Onan

Fig. 5-49. Dry-type filter element can be partially cleaned by tapping gently. New element should be installed as needed.

tapping it on a flat surface to dislodge light accumulations of dirt. However, when it will not tap clean, a treated paper filter must be replaced with a new element designed for the given engine application.

SUMMARY

The main purpose of the carburetor is to produce a mixture of fuel and air to operate the engine. The gasoline engine cannot run on "liquid" gasoline. The carburetor must vaporize the fuel and mix it with air.

The amount of air needed for combustion is far greater than the amount of fuel required. The average weight ratio is 15 parts air to 1 part fuel.

The carburetor is operated by pressure differences. It creates partial vacuum by means of a venturi. A venturi is a restriction in a passage that causes air velocity to increase and pressure to decrease. Reduction in pressure draws fuel into the airstream.

Three basic types of carburetors include the natural draft, the updraft, and the downdraft. These carburetors are named according to the direction that the air flows from their outlets to the engine manifold.

Some fuel systems are equipped with a float-type carburetors. The purpose of a carburetor float is to maintain a constant level of fuel in the float bowl.

A diaphragm carburetor does not have a float. Instead, engine vacuum pulsates a flexible diaphragm. The diaphragm draws fuel into a chamber of the carburetor from which it is readily drawn into the venturi.

A manual throttle control consists of mechanical linkage or a flexible cable that is operated manually to open and close the throttle valve.

In many small engine applications, the load on the engine changes constantly. This change would require constant throttle changes on the part of the operator. Governors work to maintain speed selected by the operator, prevent overspeeding, and limit high and low speeds.

A filtering device is used to clean incoming air. If air is not properly filtered, dirt entering the cylinder will cause rapid wear and scoring of machined parts throughout the engine.

KNOW THESE TERMS

Carburetor, Air-fuel mixture, Vacuum, Atmospheric pressure, Venturi, Venturi principle, Natural draft carburetor, Updraft carburetor, Downdraft carburetor, Float, Choke, Throttle, Load adjustment, Acceleration well, Economizer circuit, Idling circuit, Primer, Diaphragm carburetor, Manual throttle control, Governor throttle control, Hunting, Stability, Sensitivity, Air cleaner.

REVIEW QUESTIONS – CHAPTER 5

1. Give five different engine running conditions that must be met by the carburetor.
2. Normal air-fuel mixture by weight is:
   a. 12 to 1.
   b. 13 to 1.
   c. 14 to 1.
   d. 15 to 1.
3. If the barometric pressure on a standard day at 1500 ft. MSL was 29.95, then at 3500 ft. MSL the barometric pressure would be: (Refer to Fig. 5-3.)
   a. 30.95 in. Hg.
   b. 31.95 in. Hg.
   c. 29.95 in. Hg.
   d. 27.95 in. Hg.
4. In a venturi:
   a. Air pressure is greatest where velocity is greatest.
   b. Air pressure is least where velocity is greatest.
   c. Velocity is least where air pressure is least.
   d. The volume of air entering is slightly greater than the air leaving due to the restriction to flow.
5. Name the three basic types of carburetors. (Consider direction of airflow.)
6. Which type of carburetor would normally require a smaller air passage than the other two types?
7. Needle valve points in the carburetor float chamber are usually made from one of two materials. The two materials are _____ and _____.
8. The choke valve in the carburetor is always located:
   a. Nearest the intake end of carburetor.
   b. Nearest the manifold end of carburetor.
   c. In the center of the carburetor.
   d. Above the float chamber level.

9. The richest air-fuel mixture takes place during:
    a. Full throttle.
    b. Half throttle.
    c. Idle.
    d. Starting.
10. On some carburetors, the amount of fuel entering the main discharge nozzle is regulated by:
    a. The float level.
    b. A load adjusting needle.
    c. The idle adjustment needle.
    d. A spray bar needle valve.
11. The acceleration well fills when the engine:
    a. Is running at steady high speed.
    b. Is running at half throttle.
    c. Is under heavy load.
    d. Is idling.
12. The primary purpose of air bleeds is to:
    a. Increase the air-fuel ratio.
    b. Improve atomization of the fuel.
    c. Remove air bubbles that may be mixed with the fuel.
    d. Prevent vapor lock.
13. During idle and fast idle conditions, the main discharge nozzle is:
    a. Discharging a small amount of fuel.
    b. Inoperative.
    c. Acting as an air bleed.
    d. Providing most of the fuel.
14. The carburetor economizer:
    a. Reduces float bowl pressure.
    b. Reduces the amount of fuel discharged into the venturi.
    c. Operates only after the engine reaches part throttle.
    d. All of the above.
15. One main advantage of a diaphragm carburetor as compared to a float-type carburetor is its ability to _____.
16. The diaphragm in the carburetor:
    a. Forces fuel through the main discharge nozzle.
    b. Operates only at high speed.
    c. Draws fuel under spring pressure.
    d. Draws fuel during a vacuum pulse from the manifold or crankcase.
17. Governors serve three basic functions. What are they?
18. Name two basic types of small engine governors.
19. On a governor installation, the governor spring is attached to the throttle lever. The governor spring is intended to:
    a. Have no effect on the throttle valve.
    b. Close the throttle valve.
    c. Open the throttle valve.
    d. Return the throttle lever to the off position.
20. Name three types of air cleaners.

SUGGESTED ACTIVITIES

1. Make a venturi tube. Provide a connection so that air can be forced through the venturi. Install one pressure gage before the restriction and one gage in the restriction. Demonstrate what happens when the air is applied to the venturi.
2. Working with the same venturi used in activity number one, remove the pressure gage in the restriction. Place a pick-up tube in the restriction and draw water out of a beaker. Demonstrate the atomization of the water particles.
3. Make a cutaway of a float-type carburetor so that the float, needle, throttle valve, choke, and internal passages can be seen.
4. Make a working mock-up model of a variable speed centrifugal governor system that demonstrates governor principles.
5. Make a large cross section of a float-type carburetor mounted on a board. Make the choke, throttle valve, float and needle movable from the back. Paint the various parts, passages, and ports with bright colors. Give a demonstration to the class of choked, idle, part throttle, and full throttle carburetor functions.

CHAPTER 6

IGNITION SYSTEMS

After studying this chapter, you will be able to:
- ☐ List the primary purposes of the ignition system.
- ☐ Identify the components in a typical magneto system and describe the function of each part.
- ☐ Describe small engine ignition advance systems.
- ☐ List the advantages of a solid state ignition system.
- ☐ Identify the three general classifications of magneto ignition systems and explain the operation of each.
- ☐ Describe the operation of a battery ignition system.

The primary purpose of the *ignition system*, Fig. 6-1, of a small gasoline engine is to provide sufficient electrical voltage to discharge a spark between the electrodes of the spark plug, Fig. 6-2. The spark must occur at exactly the right time to ignite the highly compressed air-fuel mixture in the combustion chamber of the engine.

The ignition system must be capable of producing as many as 30,000 volts to force electrical current (electrons) across the spark plug gap. The intense heat created by the electrons jumping the gap ignites the air-fuel mixture surrounding the electrodes.

The rate, or times per minute, at which the spark must be delivered is very high. For example, a single cylinder, four cycle engine operating at 3600 rpm requires 1800 ignition sparks per minute. A two cycle engine running at the same speed requires 3600 sparks per minute. In multicylinder engines, the number of sparks per minute is multiplied by the number of cylinders.

Fig. 6-1. The major parts of this small engine magneto system are the breaker points, condenser, coil, flywheel magnets, and the spark plug.

Every spark must take place when the piston is at exactly the right place in the cylinder and during the correct stroke of the power cycle (see Chapter 3). Considering the high voltage required, the precise degree of timing, and the high rate of discharges, the ignition system has a remarkable job to do.

Most small gasoline engines use *magneto systems* to supply ignition spark, Fig. 6-1. *Magneto systems produce electrical current for ignition without any outside primary source of electricity.* They serve as simple and reliable ignition systems. Basic parts of a magneto system include:

- Permanent magnets.
- High tension coil.
- Mechanical or electronic switching device.
- Condenser (used only with mechanical switching device).
- High tension spark plug wire.
- Spark plug, Figs. 6-1 and 6-2.

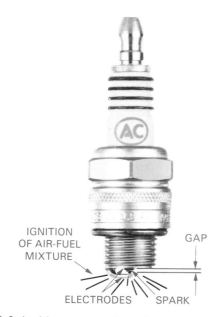

Fig. 6-2. Ignition system of small engine works hard to produce enough voltage to force electrons to jump spark plug gap.

Today, there are several types of magneto systems used on small engines. The mechanical breaker ignition (MBI) system uses mechanical breaker points to control current in the ignition coil. This type system was used exclusively until the development of the solid state ignition system. Solid state systems use electronic devices (transistors, capacitors, diodes) to control various ignition system functions. Several types of magneto systems will be discussed in detail later in this chapter.

To make it easier to understand how various magneto parts function, a review of some basic electrical principles follows.

THE ELECTRON THEORY

All matter is composed of atoms. An atom is extremely small, so small that it cannot be seen with the most powerful microscope. It is the smallest particle of an element that can exist, alone or in combination; yet it consists of electrons, protons, and neutrons.

Atoms can be broken down into types, determined by the number and arrangement of the electrons, protons, and neutrons. A few types of atoms are hydrogen, oxygen, carbon, iron, copper, and lead. There are many others, about 100 in all.

The structure of the atom determines the weight, color, density, and other properties of an element. The electrons, though varying in number, are identical in all elements. An electron from silver would be the same as an electron from copper, tin, or any other substance.

Electrons travel in orbits around the center of the atom. They are very light and their number per atom varies from one element to another. Electrons have negative (–) electrical charges, as shown in Fig. 6-3.

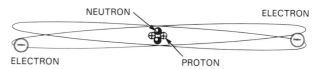

Fig. 6-3. All atoms consist of electrons, neutrons, and protons. Neutrons and protons form the nucleus. Neutrons have no electrical charge, but each proton carries a positive (+) charge. Electrons orbit the nucleus and carry negative (–) charges.

Protons are large, heavy particles when compared with the electrons. One or more protons help form the nucleus (center) of the atom and are positively charged.

A neutron is made up of an electron and proton bound tightly together. Neutrons are electrically neutral and are also located in the nucleus of the atom, Fig. 6-3. The number of electrons is equal to the number of protons in any atom. Normally, atoms are electrically neutral because the negative electrons cancel the positive force of the protons.

Actually, an atom is held together because unlike electrical charges attract each other. The positively charged protons hold the negatively charged electrons in their orbits. Since like electrical charges repel each other, the negative electrons will not collide as they spin.

The ease with which an electron from one atom can move to another atom determines whether a material is an electrical conductor or nonconductor. In order to have electric current, electrons must move from atom to atom. Materials allowing electrons to move in this way are called conductors. Examples are copper, aluminum, and silver.

In nonconductors, it is difficult, if not impossible, for electrons to leave their orbits. Nonconductors are called insulators. Some examples are glass, mica, rubber, plastic, and paper.

Electron flow in a conductor follows a path similar to that shown in Fig. 6-4. The flow of electrons will take place only when there is a complete circuit and a difference in electrical potential. A difference in potential exists when the source of electricity lacks electrons, or is positively (+) charged. Since electrons are negatively

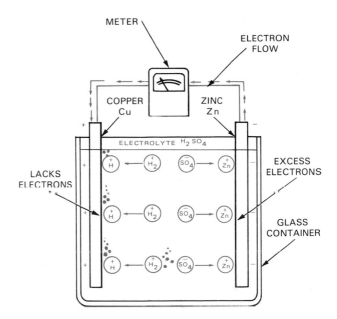

Fig. 6-5. A difference in potential exists if source of electricity lacks electrons and, therefore, is positively (+) charged. Electrons, being negatively (–) charged, are attracted to positive source.

(–) charged and unlike charges attract, the electrons move toward the positive source, Fig. 6-5.

Electrical potential is produced in three ways:
- Mechanically.
- Chemically.
- Statically.

The electrical generator is a mechanical producer of electricity and can be run by water power, steam turbines, or internal combustion engines. A magneto is a type of generator. Mechanical energy from the crankshaft is used to rotate a permanent magnet.

Electricity used in homes and factories is produced mechanically. Batteries are chemical producers of electricity. Lightning is a result of static electricity.

ELECTRICAL UNITS OF MEASUREMENT

Three basic units of electrical measurement are:
- Amperes (rate of electron flow).
- Volts (force that causes electrons to flow).
- Ohms (resistance to electron flow).

An *ampere* is a measurement of the number of electrons flowing past any given point in a specific length of time. One ampere of current is

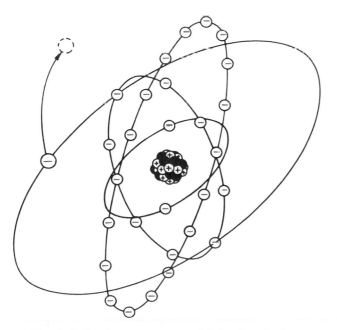

Fig. 6-4. Conductors are materials having electrons that can easily leave orbit of one atom and move to orbit of another atom. When many electrons do this, electricity is produced.

equal to 6,240,000,000,000,000,000 (6.24 x 10^{18}) electrons per second. Since electricity is generally transmitted through wires, the greater the number of electrons flowing, the larger the wire size must be.

The difference in electrical potential between two points in a circuit is measured in *volts*. Voltage is the force, or potential, that causes the electrons to flow.

Resistance to electron flow is measured in *ohms*. Some materials produce a strong resistance to electron flow; others produce little resistance. If a wire is too small for the amount of current produced at the source, the wire will create excessive resistance and will get hot.

The air gap between spark plug electrodes is highly resistant to electron flow, creating the need for high voltage to cause the electrons to jump the gap. This high resistance also creates heat, which ignites the fuel in the cylinder.

OHM'S LAW

Every electrical circuit operates with an exact relationship of volts, amps, and ohms. It is possible to work out their mathematical relationship through the application of *Ohm's Law*.

The formula for Ohm's law is $I = \dfrac{E}{R}$ where:

I = amperes
E = volts
R = ohms

If circuit voltage is 12 and resistance is 8 ohms, the current would be:

$$I = \frac{E}{R} \qquad I = \frac{12}{8} \qquad I = 1.5 \text{ amperes}$$

If amperage is 15 and voltage is 6, resistance would be:

$$R = \frac{E}{I} \qquad R = \frac{6}{15} \qquad R = .4 \text{ ohms}$$

If amperage is 3 and resistance is 10 ohms, the voltage would be:

$$E = I \times R \qquad E = 3 \times 10 \qquad E = 30 \text{ volts}$$

MAGNETISM

The molecular theory of magnetism is the one most widely accepted by scientists. Molecules are the smallest particles of matter that are recognizable as being that matter. For example, a molecule of aluminum oxide will contain atoms of aluminum and atoms of oxygen.

In most materials, the magnetic poles of adjoining molecules are arranged in a random pattern, so there is no magnetic force, Fig. 6-6. Iron, nickel, and cobalt molecules, however, are able to align themselves so that all their north poles point in one direction and south poles in the opposite direction, Fig. 6-7. The individual magnetic forces of each molecule combine to produce one strong magnetic force. In magnets, opposite poles attract each other and like poles repel each other, much in the same way that like and unlike electrical charges react.

Certain materials have good magnetic retention. That is, they retain their molecular alignment. These materials are suitable as permanent magnets. Some materials maintain their molecular alignment only when they are located within a magnetic field. When the field is removed, the

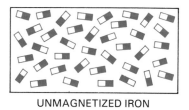

UNMAGNETIZED IRON

Fig. 6-6. An unmagnetized substance is made up of molecules whose poles are not aligned. Molecules have north and south poles, like bar magnets.

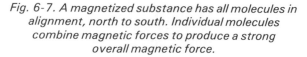

MAGNETIZED IRON

Fig. 6-7. A magnetized substance has all molecules in alignment, north to south. Individual molecules combine magnetic forces to produce a strong overall magnetic force.

molecules disarrange themselves into random patterns and the magnetism is lost. When permanent magnets are cut into pieces, each piece takes on the polarity of the parent magnet, as illustrated in Fig. 6-8.

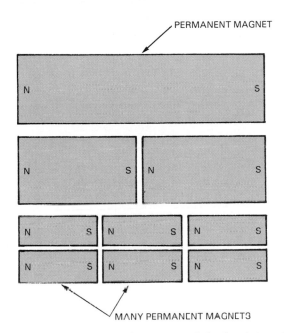

Fig. 6-8. If a permanent bar magnet is broken into subparts, each subpart has a north and south pole, like the parent magnet. If parts could be further broken into individual molecules, each molecule would be an individual magnet.

MAGNETS AND ELECTRICITY

The fact that there is a close relationship between electricity and magnetism serves as the basis for making a workable magneto system. Over 100 years ago, Michael Faraday discovered that electricity could be produced from magnetism. One of his experiments showed that if a wire is moved past a magnet, the magnetic field is cut by the wire and current will flow. See Fig. 6-9. When movement of the wire is stopped, the current also stops, Fig. 6-10. Therefore, electricity will flow when the magnetic lines of force are being cut by the wire.

An important principle used in magneto construction is that a magnetic field is developed when electrons flow through a coil of wire. Fig. 6-11 shows a simple coil of wire with current

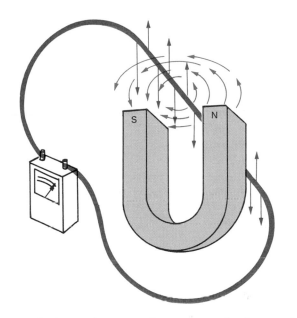

Fig. 6-9. If a conductor, such as copper wire, is moved so that it cuts magnetic lines of force, an electron flow is induced in conductor. Flow of electrons (electricity) can be measured with a sensitive meter.

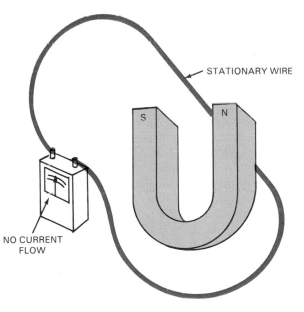

Fig. 6-10. A conductor that is not moving and not cutting magnetic lines of force will not induce electrical current.

flowing through it. In this illustration, the magnetic field is indicated by lines around the coil.

The lines of force in Fig. 6-11 come from the north pole and return to the south pole. If the direction of current is changed, the polarity also

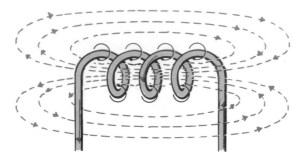

Fig. 6-11. A coil of wire with current flowing through it will produce a magnetic field around itself and around each turn of wire in coil.

changes. In magneto operation, an electric current is passed through a coil, which develops a magnetic field.

IGNITION COIL

The *ignition coil* used in a magneto system operates like a transformer. The coil contains two separate windings of wire insulated from each other and wound around a common laminated iron core. See Fig. 6-12. The primary winding is heavy gage wire with fewer turns than the secondary winding, which has many turns of light gage wire.

When electrical current is passed through the primary winding, a magnetic field is created around the iron core. When the current is stopped, the magnetic field collapses rapidly, cutting through the secondary windings. This

rapid cutting of the field by the wire in the coil induces high voltage in the secondary circuit. The high secondary voltage, in turn, causes a spark to jump the spark plug gap and ignite the air-fuel mixture.

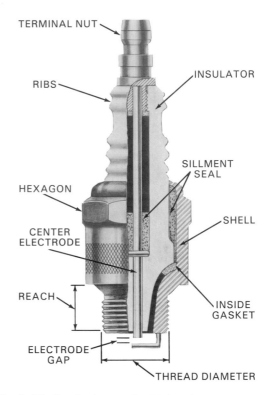

Fig. 6-13. Spark plug carries high voltage current produced by ignition system. It also must withstand the high temperatures and shock of combustion, insulate center electrode against current loss, and seal against compression leakage. (Deere & Co.)

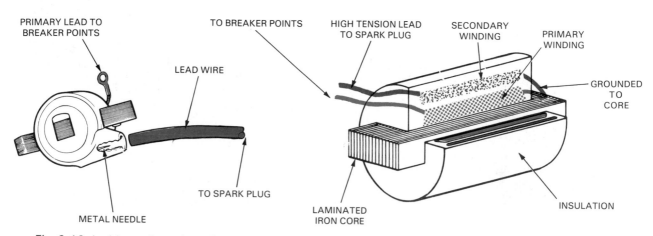

Fig. 6-12. Ignition coil consists of two windings, one inside the other. Coil functions as a step-up transformer to produce high voltage and low amperage from low voltage and high amperage.

SPARK PLUGS

At first glance, an assortment of spark plugs may look very much alike. Actually, there are many variations. Using the correct spark plug for a given engine application can greatly increase the efficiency, economy, and service life of the engine.

Fig. 6-13 shows major parts of a typical spark plug. The terminal nut is the external contact with the high tension coil. Some terminal nuts are removable, others are not.

Two common methods of *high tension lead* connections are shown in Fig. 6-14. Application A uses the exposed clip, which is satisfactory in uses where moisture, oil, or dirt will not get on the plug or can easily be wiped off. The boot type, shown at B, provides better plug protection.

The spark plug *insulator* is usually an aluminum oxide ceramic material, which has excellent insulating properties. The insulator must have high mechanical strength, good heat conducting quality, and resistance to heat shock. Generally, ribs on the insulator extend from the terminal nut to the shell of the plug to prevent "flashover." Flashover is the tendency for current to travel down the outside of the spark plug instead of through the center electrode. See Fig. 6-15.

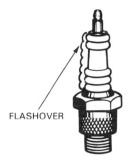

Fig. 6-15. Flashover is caused by moisture or dirt. It can also be caused by a worn out terminal boot, which allows voltage to short across outside of ceramic insulator.

The *center electrode* carries the high voltage current to the spark gap. If the electrical potential is great enough to cause the current to jump the plug gap, the grounded electrode will complete the circuit to ground. NOTE: Always refer to manufacturer's specifications for correct electrode gap, Fig. 6-16.

The sillment seal is a compacted powder that helps ensure permanent assembly and eliminates compression leakage under all operating conditions. The inside gasket also acts as a seal between the insulator and the steel shell.

Spark plug *reach* varies with type of spark plug. Some are long, others quite short, Fig. 6-17.

Fig. 6-14. Two common high tension lead connectors. A—Exposed clip type. B—Neoprene boot type. Exposed clip connector is used in conjunction with a metal strip stop switch.

Fig. 6-16. Mechanic is setting correct electrode gap from manufacturer's specifications. (Champion Spark Plugs)

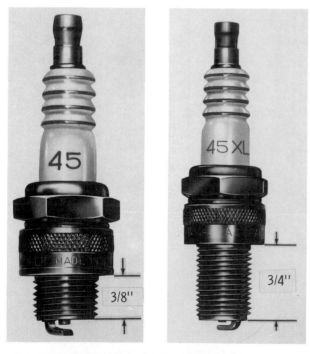

Fig. 6-17. Spark plug reach (length of thread) can vary considerably from one plug to another. Too long a reach can damage a piston. Too short a reach provides poor combustion. (AC Spark Plug Div., GMC)

Never use a spark plug that has a longer reach than specified, Fig. 6-18. Serious engine damage can result if the piston hits the plug.

Several standard thread sizes are commonly used. Threads on some spark plugs are metric sizes, usually 14 mm.

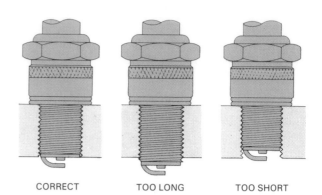

CORRECT   TOO LONG   TOO SHORT

Fig. 6-18. Spark plug reach is determined by thickness of cylinder head.

SPARK PLUG HEAT TRANSFER

Heat transfer in spark plugs is an important consideration. The heat of combustion is conducted through the plug as shown in Fig. 6-19. Spark plugs are manufactured in various **heat ranges** from "hot" to "cold," Fig. 6-20. Cold running spark plugs are those which transfer heat readily from the firing end. They are used to

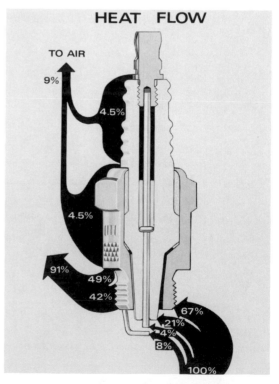

Fig. 6-19. Heat of ignition and combustion must be conducted away from critical parts of spark plug to prevent preignition and burning of electrodes.

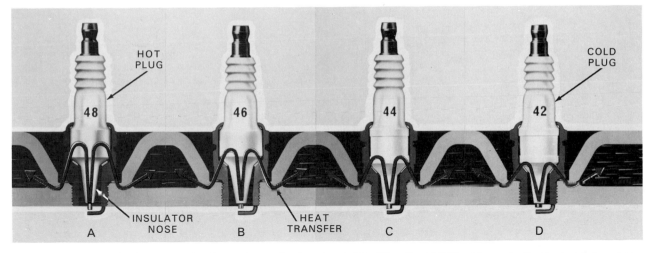

Fig. 6-20. Spark plug heat transfer determines whether plug is "hot" or "cold." Heat is controlled by insulator nose. (AC Spark Plug Div., GMC)

avoid overheating in engines having high combustion temperatures.

In figuring spark plug heat range, the length of the insulator nose determines how well and how far the heat travels. The spark plug at A in Fig. 6-20, for example, is a hot plug because the heat must travel a greater distance to the cylinder head. Spark plug at D in Fig. 6-20 is comparatively colder than A. A cold plug installed in a cool running engine would tend to foul. Cool running usually occurs at low power levels, continuous idling, or in start/stop operation.

The tip of the insulator is the hottest part of the spark plug and its temperature can be related to preignition (firing of fuel charge prior to normal ignition) or plug fouling. Experiments show that if combustion chamber temperature exceeds 1750°F (954°C) in a four cycle engine, preignition is likely to occur. If insulator tip temperature drops below 700°F (371°C), fouling or shorting of the plug due to carbon is likely to occur.

MEASURING SPARK PLUG TEMPERATURE

Specially assembled thermocouple spark plugs are used to accurately determine spark plug temperatures during actual engine operation, Fig. 6-21. These plugs have a small temperature sensing element embedded in the insulator tip and serve as a valuable aid to engineers in gathering spark plug and combustion data.

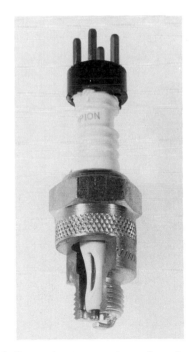

Fig. 6-21. Operating temperature of spark plug can be studied with a special spark plug having a thermocouple (heat sensor) installed in it. (Champion Spark Plug Co.)

TYPES OF ELECTRODES

Electrode configurations vary considerably. In Fig. 6-22, plug A is a retracted gap type used in some engines where clearance is a problem or protection of the firing tip is desirable. Plug B is a surface gap type, which is extremely cold.

Ignition Systems 93

Fig. 6-22. Many electrode designs are available for engine use. Surface gap spark plug is extremely ''cold'' and is finding application with capacitor discharge ignition systems. (AC Spark Plug Div., GMC)

Surface gap spark plugs are sometimes used in engines with capacitive discharge ignition systems. Plug C is a clipped gap type, in which the side electrode extends only part way across the center electrode. Plug D is a standard gap automotive spark plug.

BREAKER POINTS

Some ignition systems use mechanical *breaker points* to control primary current to the coil. The breaker points generally consist of two tungsten contacts. One contact point is stationary, the other is movable. Each contact is fastened to a bracket.

Tungsten is a hard metal with a high melting temperature. These characteristics are needed to withstand the continual opening and closing that takes place and the eroding effect of the arc that occurs when the points "break" (start to open). Fig. 6-23 shows the breaker point assembly with the dust cover removed.

The breaker point assembly is an electrical switch. When the points are closed, the magnetic field created by the flywheel magnets is being cut by the primary coil winding. This induces current in the primary circuit, and the primary winding builds its own magnetic field around the coil.

When the breaker opens, current flow in the primary circuit stops and the magnetic field collapses. Immediately upon collapse, a surge of

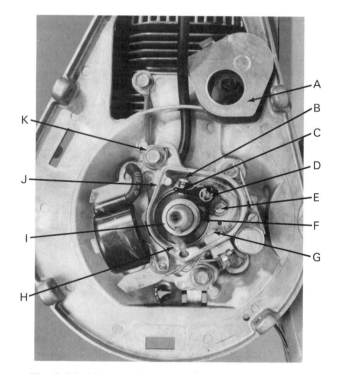

Fig. 6-23. Magneto breaker point assembly. A–Dust cover. B–Stationary point. C–Movable point. D–Pivot pin. E–Gap adjustment screw. F–Wear block. G–Breaker point spring. H–Felt lubricator. I–Cam block. J–Point housing. K–Advance adjustment screw.

high voltage induced in the secondary winding of the coil forces the electrons to jump the spark plug gap. This process is repeated every time the breaker points close and open. The spark plug sparks only at the instant the breaker points open.

CONDENSER CONSTRUCTION AND OPERATION

The *condenser* plays an important part in mechanical breaker point ignition system (MBI) operation. Its primary purpose is to prevent current from arcing across the breaker point gap as the points open. If arcing would occur, it would burn the points and absorb most of the magnetic energy stored in the ignition coil. Not enough energy would be left in the coil to produce the necessary high voltage surge in the secondary circuit.

The condenser absorbs current the instant breaker points begin to separate. Since the condenser absorbs most of the current, little is left to form an arc between the points. A condenser must be selected that has just the right capacitance to absorb the amount of energy required to produce an arc.

The construction of the condenser is quite simple. Fig. 6-24 shows a condenser partially opened up to expose laminations of aluminum foil and an insulating strip. Two strips of aluminum foil of specific length are wound together with the insulator strip between them. One foil strip is grounded, while the other is connected across the breaker points.

THE MBI MAGNETO SYSTEM

Magneto systems supply the ignition spark on most small engines. A magneto system will produce current for ignition without any outside primary source of electricity.

Major components of a typical flywheel magneto ignition system are illustrated in Fig. 6-25. Except for the spark plug, the coil, condenser and breaker points may be found inside or outside of the flywheel. This varies with engine type, but the principles of operation remain basically the same.

Magnets are usually cast into the flywheel and cannot be removed. They are strong permanent magnets made of *Alnico* (aluminum, nickel,

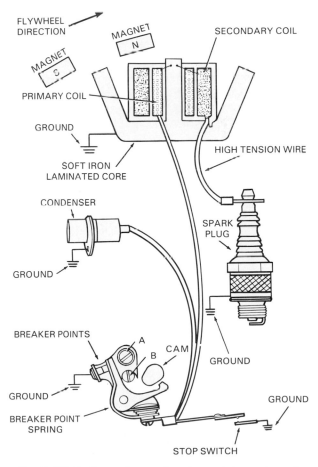

Fig. 6-25. Coil, condenser, breaker points, and spark plug make up primary and secondary circuits of this magneto system. Magnets create current flow in primary winding of coil, which induces current in secondary winding.

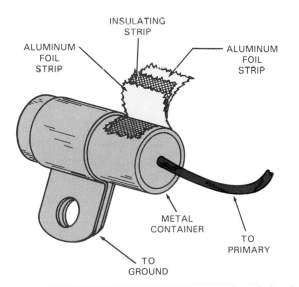

Fig. 6-24. Condenser consists of two strips of aluminum foil separated by an insulating material (dielectric). One foil strip is connected to metal container, other is attached to primary lead. (Kohler Co.)

cobalt alloy) or of a newer ceramic magnetic material.

The coil in Fig. 6-25 is cut away to show primary and secondary windings. The primary winding usually has about 150 turns of relatively heavy copper wire. The secondary winding has approximately 20,000 turns of very fine copper wire. One end of primary and one end of secondary are grounded to the soft iron laminated core which, in turn, is grounded to engine.

The odd shape of the core of the coil is designed to efficiently direct the magnetic lines of force. The spacing (air gap) between the magnets and the core ends is critical and can greatly affect the whole system. This gap can be checked with special gages or standard feeler gages. The high tension wire is heavily insulated because it carries high voltage. If the insulation deteriorates, much

of the voltage can be lost by arcing to nearby metallic parts of the engine.

In the system illustrated in Fig. 6-25, the breaker points are mechanically actuated, opened by the cam and closed by the breaker point spring. The breaker point gap is adjusted by loosening clamp screw A and turning eccentric screw B to move the stationary point. Remember that some magneto systems use solid state switching devices instead of mechanical breaker points.

THE MBI MAGNETO CYCLE

As the flywheel turns, the magnets pass over the legs of the laminated core of the coil. When the north pole of the magnet is over the center leg of the coil, the magnetic lines of force move down

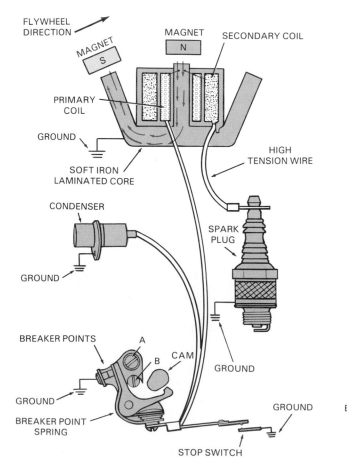

Fig. 6-26. As flywheel turns and magnets align with legs of laminated core of coil, a magnetic field is conducted through the primary winding.

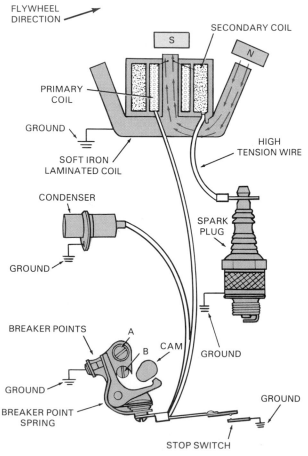

Fig. 6-27. As flywheel continues to turn, magnets realign with center and outside leg of core, causing magnetic field to reverse and induce low voltage in primary coil.

the center leg through the coil, across the bottom of the lamination and up the side leg to the south pole. See Fig. 6-26.

As the flywheel continues to turn, Fig. 6-27, the north pole of the magnet comes over the side leg and the south pole is over the center leg of the core. Now the lines of force move from the north pole down through the side leg, up through the center leg and coil, and to the south pole. At this point, the lines of force have reversed direction.

Fig. 6-28 shows the field reversal taking place in the center leg of the core and coil. The reversal induces low voltage current in the primary circuit through the breaker points. Current flowing in the primary winding of the coil creates a primary magnetic field of its own, which reinforces and helps maintain the direction of the lines of force in the center leg of the lamination. It does this until the magnets' poles move into a position

where they can force the existing lines of force to change direction in the center leg of the lamination. Just before this happens, the breaker points are opened by the cam.

Opening of the points breaks the primary circuit and the primary magnetic field collapses through the turns of the secondary winding. See Fig. 6-29. The condenser makes the breaking of the primary current as instantaneous as possible by absorbing the surge of primary current to prevent arcing between the breaker points.

As the magnetic field collapses through the secondary winding of the coil, high voltage current is induced in the secondary winding. At exactly the same time, the charge stored in the condenser surges back into the primary winding,

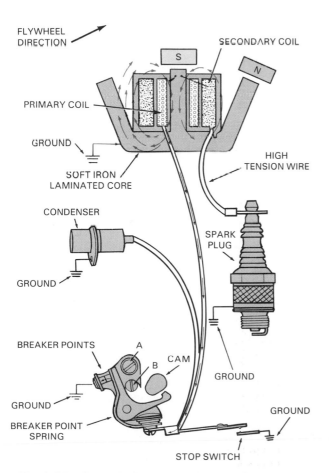

Fig. 6-28. Magnetic field reversal takes place as magnets pass from left to right. Breaker points are closed.

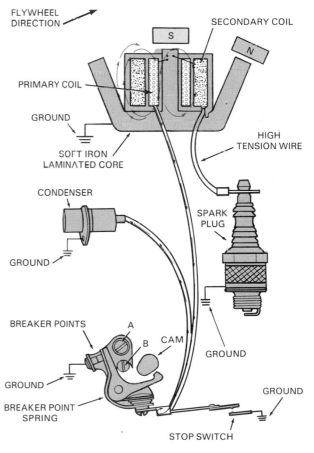

Fig. 6-29. Breaker points open, causing primary magnetic field to collapse at an extremely high rate through secondary winding. This induces high voltage required to fire spark plug. Field collapse also cuts through windings of primary coil, inducing moderate voltage that is absorbed by condenser.

Fig. 6-30, and reverses the direction of current in the primary windings. This change in direction sets up a reversal in direction of the magnetic field cutting through the secondary and helps increase the voltage in the secondary circuit. The potential of the high voltage causes secondary current to arc across the spark plug gap.

THE STOP SWITCH

The spark plug can only fire when the ignition points open the primary circuit. Using this as a basis for a stop switch, the switch is designed to ground the movable breaker point so that, in effect, the points never open. See Fig. 6-31. Therefore, the engine stops running.

Another common method of stopping single cylinder engines is by means of a strip of metal fastened to one of the cylinder head bolts. When the engine is running, the strip is suspended about 1/2 in. from the spark plug wire terminal. By depressing the strip against the plug wire, the current flows down the strip to the cylinder head to prevent a spark at the plug. There is no danger of shock to the operator. CAUTION: Do not touch the spark plug directly.

IGNITION ADVANCE SYSTEMS

Some small engines have mechanical systems that retard occurrence of spark for starting. For intermediate and high speed operation, the *ignition advance system* causes spark to occur earlier in the cycle.

One type of ignition advance system is illustrated in Fig. 6-32. Two different spark timings are provided, one for starting and one for running. For starting, the spark-advance flyweight

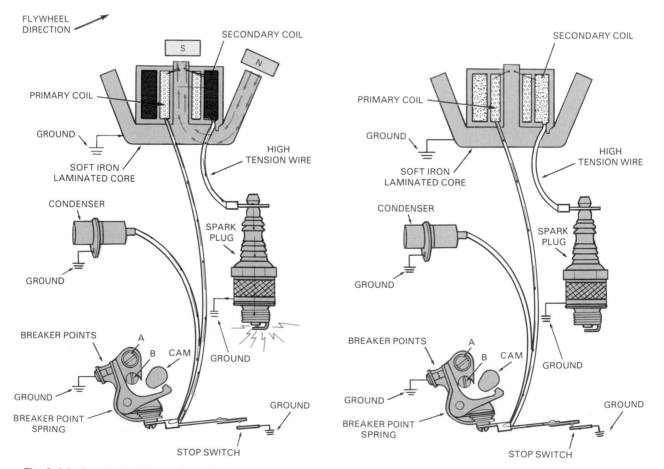

Fig. 6-30. Spark plug fires and condenser discharges voltage back into primary circuit.

Fig. 6-31. When stop switch is closed, breaker point system is grounded and engine is stopped.

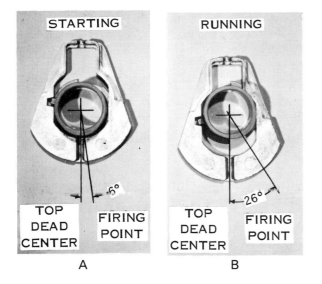

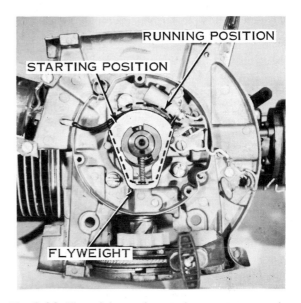

Fig. 6-32. A–For starting engine, flyweight is held in retarded position by a spring. B–When engine reaches 1000 rpm, flyweight overcomes spring and rotates cam to advanced position. (Lawn-Boy Power Equipment, Gale Products)

Fig. 6-33. Flyweight, spring, and cam rotate together as a single assembly. Outer position of flyweight is for advanced operation.

holds the cam in a position so that the ignition spark occurs at 6° of crankshaft rotation before the piston reaches top dead center (TDC). See Fig. 6-32A.

When the engine reaches a speed of nearly 1000 rpm, centrifugal force moves the flyweight out, forcing the cam to rotate. This position of the cam causes the points to open and a spark to occur at 26° before top dead center, Fig. 6-32B.

Fig. 6-33 shows the flyweight and cam assembled on the crankshaft in starting (solid lines) and running (dashed lines) positions.

DWELL AND CAM ANGLE

Dwell (cam angle) is the time the breaker points stay closed during one revolution of the cam. **Dwell** is the number of degrees measured around the cam from the point of closing to the point of opening. Fig. 6-34 shows the direct relationship between the breaker point gap setting and dwell time.

Fig. 6-34A indicates what might be considered normal dwell time. Fig. 6-34B shows a large point gap with a correspondingly short dwell. Note that the wear block must travel quite a distance to a

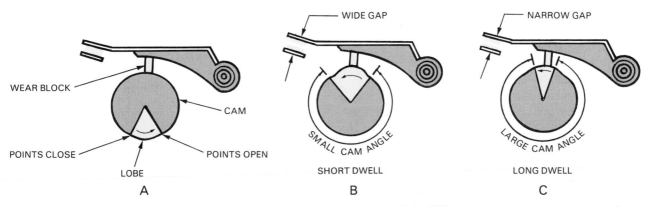

Fig. 6-34. Relationship of point gap and dwell. A–Normal gap and dwell. B–With a wider breaker point gap setting, dwell decreases. C–A narrower gap increases dwell.

lower position on the lobe before the points make contact. Then, they open with just the slightest rise in the lobe.

The narrow point gap illustrated in Fig. 6-34C causes the breaker points to close after only a slight travel down the lobe, but more distance is required up the lobe to open the points. Remember that the cam is driven directly from the crankshaft. When the breaker points open, the spark plug fires.

Obviously, then, changing the point setting can also change spark timing. The engine manufacturer specifies which gap setting is best (usually between .020 to .030 in.) and the number of degrees before top dead center (BTDC) that the spark should occur.

SOLID STATE IGNITION

Solid state is a broad term applied to any ignition system that uses electronic semi-conductors (diodes, transistors, silicon controlled rectifiers, etc.) in place of one or more mechanical ignition components.

Electronic components in solid state systems are extremely small. Since there are no moving parts, mechanical adjustments are not required. Solid state ignition systems provide many advantages:

- Eliminate ignition system maintenance.
- No breaker points to burn, pit, or replace.
- Increase spark plug life.
- Easy starting, even with fouled plugs.
- A flooded engine will start easily.
- Higher spark output and faster voltage rise.
- Spark advance is electronic and automatic. It never needs adjusting.
- Electronic unit is hermetically sealed and unaffected by dust, dirt, oil, or moisture.
- System delivers uniform performance throughout component life and under adverse operating conditions.
- Improve idling and provide smoother power under load.

OPERATION OF CAPACITIVE DISCHARGE IGNITION (CDI) SYSTEM

The *capacitive discharge ignition (CDI) system* is a solid state ignition system. It is one of the newest systems used in an internal combustion engine. It is standard equipment in many applications and has improved the reliability of modern small gasoline engines.

The CDI ignition system is breakerless. The mechanical points and accessories are replaced with electronic components. The only moving parts are the permanent magnets in the flywheel. Fig. 6-35 shows the CDI module installed on a small gasoline engine.

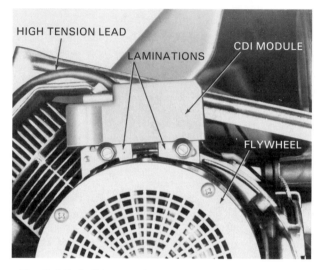

Fig. 6-35. Solid state, CDI ignition module is compact and maintenance free. Only moving parts are flywheel magnets.

The CDI module can be tested to see if it is producing a spark by grounding a known good spark plug to the engine. Turn the flywheel and watch the plug electrode gap for a spark. If there is no spark, the ignition switch or switch lead may be faulty; or the air gap may be incorrect (spacing between flywheel and CDI module). No other troubleshooting is necessary.

The ignition switch is the most vulnerable part of the solid state ignition system. The switch must be kept dry and clean.

Refer to the following illustrations to progressively trace current flow through the various electronic components in a typical CDI system.

In Fig. 6-36, flywheel magnets rotate across the CDI module laminations, inducing a low voltage alternating current (ac) in charge coil. The ac

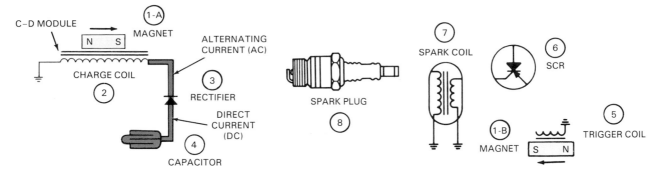

Fig. 6-36. Flywheel operation. 1A Magnets induce low voltage alternating current into charge coil at 2. 3–Rectifier changes alternating current to direct current. 4–Direct current from rectifier is stored in capacitor (condenser). See Fig. 6-39.

passes through a rectifier and changes to direct current (dc), which travels to the capacitor where it is stored.

In Fig. 6-37: The flywheel magnets rotate approximately 351° before passing the CDI module laminations and induce a small electrical charge in trigger coil. At starting speeds, this electrical charge is just great enough to turn on the silicon controlled rectifier (SCR) in a retarded firing position (9° BTDC). This provides for easy starting.

In Fig. 6-38: When the engine reaches approx-

imately 800 rpm, advanced firing begins. The flywheel magnets travel approximately 331°, at which time enough voltage is induced in the trigger coil to energize the silicon controlled rectifier in the advanced firing position (29° BTDC).

In Fig. 6-39: When the silicon controlled rectifier is triggered, the 300V dc stored in the capacitor travels to the spark coil. At the coil, the voltage is stepped up instantly to a maximum of 30,000V. This high voltage current is discharged across the spark plug gap.

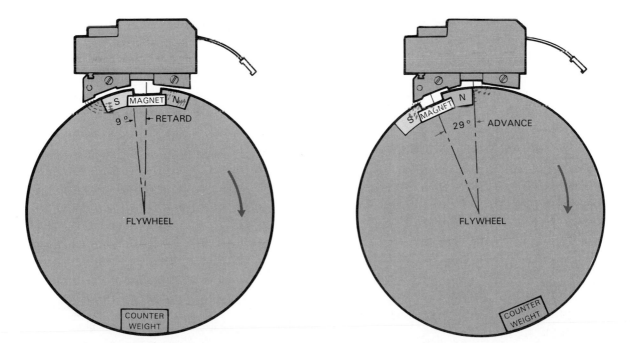

Fig. 6-37. At low speed, flywheel magnets induce a small current in trigger coil, which turns on silicon rectifier at 9° BTDC for easy starting.

Fig. 6-38. At 800 rpm, stronger trigger coil current turns on silicon rectifier at 29° BTDC for satisfactory ignition during normal engine operation.

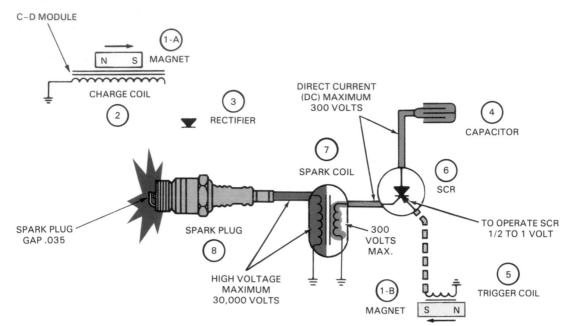

Fig. 6-39. 1B–Magnet induces a small current in trigger coil. 5–Trigger coil switches on silicon controlled rectifier. 6–Rectifier permits capacitor to discharge 300V into primary winding of spark coil. 7–Spark coil steps up voltage in secondary winding. 8–Spark plug fires.

Diode (D1, D2)	A ──▶├── K	Allows one way current only from Anode ''A'' to Cathode ''K'' as rectifier.
Flywheel		Provides magnetic flux to primary windings of ignition coil.
High-tension lead		Conducts high voltage current in secondary windings to spark plug.
Ignition coil		Generates primary current, and transforms primary low voltage to secondary high voltage.
Ignition switch		No spark across gap of spark plug when switch is at ''STOP'' position.
Resistor (R1, R2)	──/\/\/──	Resists current flow.
Spark plug		Ignites fuel-air mixture in cylinder.
Thyristor (S)	A ──◀├── K, G	Switches from blocking state to conducting state when trigger current/voltage is on gate ''G.''
Transistor (T, T1, T2)	B, C, E	Very small current in the base circuit (B to E) controls and amplifies very large current in the collector circuit (C to E). When the base current is cut, the collector current is also cut completely.

Fig. 6-40. Study components of transistor controlled ignition system.

OPERATION OF TRANSISTOR CONTROLLED IGNITION (TCI) SYSTEM

The individual components that make up the *transistor controlled ignition system* are given in a chart in Fig. 6-40. Study the function of each part carefully.

There are a variety of transistor controlled circuits. Each has its own unique characteristics and modifications. Fig. 6-41 illustrates a typical circuit for a transistor controlled ignition. Refer to this circuit as its principles are described.

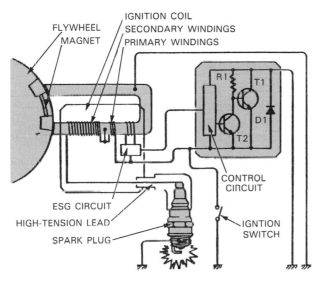

Fig. 6-41. Study how this transistor circuit is used to operate the ignition coil.

As the engine flywheel rotates, the magnets on the flywheel pass by the ignition coil. The magnetic field around the magnets induces current in the primary windings of the ignition coil.

The base circuit of the ignition system has current flow from the coil primary windings, common grounds, resistor (R1), base of the transistor (T1), the emitter of the transistor (T1), and back to the primary windings of the ignition coil.

Current flow for the collector circuit in Fig. 6-41 is from the primary windings of the coil, common grounds, collector of transistor (T1), emitter of transistor (T1), and back to the primary windings.

When the flywheel rotates further, the induced current in the coil primary increases. When the current is high enough, the control circuit turns on and begins to conduct current. This causes transistor (T2) to turn on and conduct. A strong magnetic field forms around the primary winding of the ignition coil, Fig. 6-41.

The trigger circuit for this ignition system consists of the primary windings, common grounds, control circuit, base of transistor (T2), and emitter of transistor (T2).

When transistor (T2) begins to conduct current, the base current flow is cut. This causes the collector circuit to shut off and transistor (T1) stops conducting current.

When transistor (T1) stops conducting, current stops flowing through the primary of the ignition coil. This causes the primary magnetic field to collapse across the secondary windings of the ignition coil. High voltage is then induced into the secondary to "fire" the spark plug.

The secondary circuit includes the coil secondary windings, high tension lead, spark plug, and common grounds returning to the coil secondary.

When the ignition switch is off, Fig. 6-41, the primary circuit is grounded to prevent the plug from firing.

Diode (D1) is installed in the circuit to protect the TCI module from damage.

The ESG circuit in Fig. 6-41 is used to retard the ignition timing. At high engine rpm, the ESG circuit conducts. This bypasses the trigger circuit and delays when the current reaches the base of transistor (T2).

MAGNETO IGNITION SYSTEMS COMPARED

The following are definitions for the three general classifications of magneto ignition systems.

1. A mechanical breaker ignition (MBI) system is a flywheel magneto inductive system commonly used for internal combustion engines. It employs mechanical breaker contacts to time or trigger the system.

2. A transistor controlled ignition (TCI) system is an inductive system that does NOT use mechanical breaker contacts. It utilizes semiconductors (transistors, diodes, etc.) for switching purposes.

COMPARISONS	MECHANICAL BREAKER IGNITION SYSTEM	TRANSISTOR CONTROLLED IGNITION SYSTEM	CAPACITOR DISCHARGE IGNITION SYSTEM
Abbreviation	MBI	TCI	CDI
Circuit type	Conventional	Solid state	Solid state
Energy source	Primary current of ignition coil	Primary current of ignition coil	Stored in capacitor
Trigger switch	Breaker contacts	Power transistor	Thyristor
Secondary voltage	Standard	Standard	Higher
Spark duration	Standard	Standard	Shorter
Rise time*	Standard	Standard	Shorter
Maximum operating speed	Standard	Higher	Higher
Maintenance	Regap and retime	None	None

*RISE TIME—time required for maximum voltage to occur.

Fig. 6-42. Chart compares breaker point, transistor controlled, and capacitor discharge systems. Note differences and similarities.

3. A capacitor discharge ignition (CDI) system is a solid state (no moving parts) system that stores its primary energy in a capacitor and uses semiconductors for timing or triggering the system.

Look at Fig. 6-42. It compares these three types of magneto ignition systems. Study them!

BATTERY IGNITION SYSTEMS

The *battery ignition system* has a low voltage primary circuit and a high voltage secondary circuit. Like the magneto system, it consists of a coil, condenser, breaker points (or solid state switching device), and spark plug. The basic difference is that the source of current for the primary circuit is supplied by a lead-acid battery. See Fig. 6-43.

When the ignition switch is turned on, current flows from the positive post of the battery to the ignition coil. Current traveling through the primary windings of the coil builds up a magnetic field, Fig. 6-44. During this time, the breaker points are closed. Ignition at the plug is not re-

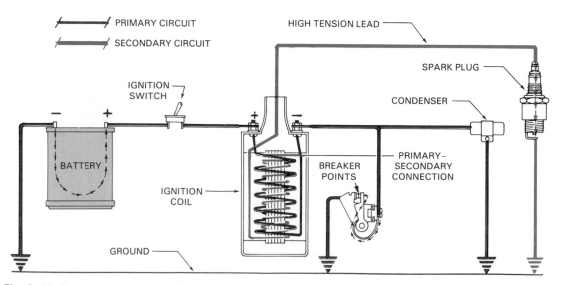

Fig. 6-43. Battery system is similar to a magnet system, except that battery replaces flywheel magnets.

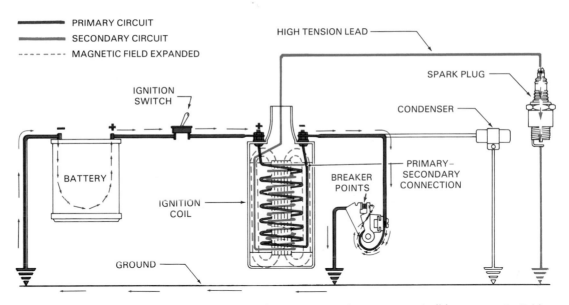

Fig. 6-44. When points close in battery ignition system, primary current builds a magnetic field around coil. (Kohler Co.)

quired, so the current returns to the battery through the common ground.

Then, at the exact time when ignition at the plug is required, the breaker points are opened by the cam. Current flow stops abruptly, causing the magnetic field surrounding the coil to collapse, Fig. 6-45. This rapid change of magnetic flux causes voltage to be induced in every turn of the primary and secondary windings. Voltage in the primary winding (about 250 volts) is quickly absorbed by the condenser.

Without the condenser, the current would arc at the breaker point gap, burning the points. The condenser acts as a reservoir for the sudden surge of power in the primary windings of the coil. The condenser holds the current for an instant, then releases it to the primary circuit, as shown in Fig. 6-46. Note that many battery ignition systems

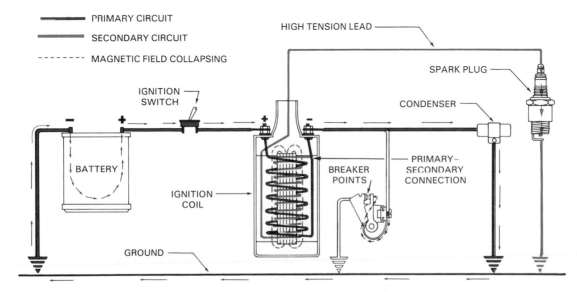

Fig. 6-45. Breaker points open and field collapses, inducing high voltage in secondary winding of coil. Condenser absorbs voltage surge in primary winding.

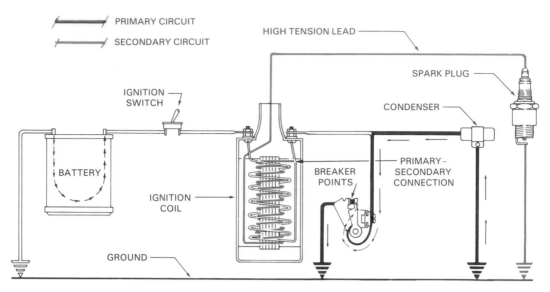

Fig. 6-46. Condenser discharges back into primary circuit.

employ solid state ignition components instead of mechanical breaker points.

HIGH VOLTAGE SECONDARY CURRENT

The voltage built up in the secondary winding of the coil can become as high as 25,000 volts. It has approximately 100 times as many turns of wire as the primary, Fig. 6-47. Normally, the voltage does not reach this value. Once it becomes great enough to jump the spark plug gap, the voltage drops. Usually, the amount required to jump the gap is between 6000 and 20,000 volts. The actual amount of voltage required depends upon variables such as compression, engine speed, shape and condition of electrodes, spark plug gap, etc.

AUTO-TRANSFORMER TYPE IGNITION COIL

The *auto-transformer* type ignition coil used on some small gasoline engines serves as a step-up transformer. It increases low voltage primary current to the high voltage required to bridge the spark plug gap. The primary and secondary windings are connected, and the common ground of the battery and primary circuit is used to complete the secondary circuit.

With this type of coil, very little primary current can flow into the secondary circuit because

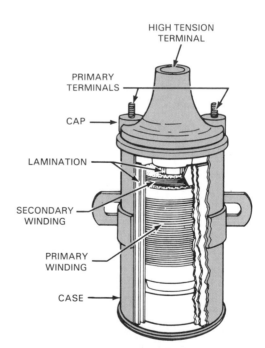

Fig. 6-47. Auto-type ignition coil has about 100 times as many turns in secondary winding as in primary winding. Coil can produce 25,000V, if needed.

the secondary circuit is normally open at the spark plug gap. Primary current is just not great enough to jump the gap. Therefore, the two circuits function separately.

The primary winding of the coil consists of about 200 turns of heavy copper wire. The secondary winding has approximately 20,000 turns of very fine copper wire. Because the magnetic field collapses through such a great number of conductors in the secondary, a very high voltage (electrical potential) is developed. The amperage (rate of electron flow), however, is proportionately low. This is a typical characteristic of any transformer.

Laminated iron is used as the center core of the coil, Fig. 6-47. It also forms the outer shell of the inner assembly, providing maximum concentration of the magnetic field. The inner assembly is sealed in a coil case, and the remaining space inside is filled with a special oil to minimize the effects of heat, moisture, and vibration.

The top of the coil is provided with two primary terminals. They are marked positive (+) and negative (-). The positive terminal must be connected to the positive side of the battery. The negative terminal connects to the breaker points, Fig. 6-46. The center tower of the coil contains the high tension terminal.

THE LEAD-ACID BATTERY

The battery is the sole source of energy for the battery ignition system of a small gasoline engine. A generator is used to replenish energy in the battery. However, the generator does not supply energy directly to the ignition system.

Lead-acid type batteries are used in battery ignition systems. The cell plates are made of lead, and a sulfuric acid and water solution serves as the electrolyte. "Wet" or "dry charged" types are available. Wet batteries are supplied with the electrolyte in them, ready for use if the charge has been kept up. Dry charged batteries must have electrolyte installed after purchase. Both types of batteries function in the same way.

BATTERY CONSTRUCTION

The typical 12V battery is constructed with a hard rubber case and six separate compartments called *cells*, Fig. 6-48. There is a specific number of negative and positive plates in each cell. The greater the number of plates per cell, the higher the ampere-hour rating (capacity to provide current for a specific length of time) of the bat-

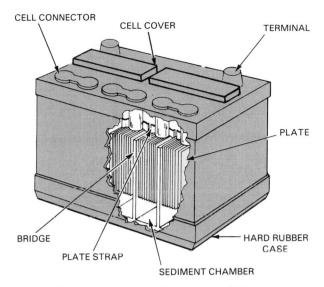

Fig. 6-48. Battery provides all current for battery ignition system. A 12V battery has six cells each producing 2V.

tery. The positive plates have a lead oxide covering. The negative plates have a porous or spongy surface.

BATTERY VOLTAGE

A chemical reaction causes each negative plate to lose electrons and each positive plate to gain electrons when surrounded by electrolyte. The plates, therefore, develop an electrical potential between them. All plates of a like charge are electrically connected, causing accumulative charges to be present at the positive and negative battery terminals.

Each cell of a battery in good condition contributes approximately 1.95 to 2.08V. Six fully charged cells will produce at least 12V. If not, the battery must be recharged or replaced.

A DISCHARGING BATTERY

When a battery discharges without replacement of energy, the sulfuric acid is chemically withdrawn from the electrolyte, specific gravity goes down, and lead sulfate deposits accumulate on the plates. If sulfate deposits become too great or the level of the electrolyte falls lower than the top of the plates, permanent damage may be done to the battery.

When a battery is recharged, a controlled direct current is passed through the battery in reverse direction from normal operation. This causes a reversal in chemical action and restores the plates and electrolyte to active condition.

SUMMARY

The primary purpose of the ignition system is to provide sufficient voltage to discharge a spark between the electrodes of a spark plug. Many small engine use magneto systems to supply ignition spark. Magneto systems are self-contained systems that produce electrical current for ignition without an outside primary source of electricity. Basic magneto system parts include: permanent magnets; high tension coil; breaker points, transistor, or capacitor; high tension spark plug wire; and spark plugs.

Using the correct spark plug can greatly increase engine efficiency and service life. Reach, heat range, and electrode consideration must all be considered.

There are three general types of magneto ignition systems. The mechanical breaker ignition system is a flywheel magneto inductive system. It employs mechanical breaker points to time triggering the ignition system. Some small engines are equipped with mechanical systems that retard and advance timing. Dwell (cam angle) is the amount of time that the breaker points stay closed during one revolution of the cam.

Many small engines are equipped with solid state ignition systems that use electronic devices in place of one or more mechanical ignition components. Most solid state systems have no moving parts and do not require mechanical adjustments. The two most common solid state systems are the capacitive discharge ignition (CDI) system and the transistor controlled ignition (TCI) system.

The CDI system stores primary energy in a capacitor and uses semiconductor devices to trigger the ignition system. The TCI system is an inductive system that utilizes semiconductor devices (transistors, diodes, etc.) for switching purposes.

Instead of a magneto, some ignition systems use a lead-acid battery to supply primary current. These systems generally employ an auto-type ignition coil. Because the battery is the only source of energy for battery ignition systems, a generator is used to replenish energy in the battery.

KNOW THESE TERMS

Ignition system, Magneto system, Electron theory, Amperes, Volts, Ohms, Ohm's law, Magnetism, Ignition coil, Spark plug, High tension lead, Spark plug insulator, Flashover, Center electrode, Reach, Heat range, Breaker points, Condenser, Tungsten, Stop switch, Ignition advance, Dwell (cam angle), Solid state, Capacitive discharge ignition, Transistor controlled ignition, Battery ignition systems, Auto-transformer type coil, Lead-acid battery, Cells.

REVIEW QUESTIONS–CHAPTER 6

1. The two major tasks performed by an ignition system are to _____ and _____.

2. If a four cycle engine runs at 3600 rpm, the number of sparks per minute required at the spark plug would be _____ per minute.

3. Name the six main electrical components that make up the magneto ignition system.

4. Regarding atoms:
   a. Electrons and protons are tightly bonded together.
   b. Electrons are positively charged particles.
   c. Neutrons are negatively charged, protons are positively charged.
   d. Electrons do not collide because they are negatively charged.

5. The electron theory states that:
   a. Like charges attract each other.
   b. Like charges repel each other.
   c. Only negative charges repel each other.
   d. Electrons orbit the nucleus because the protons repel the electrons.

6. Substances that have electrons which can move freely from atom to atom are said to be good:
   a. Conductors.
   b. Dielectrics.
   c. Insulators.
   d. Nonconductors.

7. List three good conductors of electrical current.
8. List three good insulating materials.
9. A source of electricity:
   a. Has an excess of electrons and is positively charged.
   b. Has an excess of electrons and is negatively charged.
   c. Lacks electrons and is negatively charged.
   d. Lacks electrons and is positively charged.
10. Electrical current in a battery flows from:
    a. Negative to positive.
    b. Positive to negative.
11. Electrical current in the primary circuit of a battery ignition system flows from:
    a. Negative to positive.
    b. Positive to negative.
12. What are the three ways in which an electrical potential can be produced?
13. Match the following units of electrical measurement:
    a. _____ Rate of electron flow.   Ohms.
                                            Volts.
    b. _____ Resistance to electron flow. Amperes.
    c. _____ Electrical potential.
14. In a circuit of 5 amperes and 2 ohms, the voltage would be _____.
15. In a circuit with 12 volts and 5 amperes, the resistance would be _____ ohms.
16. Soft iron used in the laminations of a coil or transformer has:
    a. Good magnetic retention.
    b. Poor magnetic retention.
    c. Nonmagnetic properties.
17. What happens when a coil of wire is passed through a magnetic field?
18. What happens when electric current is passed through a coil of wire?
19. In the ignition coil, the primary winding has:
    a. Many turns of fine wire.
    b. Few turns of fine wire.
    c. Few turns of heavy wire.
    d. Many turns of heavy wire.
20. The coil acts as a transformer that:
    a. Steps down the voltage and increases the output amperage.
    b. Steps up the voltage and amperage.
    c. Steps down the voltage and amperage.
    d. Steps up the voltage and decreases the output amperage.
21. The greatest resistance to the highest voltage in the ignition circuit occurs at:
    a. Spark plug.
    b. Breaker points.
    c. Condenser.
    d. Secondary of the coil.
22. Would a "cool" spark plug have a short or long insulator nose?
23. Spark plugs of the "cool" type should always be matched with cool running engines. True or False?
24. The device used by engineers and technicians to measure spark plug temperatures under running conditions is a _____ _____.
25. Breaker point contacts are made of a very hard material called _____.
26. When the breaker points in the magneto are closed:
    a. Current is induced in the primary circuit by the flywheel magnets and coil.
    b. The spark plug fires.
    c. A high voltage is induced in the secondary circuit.
    d. The condenser absorbs current from the primary circuit for later use in the secondary.
    e. All of the above are correct.
27. The spark plug fires when:
    a. Magnetic field collapses through the secondary windings of the coil.
    b. Piston is nearing top of compression stroke.
    c. Breaker points open.
    d. Voltage potential is high enough to bridge the electrodes.
    e. All of the above.
28. The condenser acts as a voltage reservoir or shock absorber for the primary circuit. Its primary purpose is to prevent _____.
29. When the breaker points open, voltage in the primary circuit may surge to as much as _____.

30. The stop switch grounds the _____.
31. When connecting the auto-type ignition coil in the circuit, the positive terminal of the battery must be connected to:
    a. The positive terminal of the coil.
    b. The negative terminal of the coil.
    c. Either terminal of the coil.
32. When breaker points are set with a wider gap, the dwell:
    a. Becomes greater.
    b. Becomes less.
    c. Does not change.
33. Name five advantages of a solid state ignition system.

SUGGESTED ACTIVITIES

1. Experiment with a coil of wire and a magnet to induce a current. Use a galvanometer to show the current flow.
2. With a dry cell, iron filings, copper wire and a piece of paper, demonstrate the magnetic flux produced when the current flows through the wire.
3. Using copper wire, form it into a loose coil. Use a dry cell to pass a current through the wire and determine the polarity with a permanent magnet. Reverse the polarity.
4. Using the coil of wire from No. 3, wrap insulating tape around an iron bar and place it in the coil. Demonstrate how this can improve the magnetic strength of the coil when current is flowing.
5. Make a visible magneto mounted on a display board or built into a clear acrylic box so that it can be manually turned with a crank. Old, but usable, engine parts can be used.
6. A workable battery ignition system can be built and mounted as a display board. Demonstrate the operation and principles involved in this system.
7. Make a large Ohm's law triangle and display it until it has been learned. Make up some problems that require use of the formulas.

8. Make a collection of various kinds of spark plugs.
9. Section a spark plug, so that the inner construction can be studied.
10. Cutaway a "dry" type battery that has not had electrolyte added to it and discuss how it works.
11. Section an old ignition coil to show the primary and secondary windings around the core.
12. Carefully open a condenser to display the lamination of aluminum foil and insulation.
13. Collect several engines with different types of ignition advance mechanisms. Demonstrate them to the class.
14. Disassemble a magneto and demonstrate how it works.

CHAPTER 7

ENGINE LUBRICATION

After studying this chapter, you will be able to:
- ☐ Define friction and explain how it affects internal engine components.
- ☐ List the functions of lubricating oil.
- ☐ Differentiate between the lubrication systems in two cycle engines and four cycle engines.
- ☐ Explain the operation of ejection pumps, barrel pumps, and positive displacement pumps.
- ☐ Explain the function of oil filter systems and differentiate between the three main types.

Lubrication is the process of reducing friction between sliding surfaces by introducing a slippery or smooth substance between them, Fig. 7-1. Lubricants come in dry (powdered), semi-dry (grease), and liquid (oil) forms. Oil is the most important lubricant for small engine use, simply because it is often the only lubrication the engine needs, Fig. 7-2.

FRICTION

Friction is the resistance to motion created when one dry surface rubs against another. Even highly polished metal surfaces have irregularities

Fig. 7-2. Oils, grease, and graphite are commonly used in engine lubrication.

(when studied under a microscope) that will create a great deal of friction if rubbed together. The microscopic roughness will resist movement and create heat.

As the relatively rough projections on the contact surfaces rub across each other, they eventually break off and become loose particles. These particles, in turn, work between the contact surfaces and gouge grooves in the metal. Then, as friction and heat increase, the metal parts expand, causing greater pressure between the surfaces and creating even greater friction. This condition of wear exists until the parts either weld themselves together or seize (expand so much that mating parts cannot move).

In some cases, the excessively worn parts lose so much material from their contact surfaces that they become too loose to function properly. When this happens, the scored part should be replaced.

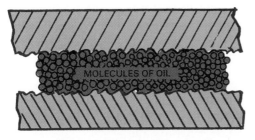

Fig. 7-1. A film of engine oil serves to separate and lubricate machined surfaces. When parts are in motion, oil molecules roll over one another like microscopic ball bearings.

PREVENTING WEAR DUE TO FRICTION

In designing a small gasoline engine, the manufacturer selects suitable materials for parts that will be in moving contact with one another. For example, precision insert bearing shells are used in connecting rods and caps and in main bearing saddles and caps, Fig. 7-3. The steel backing has a cast babbitt surface. Babbitt, an alloy of tin, copper and antimony, has good anti-friction qualities.

Rod and main bearing inserts must withstand reciprocating and rotational forces while fitting closely to the crank journal or throw. Notice that the oil hole in the insert in Fig. 7-3 permits oil to enter the oil groove. With the crankshaft in motion, the oil groove distributes oil all the way around the insert. Then, oil travels outward to bearing edges and back into crankcase.

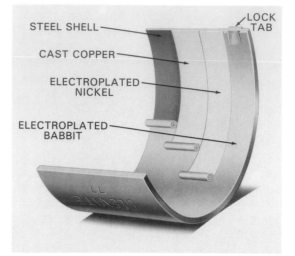

Fig. 7-4. Some precision inserts have several layers of material between steel shell and babbitt. (Clevite Corp.)

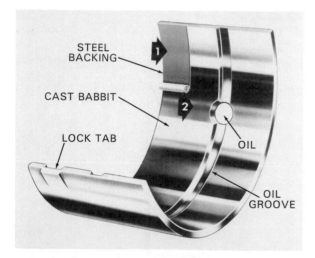

Fig. 7-3. Precision insert type of engine bearing has a steel shell with a cast babbitt inner surface. Oil hole and groove carries pressurized oil to bearing surfaces.

Some bearing inserts have laminations of other metals to back up the babbitt. Refer to Fig. 7-4. This construction enhances the properties of the insert, making it more heat resistant and durable.

The lock tab illustrated in Fig. 7-4 is used to prevent the bearing insert from rotating in the connecting rod or the main bearing cap. An insert with the tab in place in a rod assembly is shown in Fig. 7-5.

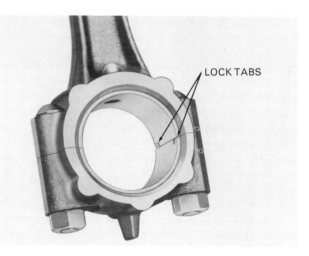

Fig. 7-5. Lock tabs prevent bearing insert from rotating in connecting rod. Tabs must be seated in recesses provided in rod. (Wisconsin Motors Corp.)

Regardless of the quality of the material used, all bearing surfaces in small gasoline engines must have oil separating moving parts that are in close contact. See Fig. 7-6. A thin film of engine oil must coat the area between: piston/piston rings and cylinder wall, piston pin and piston or connecting rod, valve stem and guide, valve tappet and guide, main and rod bearing inserts and crankshaft journals, etc.

The thin film of oil between the close fitting parts may be only a few molecules thick, yet this is enough to prevent the two metal surfaces from

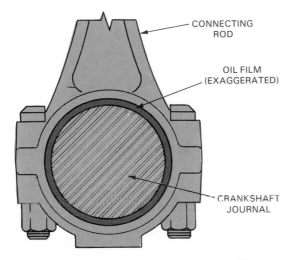

CONNECTING
ROD

OIL FILM
(EXAGGERATED)

CRANKSHAFT
JOURNAL

Fig. 7-6. Oil film between close fitting metallic parts prevents actual contact. Oil provides a relatively frictionless movement of parts.

actually touching. The molecules of oil then roll over one another, acting like microscopic ball bearings between the surfaces, Fig. 7-1.

LUBRICATING OIL

Modern motor oil is a highly specialized product, which has been developed by engineers and chemists to perform many essential functions in an engine. In order to operate efficiently, engines depend on motor oil to do the following:

- Permit easy starting.
- Lubricate engine parts.
- Protect against rust and corrosion.
- Keep engine parts clean.
- Cool engine parts.
- Seal combustion pressures.
- Prevent foaming.
- Aid fuel economy.

PERMITS EASY STARTING

The proper oil must be used if an engine is to start easily. If the oil is too thick, it will create so much drag between moving parts that the engine will not crank fast enough to start quickly and keep running.

Cold temperatures thicken oil. Oil for winter use must be thin enough to permit adequate cranking speeds at the lowest anticipated temper-

ature. Once the engine is started, the oil must be fluid enough to flow quickly to the bearings to prevent wear. On the other hand, the oil must also be thick enough to provide adequate protection when the engine reaches normal operating temperatures.

LUBRICATES AND PREVENTS WEAR

Once an engine is started, oil must circulate quickly to prevent the metal-to-metal contact that can cause wear, scoring, or seizure of engine parts, Fig. 7-1. Bearings and cylinder walls are particularly sensitive to movement, pressure, and oil supply. Oil supplies to these components must be continually replenished by adequate flow and distribution.

Once the oil reaches the moving parts, it must lubricate and prevent wear of the moving surfaces. The oil is expected to establish a complete, unbroken film between surfaces. Lubrication engineers call this full-film or *hydrodynamic lubrication.*

Under some conditions, it is impossible to maintain a continuous oil film between moving parts, and there is intermittent metal-to-metal contact between the high spots on sliding surfaces. Lubrication engineers refer to this *boundary lubrication.* When this occurs, the friction generated by the contact can produce enough heat to cause the metals to melt and weld together. Unless counteracted by proper additive treatment, the result is either part roughening or tearing.

Boundary lubrication always exists during engine starting and often exists during the operation of a new or rebuilt engine. Boundary lubrication is also found around the top piston ring where oil supply is limited, temperatures are high, and piston motion is reversed.

PROTECT AGAINST RUST AND CORROSION

Normally, burning fuel forms carbon dioxide and water. Gasoline engines, however, do not burn all of their fuel completely. Some of the partially burned gasoline undergoes complex chemical changes during combustion and, under some conditions, forms soot or carbon. Some of this soot and partially burned fuel escapes through the exhaust in the form of black smoke. Part of the soot and fuel escapes past the rings and into the crankcase. They tend to combine with water to form sludge and varnish deposits on critical engine parts. Sludge buildup can clog oil passages, reducing oil flow. Varnish buildup interferes with proper clearances, restricts oil circulation, and causes vital engine parts to stick, resulting in rapid engine failure.

Water causes a considerable problem in the engine. For each gallon of fuel burned, more than one gallon of water is formed. Although most of this water is in vapor form and escapes with the exhaust, some condenses on the cylinder walls or escapes past the piston rings and is trapped temporarily in the crankcase. This occurs most frequently in cold weather, before the engine reaches its normal operating temperature.

In addition to water, other corrosive combustion gasses get past the rings and are condensed or dissolved in the crankcase oil. Add to this the acids formed by the normal oxidation of oil, and the potential for rust and corrosive engine deposits becomes significant.

Engine life depends in part on the ability of motor oil to neutralize the effects of these corrosive materials. Due to extensive research by oil chemists, effective, oil-soluble, chemical compounds have been developed and are added to motor oil to provide protection for engine parts.

KEEPS ENGINE PARTS CLEAN

Engines are unable to tolerate excessive amounts of sludge and varnish on critical parts. Sludge can collect on oil pump screens and limit the flow of oil to vital engine parts. Varnish accumulation can cause piston rings to drag, preventing the engine from developing full power. Plugged oil-control rings prevent the removal of excess oil from the cylinder walls, resulting in excessive oil consumption.

Straight mineral oils have very limited ability to keep contaminants from forming masses of sludge in the engine. Therefore, various *detergent/dispersant additives* have been blended into

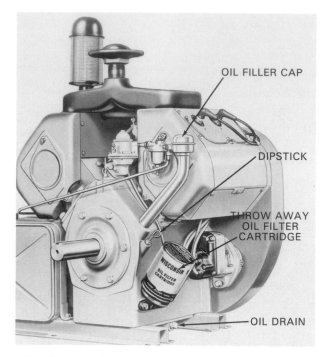

Fig. 7-7. An oil filter traps contaminants picked up by engine oil. Most disposable filters are located on side of engine. (Wisconsin Motors Corp.).

modern motor oils. These additives keep engine parts clean by suspending fine particles of oil contaminants until they can be trapped by the oil filter, Fig. 7-7.

Detergent/dispersants are also effective at preventing varnish deposits within an engine. Varnish-forming materials react chemically with the oxygen in the crankcase to form complex chemical compounds. These compounds, which react with each other and with oxygen, are baked into a hard coating on hot engine parts. Piston rings and bearings are particularly sensitive to varnish deposits. When deposits are formed on these components, engine operation can be severely impaired.

COOLS ENGINE PARTS

The crankshaft, camshaft, timing gears, bearings, pistons, and other components in the lower part of the engine are directly dependent on the motor oil for cooling. Each of these parts has definite temperature limits that must not be exceeded. Some can tolerate fairly high temperatures, while others, such as the main and connecting rod bearings, must run relatively cool to avoid failure. These parts must get an ample supply of cool oil to absorb heat and transfer it back to the crankcase.

SEALS COMBUSTION PRESSURES

The surfaces of the piston rings, ring grooves, and cylinder walls are not completely smooth. When examined under a microscope, these surfaces consist of minute hills and valleys. Therefore, the rings cannot completely prevent high combustion and compression pressures from escaping into the low pressure area of the crankcase. This results in a reduction in engine power and efficiency. Motor oil fills in the hills and valleys between ring surfaces and cylinder walls and helps seal compression and combustion pressures. Because the oil film at these points is quite thin, it cannot compensate for excessively worn rings, ring grooves, or cylinder walls. When such conditions exist, oil consumption may be high. New and rebuilt engines may also consume excessive amounts of oil until the hills and valleys on the surfaces have smoothed out enough to allow the oil to form a good seal.

PREVENTS FOAMING

Modern engine oil has anti-foaming additives to prevent it from whipping into bubbles that do not readily collapse. Oil foam does not cool well and does not provide adequate lubrication.

AIDS FUEL ECONOMY

Oils are available that have been formulated to improve fuel economy in gasoline-fueled engines. The fuel economy benefits are achieved by various means, including the use of friction-modifier additives.

OIL SELECTION

The cost of oil for a small gasoline engine is relatively low. However, the particular oil selected for use in a small engine is extremely important to the life of the engine, Fig. 7-8.

Fig. 7-8. Use a sealing cap to keep engine oil clean during storage when a full can is not used.

The oil recommended for use in a given engine may be shown on the engine nameplate or on a special label attached to the engine. The operator's manual also carries the manufacturer's recommendation, as does the lubrication guide provided by major oil companies.

Specifications for engine oils are given in two ratings:
1. SAE (Society of Automotive Engineers) Viscosity, referred to as "viscosity grade."
2. API (American Petroleum Institute) Service Classification, often referred to as the "type" of oil.

SAE VISCOSITY GRADE

Viscosity must be considered when selecting engine oil. *Viscosity* is a measure of the oil's resistance to flow. This resistance keeps the oil from being squeezed out from between engine surfaces as they move under load or pressure. The resistance to flow is a function of the molecular structure of the oil. Because it is this resistance that causes most of the drag during starting, it is important to use an oil with viscosity characteristics that ensure satisfactory cold cranking, good oil circulation, and adequate temperature protection.

The Society of Automotive Engineers (SAE) has established a viscosity range classification system for engine lubricating oils. All motor oils are classified according to this system, which is used worldwide. Each oil is assigned an SAE grade (or grades) that signifies the range into which it falls. Single grade motor oils commonly used today are SAE 5W, 10W, 15W, 20W, 20, 30, 40, and 50. Thick, slow-flowing oils have high numbers. Thin, free-flowing oils have low numbers. The "W" denotes oils suitable for use at low ambient (encompassing) temperatures. SAE numbered oils that do not have "W" designations are measured for viscosity at 212°F (100°C) to ensure adequate viscosity at normal engine operating temperatures.

Fig. 7-9 compares the viscosity recommendations of five manufacturers for their four-cycle engines at various operating temperatures. Note that the higher viscosity oils (more resistant to flow) are recommended for higher temperatures. A thick oil in low temperature operation makes a cold engine very difficult to start and may deprive critical parts from adequate lubrication while the oil is gaining heat from combustion. "Cold running" can result in scored cylinder walls and engine bearings.

The temperature effect on viscosity varies widely with different types of oils. A standard has been developed for measuring the relationship between viscosity and temperature. This standard is called the viscosity index (V.I.). Oil with a high viscosity index shows little change in viscosity over a wide range of temperatures, Today, through the use of selective crude oil stocks, new refining methods, and special chemical additives, there are many high viscosity index oils that are light enough to provide easy cranking at low temperatures and heavy enough to perform satisfactorily at high temperatures. These oils, which meet the viscosity requirements of two or more SAE grades, are known as *multigrade* or *multi-viscosity oils.* Examples are oils labeled as SAE 5W-20, SAE 5W-30, and SAE 10W-30.

Some of the oils listed in Fig. 7-9 are single viscosity grade oils, such as SAE 20. Others are multi-viscosity grade oils such as SAE 5W-20. Although multi-viscosity oils can be substituted for single viscosity grades in four cycle engines, they should not be used in two cycle engines.

API ENGINE SERVICE CLASSIFICATION SYSTEM

The *API Engine Service Classification System* is a dynamic arrangement that allows new categories to be added as engine designs change, placing more demands on motor oil. Currently, the classification system defines 13 categories of engine oil service. Of these categories, only SE, SF, SG,

FOUR-CYCLE CRANKCASE LUBRICATION (VISCOSITY-GRADE) RECOMMENDATIONS OF MANUFACTURERS

MANUFACTURER	Above 40°F.	Above 32°F.	Below 5°F.	Below 0°F.	Below -10°F.
Briggs and Stratton	SAE 30 or 10W-30	5W-20 or 10W			
Clinton	SAE 30			10W	5W
Kohler	SAE 30		SAE 10W	5W or 5W 20	
Wisconsin	SAE 30	SAE 20 or SAE 20W	10W		
Tecumseh	SAE 30		10W-30		

Fig. 7-9. A comparison of viscosity-grade recommendations by five engine manufacturers. Recommendations are for specific models only, not for full-line coverage.

I. Passenger Cars, Vans, and Light-Duty Trucks

SA	Obsolete
SB	Obsolete
SC	Obsolete
SD	Obsolete
SE	Obsolete
SF	Active
SG	Active (Primary recommendation)

II. Heavy-Duty Diesel (Commercial and Fleet Service)

CA	Obsolete
CB	Obsolete
CC	Active
CD	Active
CD-II	Active (Primary 2 cycle diesel recommendation)
CE	Active (Primary 4 cycle heavy duty diesel recommendation)

Fig. 7-10. American Petroleum Institute (API) has published oil service classifications in which oils are recommended for specific service conditions.

CC, CD, CD-II, and CE are considered suitable for any road vehicle in operation today. All other categories are considered obsolete for automotive and heavy-duty truck service. They are shown for historical purposes only. Category SE became obsolete after January 1, 1989. See Fig. 7-10. The typical applications of the active categories include:

- **SF** (1980 gasoline engine warranty maintenance service).
  Service typical of gasoline engines in passenger cars and some trucks beginning with 1980 models operating under engine manufacturer's warranties.
- **SG** (1989 gasoline engine warranty maintenance service).
  Service typical of present gasoline engines in passenger cars, vans, and light duty trucks operating under manufacturer's recommended maintenance procedures.
- **CC** (diesel engine service).
  Service typical of certain naturally aspirated, turbocharged, or supercharged diesel engines operated in moderate to severe service. Also for use in certain heavy-duty gasoline engines.
- **CD** (diesel engine service).
  Service typical of certain naturally aspirated, turbocharged, or supercharged diesel engines where highly effective control of wear and deposits is vital. Also used when using a wide quality range of fuels including sulfur fuels.

- **CD-II** (severe duty two-stroke cycle diesel engine service).
  Service of typical two-stroke cycle diesel engines requiring highly effective control over wear and deposits.
- **CE** (1983 diesel engine service).
  Service typical of certain naturally aspirated, turbocharged, or supercharged heavy-duty diesel engines manufactured since 1983 and operated under both low-speed, high-load and high speed, high load conditions.

THE API ENGINE OIL SERVICE CLASSIFICATION SYMBOL

The API engine oil service classification symbol provides the consumer with information about an oil's characteristics and applications, Fig. 7-11. The symbol typically appears on the

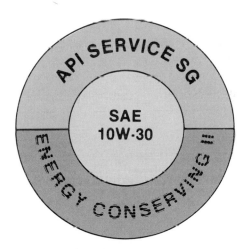

Fig. 7-11. Typical API engine oil service classification symbol for one type of SAE 10W-30 oil.

label of an oil container. It may also be found on the engine's oil fill cap or in the owner's manual. The symbol is divided into three parts including:

1. The top portion specifies the oil's service classification or recommended applications.
2. The center describes the oil's viscosity.
3. The bottom section is reserved for information on the oil's demonstrated fuel saving properties. Oils labeled "energy conserving II" offer better fuel-saving properties than oils labeled "energy conserving."

ENGINE LUBRICATION

The way that moving parts are lubricated differs in two cycle and four cycle small gasoline engines. In preparing a two cycle engine for use, a specified amount of two cycle engine oil is mixed with each gallon of gasoline to provide fuel for the engine. This fuel is thoroughly mixed in a separate container before being poured into the fuel tank, Fig. 7-12. In operation, an oil mist is created that lubricates the cylinder wall and all internal engine parts.

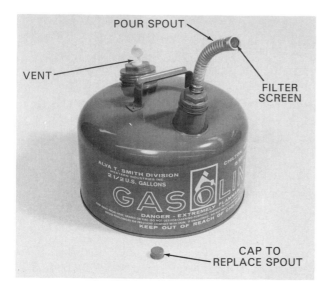

Fig. 7-12. Fuel for small gasoline engines should be stored in a clean, properly marked can with a vent cap and pouring spout.

In readying a four cycle engine for use, the fuel tank is filled with fresh gasoline. In addition, a specified amount of four cycle engine oil (type and viscosity class recommended by the manufacturer) is poured into the crankcase of the engine. In operation, the gasoline-air mixture is "fired" in the combustion chamber. At the same time, the oil sump in the crankcase supplies lubrication for the cylinder wall and all of the internal engine parts.

TWO CYCLE ENGINE LUBRICATION

Air-cooled engine operation covers a wider range of varying speeds with much higher combustion chamber temperatures than water-cooled engines. Therefore, automotive engine oils are NOT suitable for two cycle, air-cooled engines. Likewise, multi-viscosity and detergent type oils may contain additives not intended for two cycle engine operation and should not be used. *Always use the type and viscosity grade of oil recommended by the engine manufacturer.*

Certain manufacturers recommend the use of a specific SAE viscosity, diluted, two cycle engine oil. The diluent is added to make the oil pour more freely, particularly at low temperatures. As engine oil gets colder, one of its ingredients forms interlocking crystalline structures, which bind together until the oil finally solidifies. The diluent blocks the formation of these crystals, and the oil stays fluid.

Special additives are often recommended for two cycle engine use. Since two cycle engines are lubricated by mixing oil with the fuel, the oil eventually enters the combustion chamber and is burned. Some regular engine oils have additives that do not burn completely and leave a residue that fouls spark plugs and clogs exhaust ports.

Fig. 7-13. Spark plugs A, B, C, D were used in identical test engines using different oils. Plug A was taken from engine using oil recommended by engine manufacturer.

The spark plugs shown in Fig. 7-13 were used in four test engines. Each engine ran for the same length of time, and identical preventive maintenance and adjustments were performed. The only exception in the test procedures was in the brand of engine oil used. Spark plug A in Fig. 7-13 was taken from an engine that used the oil recommended by the manufacturer. The other three spark plugs were removed from test engines operating with other brands of engine oil.

Not all engine oils will produce deposits of the type shown in Fig. 7-13, but the importance of using the recommended oil should be quite clear from this comparison. Special additives for two cycle oils must be selected to avoid or prevent unburned deposits. Oils containing these additives generally are sold under the brand name of the engine manufacturer.

FOUR CYCLE ENGINE LUBRICATION

Four cycle engines must be operated with the proper "type" of oil. For this reason, always use engine oil service classifications that have been established by the Society of Automotive Engineers and the American Petroleum Institute.

Most four cycle engine manufacturers recommend oils supplemented by additives. These chemicals are added to improve the quality of the oil. They may prevent corrosion, provide a better cushioning effect between moving parts, help prevent scuffing, and/or reduce wear.

Detergent/dispersants, commonly called "detergents," are added to some oils. Detergents suspend the dirt and sludge in the oil, where the contaminants can be trapped by a filter or readily drained before fresh oil is put in. Basically, if the recommended oil is used in an engine, special oil treatments should not be required.

SPLASH LUBRICATION SYSTEM

Small, four cycle, gasoline engines generally use some type of *splash lubrication system* to lubricate internal machined surfaces. The splash lubrication system shown in Fig. 7-14 features an oil dipper arm on the connecting rod cap. The *dipper* is designed to pick up oil from the crankcase on every revolution of the crankshaft, splashing oil on the various moving parts as it is carried around by the crank throw.

With the splash system, the cylinder wall receives a generous amount of oil. To avoid oil-burning problems, the oil control ring on the piston removes excess oil, returning it to the crankcase as illustrated in Fig. 7-15. The connecting rod bearings and piston pin receive lubrication through oil passage holes, Fig. 7-16.

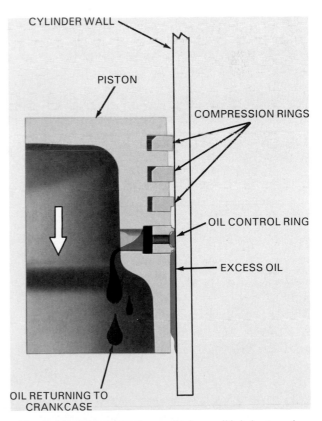

Fig. 7-15. Oil splashed on cylinder wall lubricates piston and piston rings. Excess oil is scraped from wall by oil control ring.

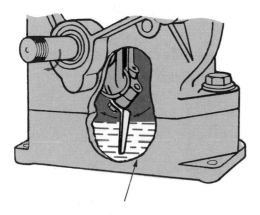

Fig. 7-14. With splash system, some oil dippers are cast onto connecting rod, others are bolted on. Oil level must be high enough for dipping action. (Briggs and Stratton Corp.)

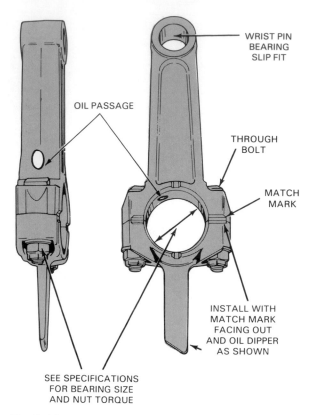

Fig. 7-16. Holes drilled in connecting rod, bearing insert and piston pin boss provide passageways for lubricating oil.

CONSTANT LEVEL SPLASH SYSTEM

The *constant level splash system* provides three major improvements over the simple splash system:

1. A lubrication oil pump.
2. A splash trough.
3. A strainer.

The cam-operated pump supplies oil to the trough where the oil dipper picks it up and distributes it to the cylinder wall and moving parts. The term "constant level" is used because the pump can supply more oil than the dipper can remove. Therefore, the trough is always full. Oil returning to the crankcase must pass through the filter before it can be pumped back to the trough. This keeps large contaminants in the crankcase.

The constant level splash system will provide adequate lubrication as long as there is enough oil to supply the pump. However, if the oil level is low, the cooling effect of the oil is reduced.

EJECTION AND BARREL PUMPS

The *ejection pump* forces oil under pressure against the rotating connecting rod. Some oil enters the connecting rod bearings, while the remaining oil is deflected to other parts in the crankcase. The ejection pump system is similar to the splash system, but it provides a more forceful spray of oil.

The *barrel pump*, Fig. 7-17, is a cylinder and plunger type of lubrication pump. By design, an eccentric on the camshaft moves the plunger in and out of the pump cylinder. The camshaft is hollow and has holes from the center of the shaft to the eccentric.

Fig. 7-17. Plunger of barrel pump is actuated by an eccentric shaft driven by valve camshaft. Ball-shaped plunger end is held in a socket so that it can pivot as shaft turns. (Deere & Co.)

In operation, the plunger is drawn out until a hole in the eccentric aligns with a hole in the plunger. This allows the cylinder to fill with oil. When the plunger is forced in, a different hole in the eccentric aligns with the plunger, and oil is forced through passages to the main bearings and crankshaft connecting rod journal.

POSITIVE DISPLACEMENT PUMPS

Several types of *positive displacement oil pumps* are used in pressurized lubrication sys-

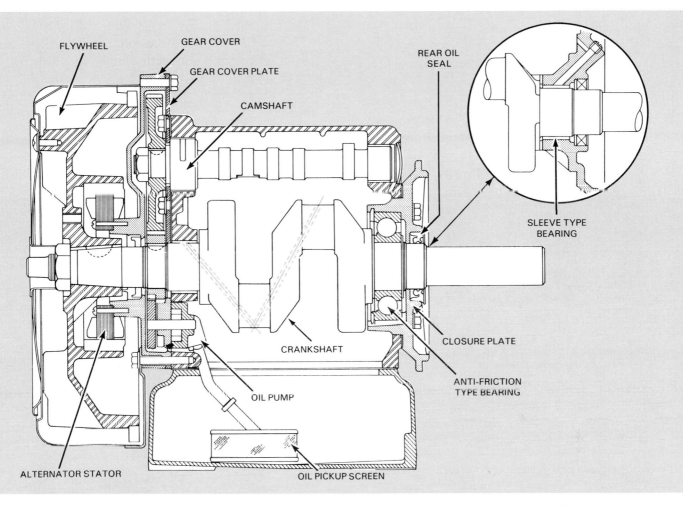

FLYWHEEL GEAR COVER

GEAR COVER PLATE

CAMSHAFT

REAR OIL SEAL

SLEEVE TYPE BEARING

CRANKSHAFT

CLOSURE PLATE

ANTI-FRICTION TYPE BEARING

OIL PUMP

ALTERNATOR STATOR

OIL PICKUP SCREEN

Fig. 7-18. Oil pump used in two cylinder engine supplies oil to moving internal engine parts. Note drilled crankshaft.

tems. See Fig. 7-18. One common type is the gear pump shown in Fig. 7-19. The end cover has been removed to expose two meshed gears. One gear is shaft-driven from the engine. It drives the second gear.

Note that the driving gear in Fig. 7-19 is keyed to the driving shaft. As the gears turn, oil fills the spaces between the teeth and is carried around to the oil outlet. No oil passes between the gears where the teeth are meshed, because of the tight fit.

If, for some reason, oil flow is restricted somewhere in the engine, the increase in pressure would raise the ball against the spring in the pressure relief valve. When this happens, oil will pass through the valve and recirculate through the pump. Recirculation of the engine oil continues until the restriction to flow ceases and pressure

declines, allowing the ball to seat and the relief valve to close. Without a pressure relief valve in the system, pressures would become excessively high during high engine speeds.

FULL PRESSURE LUBRICATION SYSTEM

A *full pressure lubrication system* is the type used in automobile engines. On some of the larger small engines, an almost completely pressurized system is used, including a positive displacement gear or rotor pump. Passages for oil flow are drilled to all critical points, such as camshaft bearings, main bearings, connecting rod bearings, and piston pins. A splash system is used in conjunction with the pressure system, particularly for lubricating cylinder walls.

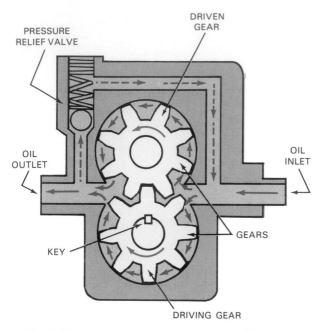

Fig. 7-19. In gear type pump operation, oil is carried between teeth of matching gears. If oil pressure is too high, relief valve recirculates oil through pump.

OIL FILTER SYSTEMS

Oil filters similar to the one shown in Fig. 7-7 are used on some small engines. *Filters trap dirt, carbon, and other harmful materials, preventing them from circulating through the engine.* The oil filter prevents very fine particles from circulating. The oil strainer, Fig. 7-20 (usually attached to intake side of oil pump), prevents large particles from entering the filter.

Three basic types of oil filter systems are in common use: bypass, shunt, and full-flow. Generally, the filter element is replaceable and can be discarded when dirty.

BYPASS SYSTEM

The *bypass filter system*, Fig. 7-20, pumps part of the oil through the filter, while the remaining oil is pumped to the engine bearings. Oil pumped through the filter is returned directly to the crankcase. The primary purpose of the filter in the bypass system is to keep a clean supply of oil in the crankcase.

The pressure relief valve (regulating valve) in the bypass system controls the maximum allowable pressure in the system. If there is a restriction to oil flow, pressure buildup will overcome relief valve spring tension and the valve will open. When this occurs, oil pressure will be relieved and oil will flow through the valve, back to the crankcase.

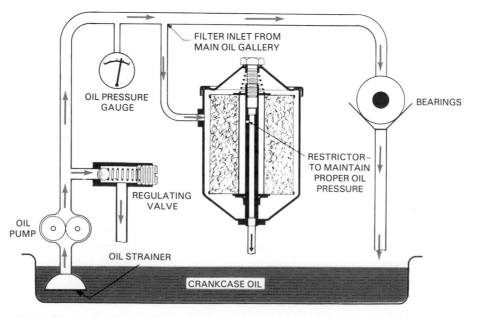

Fig. 7-20. Bypass filter system pumps some engine oil through filter. Remaining oil goes to engine bearings. (Wix Filters)

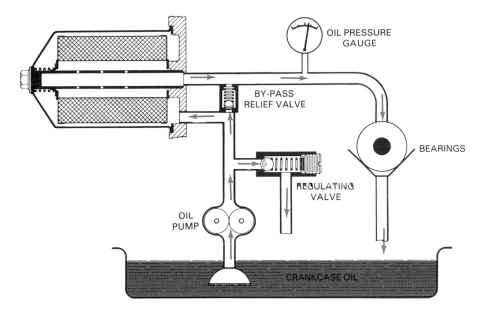

Fig. 7-21. Full-flow filter system directs all engine oil through filter. Relief valve opens if filter becomes clogged.

Pressure relief valves are installed in all pressurized lubrication systems. Often they are an integral part of the oil pump.

SHUNT FILTER SYSTEM

In the *shunt filter system,* part of the oil delivered by the pump is filtered and directed to the engine bearings. Some of the oil is shunted past the filter. The remaining oil is circulated through the pressure relief valve and back into the crankcase.

FULL-FLOW FILTER SYSTEM

The *full-flow filter system,* Fig. 7-21, directs the entire volume of pumped oil through the filter to the bearings. If the filter element becomes clogged with dirt, oil pressure will increase. The added pressure will open the relief valve, permitting the oil to flow. If the filter did not have a relief valve and the filter became clogged, serious engine damage would result.

The filter cartridge must correspond to the filter system on the engine. For example, a full-flow cartridge used in a partial-flow system will give longer service life, but initial efficiency will be poor due to its high flow rate. One the other hand, a partial-flow cartridge used in a full-flow system would drastically reduce oil pressure. Proper oil filtration in modern gasoline engines cannot be overemphasized.

SUMMARY

Lubrication is the process of reducing friction by introducing a slippery substance between sliding surfaces. Friction is the resistance to motion created when one dry surface rubs against another.

All bearing surfaces in small gas engines must have oil separating moving parts that are in close contact. The oil film between close fitting parts may only be a few molecules thick, but it is enough to keep the metal surfaces from actually touching.

In addition to lubricating, oil cools and cleans the engine. It also provides a seal between the piston rings and the cylinder wall.

When selecting oil for use in a small engine, service classification and viscosity grade must be considered.

The way that moving parts are lubricated differs in two cycle and four cycle engines. Oil in a two cycle engine is mixed with gasoline and poured into the fuel tank. In operation, an oil mist is created that lubricates all internal parts. In a four cycle engine, oil is poured into the crankcase. The oil sump in the crankcase supplies lubrication to all internal engine parts.

Oil filters are used on many small engines. These filters trap dirt, carbon, and other harmful materials, preventing them from circulating through the engine. Three common types of oil filter systems include the bypass, shunt, and full-flow.

KNOW THESE TERMS

Lubrication, Friction, Oil, API service classification, Viscosity grade, Multi-viscosity oils, Detergent/dispersants, Splash system, Constant level splash system, Dipper, Ejection pump, Barrel pump, Positive displacement pump, Full pressure lubrication system, Oil filter, Bypass filter system, Shunt filter system, Full-flow filter system.

REVIEW QUESTIONS–CHAPTER 7

1. Give three general types of lubricants.
2. Name four important jobs performed by a good engine lubricant.
3. Which engine lubrication system utilizes a trough in the crankcase?
   a. Barrel type lubrication system.
   b. Ejection pump system.
   c. Constant level system.
   d. Splash system.
4. Which small engine lubrication system is similar to the system used on automobiles?
   a. Ejection pump system.
   b. Constant level splash system.
   c. Barrel type pump system.
   d. Full pressure system.
5. Babbitt metal used as a bearing material is an alloy of three metals. Name them.
6. The type of bearing shell shown in Fig. 7-3 is referred to as a _____ _____ bearing.
7. The lock tabs on the bearing in Fig. 7-3 are for the purpose of _____.
8. The lubricating oil recommendation for a

particular make and model engine would be found on a special label on the engine. True or False?
9. API stands for _____.
10. SAE stands for _____.
11. "Type" of oil refers to its:
    a. Service classification.
    b. Viscosity.
    c. Grade.
    d. Maximum operating temperature.
12. Most four cycle engine manufacturers recommend oils supplemented by additives. True or False?
13. What is the purpose of detergent-dispersants in oils?
14. Why are some oils with additives not suitable for two cycle engine use?
15. A high viscosity oil would:
    a. Be thin.
    b. Be suitable for high temperature use.
16. Several types of positive displacement oil pumps are used in pressurized lubrication systems. Yes or No?
17. Name three types of oil filter systems in use.
18. Generally, the oil filter element is replaceable. True or False?

SUGGESTED ACTIVITIES

1. Demonstrate friction to the class by first rubbing two sheets of wasted paper together, then rub two sheets of coarse abrasive together. Discuss how a lubricant could reduce friction in each case.
2. Make viscosity measurements of several engine oils with a viscosimeter.
3. Compare a multi-viscosity oil with single viscosity oils, using a viscosimeter. Test the oils at various temperatures and make a chart of your results.
4. Compare a detergent oil with a non-detergent oil by placing equal amounts of carbon black in each. Shake them up, then let them set for several days. Explain any observed differences in the oils.
5. Produce condensation in a glass beaker partially filled with gasoline. Place the loosely covered breaker in a warm place or in direct

sunlight. When the air in the beaker is warm, pack ice around the outside. Repeat the process until a quantity of water can be seen in the bottom of the beaker. A hot humid day will produce best results.

6. Create heat with friction. Place a dull, unwanted drill in a drill press. Try to drill a bar of metal. Show what happens to drill as a result of friction.

7. Demonstrate to the class how oil cleans. With engine grease and grime on your hands, wipe them clean in a container of clean engine oil. Compare the cleaning power with soap or a detergent.

8. Remove the end cover from a gear-type oil pump and demonstrate how it works. Reinstall the cover, place the pump in oil, and operate it by hand.

This lawn mower is equipped with a 5.5 hp, overhead valve, air-cooled engine. Air-cooled engines rely on the flywheel fan for air circulation. They also depend on the cooling fins and the cylinder head for heat dissipation. (Kubota Tractor Corp.)

CHAPTER 8

ENGINE COOLING

After studying this chapter, you will be able to:
☐ Explain how air cooling, exhaust cooling, and water cooling work to lower engine operating temperatures.
☐ Define the basic function of a water pump and give examples of several common types.
☐ Describe the basic operation of outboard water circulation systems.
☐ Explain the function of a thermostat and a radiator.

The efficiency and life of an engine depend upon how well it is cooled. The average temperature of burned gases in the combustion chamber of an air-cooled engine is about 3600°F (1982°C).

Fig. 8-1. The air cooling system of a small gasoline engine consists of the shroud, screen, flywheel, baffles, and cooling fins. Airflow follows the path shown by arrows. (Wisconsin Motors Corp.)

About a third of the heat is carried away by the cooling system, Fig. 8-1. Various sheet metal parts surrounding the air-cooled engine direct the flow of cooling air as shown in Fig. 8-2. The exhaust system carries away another third of the heat. That which remains is used to produce engine power.

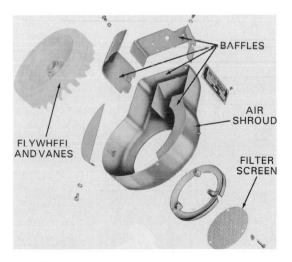

Fig. 8-2. A number of sheet metal parts enclose flywheel and cylinder. Baffles direct airflow to critical areas.

Loss of heat through the cylinder walls to the cooling system reduces the temperature from 1200°F (649°C) to approximately 350°F (177°C). The temperature drops to 100°F (38°C) by the time it reaches the outer edges of the cooling fins. Fig. 8-3 illustrates cylinder wall temperatures at various locations.

There is a reason for the comparatively low temperature along the inner cylinder wall. A *boundary layer* of stagnant gas lies next to the

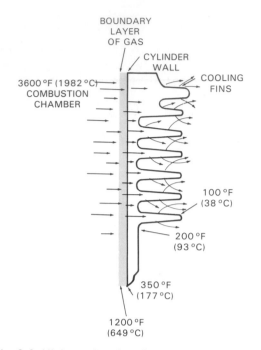

BOUNDARY
LAYER
OF GAS

CYLINDER
WALL

COOLING
FINS

3600°F (1982°C)
COMBUSTION
CHAMBER

100°F
(38°C)

200°F
(93°C)

350°F
(177°C)

1200°F
(649°C)

Fig. 8-3. High combustion chamber temperature is reduced by exhaust gases and cooling system. Large area
of cooling fins controls heat dissipation from cylinder.

wall and acts as an insulator. Therefore, less heat travels through the cylinder wall than you might expect from the 3600°F (1982°C) produced by the burning fuel.

HOW AIR COOLING WORKS

Operating temperature is lowered to about 200°F (93°C) as heat passes through the cylinder wall to the outer surfaces of the cylinder. Air forced over the cooling fins and directed by the sheet metal baffles rapidly dissipates the heat. The cooling system, then, carries heat away from the engine.

The heat of combustion (rapid burning and expansion of gas) travels from the cylinder through the cylinder walls by conduction. *Conduction* is heat transfer through a solid material.

When the heat reaches the outer surfaces of the cylinder, air forced over the surface carries it away by convection. *Convection* occurs when heat transfers through movement of a gas–in this case, air.

The flywheel has fins, which blow air around the cylinder wall and cooling fins. The flow of air

is controlled and directed by a sheet metal shroud and baffles surrounding the flywheel and cylinder. An engine should never be run without the shroud in place or it will quickly overheat. For safety, the flywheel is covered with a screen or perforated plate, which allows air to be drawn through it, Fig. 8-4. *The screen should be kept clean to permit unrestricted airflow.*

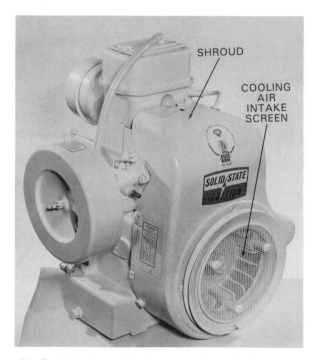

SHROUD

COOLING
AIR
INTAKE
SCREEN

Fig. 8-4. Cooling air intake screen must be kept clean
for unrestricted airflow. Screen must be kept in
place to prevent clogging of cooling fins.
(Tecumseh Products Co.)

Thin *cooling fins* increase the surface area around the outside of the cylinder, Fig. 8-5. The greater the surface area in contact with cool air, the more rapidly the heat can be carried away. Cooling fins are necessary on air-cooled engines but not on water-cooled engines. Water is four times more effective than air for engine cooling.

HOW EXHAUST COOLING WORKS

As noted earlier, the exhaust system carries away approximately a third of the engine heat. If the exhaust system is restricted in any way, part of the 1200°F (649°C) temperature will remain in

Fig. 8-5. Cooling fins are designed and placed to provide adequate cooling for each part of the cylinder. Thickness, surface area, and spacing are important considerations.

the metals of the engine. This heat buildup will increase friction. In turn, maintenance costs will increase on moving parts and engine life will be shortened. Dirt, grass clippings, leaves, straw, or other materials lodged between the cooling fins

will tend to insulate the cylinder, causing "hot spots" and engine overheating. *Keeping an engine clean, cool, and properly lubricated will significantly extend its life.*

HOW WATER COOLING WORKS

Water is an excellent medium for cooling engines. It is inexpensive, readily available, and absorbs heat well.

Some small engines are water cooled because they are used in or around a water source. The outboard engine in Fig. 8-6 is typical of a relatively small water-cooled engine. Notice the absence of cooling fins, which are found only on air-cooled engines.

Water-cooled engines are generally made with coolant passages surrounding the cylinder, Fig. 8-7. These passages are called *water jackets.* A small pump keeps water circulating through the jackets. The water absorbs the heat of combustion and carries it away from the engine. In cold weather, *antifreeze* solutions are added to the water to prevent it from freezing.

Fig. 8-6. Outboard engines are frequently water cooled. They are identified by the absence of cooling fins. (Evinrude Motors)

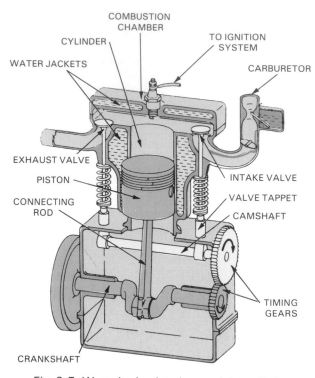

Fig. 8-7. Water is circulated around the cylinders through the water jacket, where it absorbs combustion heat.

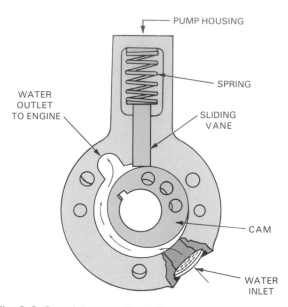

Fig. 8-8. Revolving cam in sliding vane water pump fills large volume of space, and water is pushed through outlet to engine. Sliding vane prevents water from revolving with cam.

WATER PUMPS

Water pumps of many different designs are used to move the water through the cooling system. In outboard engines, the pump is generally located in the lower unit near or below the water line. The main pump member is driven by the vertical drive shaft or horizontal propeller shaft. The pump must have an opening for water to enter and one for it to exit. These openings are called the inlet and outlet.

A *sliding vane pump* is illustrated in Fig. 8-8. As the eccentric cam (an off-center enlargement on a shaft) is rotated in the sliding vane pump, the volume of space between the cam and pump housing is constantly changing. Water enters the inlet and fills this space. The cam rotates and closes the inlet, pushing the water ahead of it toward the pump outlet.

The outlet is connected to the water jacket by suitable tubing. A second tube carries the heated water out of the engine. The sliding vane is kept in close contact with the cam by a spring. The vane provides a seal so that water cannot continue past the outlet. However, during high-speed operation, water pressure may become high enough to lift the vane against the spring, allowing some water to recirculate through the pump. This action prevents too much pressure from building up in the cooling system.

The *rotor-type pump* operates much like the sliding vane pump. The vane and rotor are one piece, and the eccentric gyrates the rotor (rotates the rotor with a bobbing motion), causing a pumping action.

The *plunger pump*, Fig. 8-9, has a cylinder and plunger. The plunger is raised and lowered in the cylinder by an eccentric on the propeller shaft. A spring keeps the plunger close to the eccentric.

When the spring forces the plunger down in the cylinder, water is drawn in through the inlet ball check valve while the outlet ball check valve is closed. As the eccentric lifts the plunger, the inlet check valve is closed by the increasing pressure in the cylinder and the outlet check valve is opened. Water in the cylinder is forced through tubing to the engine.

Vari-volume pumps use a synthetic rubber impeller. See Fig. 8-10A. Since the impeller housing

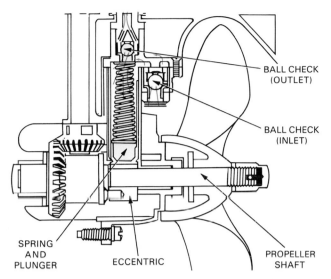

Fig. 8-9. Plunger pump draws water in through one ball check valve and forces it out through a second valve. Plunger is operated by an eccentric.

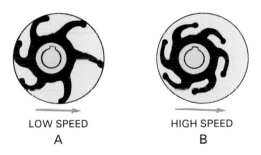

LOW SPEED
A

HIGH SPEED
B

Fig. 8-10. Vari-volume pump has a flexible synthetic rubber impeller that is offset from center of impeller housing. A--Space between impeller arms varies as they rotate. B--At high speed, increased water pressure forces impeller arms inward to prevent overcooling of engine.

is off-center with the drive shaft, the impeller arms must flex as they revolve. The volume between the arms increases and decreases with rotation.

The inlet port or opening is located where the volume is increasing and the water is drawn into the housing. The outlet is on the side where volume is decreasing and the water is forced out into the water jacket.

When a vari-volume pump is driven at high speeds, the back pressure of the water in the system becomes great enough to force the impeller arms inward, Fig. 8-10B. The pump loses some of its effectiveness and moves only enough water

through the engine to maintain proper operating temperature in the cylinders. Overcooling of the cylinders of an outboard engine can happen at high engine speeds unless some sort of flow control, such as this, is designed into the cooling system.

OUTBOARD WATER CIRCULATION SYSTEMS

Some small outboard engines use a simple cooling device called a *pressure-vacuum water flow system*, Fig. 8-11. The water flow from the propeller tips creates pressure against the intake port and vacuum at the outlet port, which is located immediately forward of the propeller. Propeller action and the forward motion of the boat provide water circulation.

At slow speeds, like those used for trolling, pressure of water from the propeller tips may not be enough to force water through the cooling system. Sufficient cooling is still maintained, however, by the siphon (sucking) effect of the discharge channels as the boat moves through the water. When starting an engine with a pressure-vacuum system, the engine should be given a short burst of speed to fill the channels and water jacket with water, Fig. 8-11.

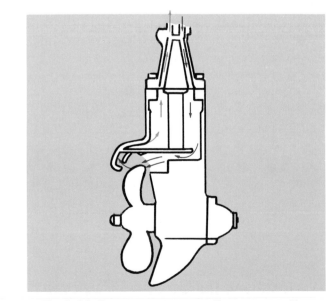

Fig. 8-11. Pressure-vacuum cooling system utilizes thrust of propeller tips and venturi (narrowed section of a passage) vacuum principle created by forward motion of engine to circulate water.

Another type of pressure-vacuum water flow system has the discharge ports located in the propeller blades. The centrifugal force created by the turning propeller aids in discharging the water. *Centrifugal force* is the tendency of spinning matter to move away from the center of its path. Since there are no moving parts, except for the propeller, the system will function as long as the water channels and jackets remain unobstructed.

With this system, vacuum must be maintained, particularly at low speeds. Therefore, all water connections in the system are airtight. Air seepage into the cooling system would destroy the slow speed siphoning effect and cause overheating.

Worn propeller blades can also cause poor circulation. The propeller's reduced diameter puts the tips farther from the water scoop opening. This reduces water pressure. Salt water corrosion, marine growth, and mud-clogged water channels are all causes of faulty water circulation in this type of system.

Other outboard water circulation systems are in use. The basic principles are similar for most types. Some examples follow.

Fig. 8-12 shows a pump-driven cooling system. Water is drawn into the intake, pumped through the water jacket surrounding the cylinders, and discharged. This is a relatively simple system. Notice that the pump is driven by the drive shaft.

In a similar system, the water pump is driven by the propeller shaft. When this pump location is used, the water is drawn in through ports in the propeller hub.

In some engines, water temperature is carefully controlled by a thermostat. The *thermostat* is simply a valve arrangement that stops the circulation of engine coolant until it reaches the proper operating temperature.

There are two types of thermostats, each opening and closing the valve at predetermined temperatures. Both work on the principle of the expansion of heated materials.

In the old bellows-type thermostat, the valve is attached to a gas-filled, sealed bellows made of thin copper. Heat causes the gas to expand, opening the valve.

The more popular pellet thermostat operates in the same way but depends upon a small, wax-

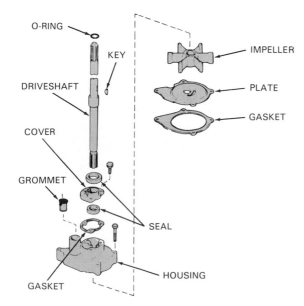

Fig. 8-12. Water pump for outboard engine is driven by vertical drive shaft. Some pumps are mounted on propeller shaft.

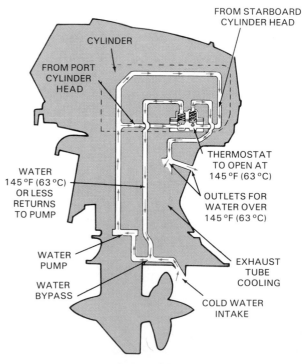

Fig. 8-13. Some outboard engines have a thermostatically controlled cooling system. Water recirculates until thermostat opens, discharging heated water and allowing more cold water to enter intake. This thermostatically controlled cooling system employs a double feed bypass.

filled, copper cylinder. When heated, the wax pushes against a rubber diaphragm (disc). This moves a piston that is attached to the valve. In some cases, the piston remains stationary and the cylinder moves away to open the valve.

A thermostatically controlled engine, Fig. 8-13, will be kept at a constant temperature regardless of speed or outside temperature. Cold water enters the intake and passes through the pump to the water jacket. If the temperature of the water from the jacket is high enough to open the thermostat, the water is discharged from the engine. If the water is too cool to open the thermostat, it is recirculated through the pressure control valve and back to the pump. When water is recirculated, it retains some of its heat and eventually brings the cylinder head temperature up to thermostat temperature. This type of system can maintain a constant operating temperature and will automatically compensate for even slight changes in cooling water temperature. A thermostat in a cold engine will remain closed until it reaches thermostat opening temperature.

OBSERVING THERMOSTAT OPERATION

With the thermostat removed from the engine, it is easy to see how it works. Place the thermostat in a beaker of cool water. Put the beaker on a hot plate and slowly heat the water. When the water has reached the proper temperature, the thermostat will open. By placing a thermometer in the beaker, Fig. 8-14, you can determine the temperature rating of the thermostat.

RADIATORS

Radiators are water reservoirs, which are made from many thin copper or aluminum tubes. These tubes are connected at the top and bottom (or each side) to water tanks. The tubes are held in place by thin metal fins. The fins increase the cooling surface area of the tubes. The tube and fin assembly shown in Fig. 8-15 is called the *radiator core.*

The water, which is heated by the engine, is pushed into the top or side of the radiator core by the water pump. The hot water travels downward

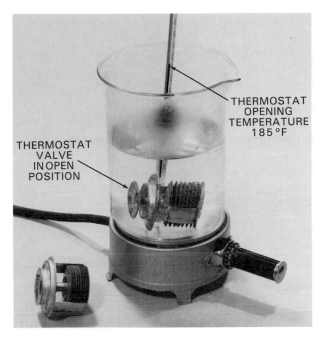

Fig. 8-14. A thermostat can be observed ''in action'' by heating it in a beaker of water. Thermometer registers temperature at which thermostat opens.

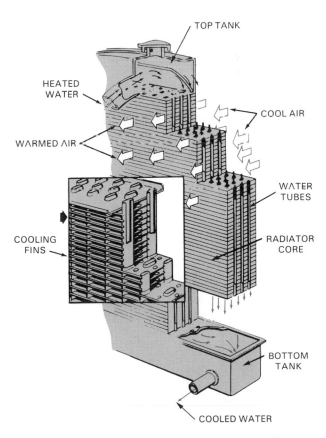

Fig. 8-15. Radiator carries water through tubes in core, where it is cooled by air forced across cooling fins.

Engine Cooling   133

or across through the tubes to the bottom tank. Cool air is forced between the fins in the core by an engine-driven fan. The air cools the water by conduction, radiation, and convection. The cooled water passes from the bottom tank to the engine water jacket, ready to absorb more heat. Engines cooled by radiators also have thermostats, which open when the proper operating temperature is reached. When engine temperature is below normal, the thermostat remains closed and the water is directed through a bypass channel. The water bypassing the radiator retains its heat and returns directly to the engine. This action retards forced cooling until the engine reaches its normal operating temperature.

The fine balance of cooling maintained by the radiator and thermostat assures that the engine will not overheat under loads in the hot summer or run too cold in the winter.

OIL COOLING

Most of this chapter has been devoted to engine cooling by air or water. Nevertheless, engine oil also plays an important role in the cooling process. *Oil circulating through the engine also acts as a coolant.* The oil absorbs heat as it contacts the various metal parts. It then draws off the heat as it leaves the engine (two cycle engine) or transfers the heat through the crankcase to the outside air (four cycle engine).

SUMMARY

The efficiency and life of an engine depends on how well it is cooled. Combustion heat is carried away from the engine by conduction and convection.

Approximately a third of engine heat is carried away by the cooling system, another third is carried away by the exhaust system, and the remaining heat is used to produce engine power.

The flywheel on a small engine has fins that blow air around the cylinder wall and cooling fins. The cooling fins increase the surface area around the outside of the cylinder. The greater the surface area in contact with air, the more rapidly heat can be carried away. In order to effi-

ciently carry away heat, the exhaust system must be free from restrictions. Keeping an engine clean, cool, and properly lubricated will significantly extend its life.

Some small engines are water cooled. Water is an excellent medium for cooling engines.

Some outboard engines use a pressure-vacuum water flow system. Water flow from the propeller tips creates pressure against an intake port and vacuum at an outlet port. Another type of pressure-vacuum system has discharge ports located in the propeller blades. The centrifugal force created by the turning propeller aids in discharging the water.

In some engines, the temperature of the cooling water is controlled by a thermostat, which stops circulation of coolant until the engine reaches a specific operating temperature.

Some small engines are equipped with radiators. Radiators are water reservoirs made of many copper or aluminum tubes, which are connected at the top and bottom to water (coolant) tanks. The tubes are held in place by thin metal fins. The fins increase the cooling area of the tubes. If a radiator system is used, antifreeze must be added to water to prevent freezing. Water pumps are used to move water through the cooling system.

Oil circulating through an engine also acts as a coolant. Oil absorbs heat as it contacts various metal parts.

KNOW THESE TERMS

Boundary layer, Conduction, Convection, Cooling fins, Water Jacket, Antifreeze, Sliding vane pump, Rotor-type pump, Plunger pump, Vari-volume pump, Pressure-vacuum cooling system, Centrifugal force, Thermostat, Radiator.

REVIEW QUESTIONS–CHAPTER 8

1. The average temperature of the burned gases in the combustion chamber is:
   a. 2400°F (1315.5°C).
   b. 3000°F (1649°C).
   c. 6200°F (3427°C).
   d. 3600°F (1982°C).

2. The heat of combustion is carried away in three ways. One third is removed by the cooling system and one third by the exhaust system. That which remains is used to produce _____ _____.

3. Why is the temperature of the inside cylinder wall relatively cool compared to the combustion temperature?

4. The heat that reaches the cooling fins is carried away mainly by:
   a. Convention
   b. Conduction.
   c. Radiation.

5. What are two ways to prevent overheating of air-cooled engines?

6. When used for engine cooling, water is about _____ times as effective as air.

7. Cylinders of water-cooled engines are surrounded by:
   a. Insulation material.
   b. A water reservoir.
   c. A water jacket.
   d. Cooling water tubes.

8. Name four types of water pumps discussed in this chapter.

9. The _____ _____ water flow system in outboard engines utilizes the tips of the propeller to circulate water.

10. A heat sensing device that maintains a constant engine operating temperature is the:
    a. Check valve.
    b. Thermostat.
    c. Rheostat.
    d. Flexible impeller.

11. In cold weather, an _____ should be added to water-cooled engines.

12. What cooling system defect can cause "hot spots?"

13. What two fluids are used to cool engines?

SUGGESTED ACTIVITIES

1. Place a thermostat and thermometer into a beaker of water. Heat the water and check thermostat operation.

2. Operate an air-cooled engine and observe the flow of air through the shroud and around the cooling fins.

3. Inspect a cutaway of a radiator and explain how it cools water.

4. Disassemble the lower unit of an outboard engine and study the pumping and circulation system.

5. Place the outboard engine in a tank and operate it to observe the cooling system at work.

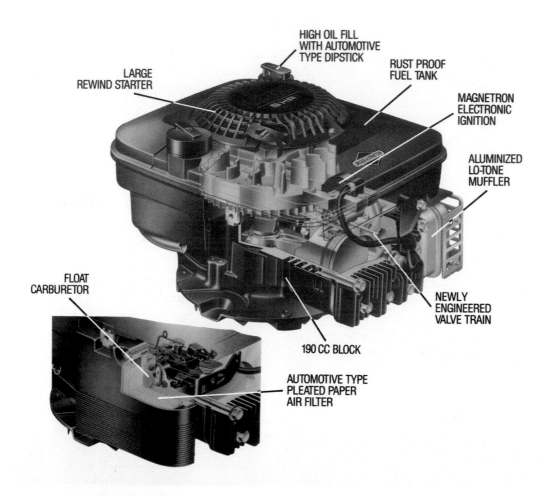

HIGH OIL FILL
WITH AUTOMOTIVE
TYPE DIPSTICK

RUST PROOF
FUEL TANK

LARGE
REWIND STARTER

MAGNETRON
ELECTRONIC
IGNITION

ALUMINIZED
LO-TONE
MUFFLER

FLOAT
CARBURETOR

NEWLY
ENGINEERED
VALVE TRAIN

190 CC BLOCK

AUTOMOTIVE TYPE
PLEATED PAPER
AIR FILTER

Cutaway view of a 5 hp, vertical shaft, four cycle engine. This engine is equipped with electronic ignition, a paper air filter element, and a float-type carburetor. (Briggs and Stratton Corp.)

CHAPTER 9

MEASURING ENGINE PERFORMANCE

After studying this chapter, you will be able to:
☐ Define engine performance.
☐ Define and compute bore, stroke, displacement, compression ratio, force, work, power, energy, and horsepower.
☐ Differentiate between the various types of horsepower.
☐ Explain the function of a Prony brake and a dynamometer.
☐ Define and calculate torque.
☐ Explain volumetric efficiency, practical efficiency, mechanical efficiency, and thermal efficiency.

The small gasoline engine belongs in the "heat engine" category. Other heat engines include automotive reciprocating piston engines, gas or steam turbines, steam engines, diesel engines, rotary combustion engines, rocket engines, and jet engines. Only the small, one- and two-cylinder piston engines will be discussed here.

The small gasoline engine is called an internal combustion engine because an air/fuel mixture is ignited (fired) and burned inside the engine, Fig. 9-1. The heat from the burning mixture causes the gases to expand rapidly within the closed cylinder. The expanding gases apply strong force and push out in all directions within the cylinder, but only the piston can move.

The piston is pushed away from the center of combustion. If it were not fastened to the crankshaft, it would come out of the cylinder the way a bullet comes out of a gun. The crankshaft and connecting rod keep the piston under control and allow it to travel only a short distance, Fig. 9-2.

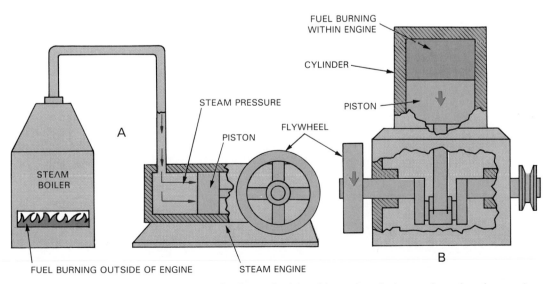

Fig. 9-1. A–External combustion engine burns fuel outside engine. B–Internal combustion engine burns fuel within engine.

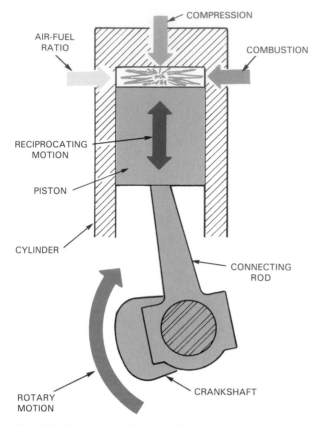

Fig. 9-2. Piston travel is controlled by connecting rod and crankshaft. Notice change from reciprocating (up and down) motion to rotary motion. (Deere & Co.)

When the piston has moved downward as far as it can go, a port is opened. This allows burned gases to escape as the piston returns upward. The burned, escaping gases are called exhaust and come out through the exhaust pipe or manifold. A manifold is a chamber that collects the exhaust and directs it to the exhaust pipe.

The piston does not pause long at the end of its stroke before the movement of the crankshaft and flywheel carry it back to the top of the cylinder. As you may have learned from your science classes, bodies in motion tend to continue in motion. This tendency is called inertia. It is what keeps the crankshaft and flywheel moving in the engine.

When the piston reaches the top of its stroke, it is ready to be forced back down again. Because the piston continues this back and forth or up and down motion, piston engines are often called "reciprocating engines."

BASIC TERMINOLOGY

To better understand how a gasoline engine works and to appreciate the power it provides, you must learn certain basic terms. These terms will be defined here only as far as necessary to provide a background for further discussion of measuring engine performance. *Performance* can be defined as the work engines do and how well they do it.

ENGINE BORE AND STROKE

Engine *bore* is the diameter or width across the top of the cylinder. *Stroke* is the up or down movement of the piston. Length of stroke is determined by the distance the piston moves from its uppermost position (top dead center or TDC) to its lowest position (bottom dead center or BDC), or in reverse order.

The amount of crank offset determines the length of the stroke. *Crank offset* is the distance from the centerline of the connecting rod journal

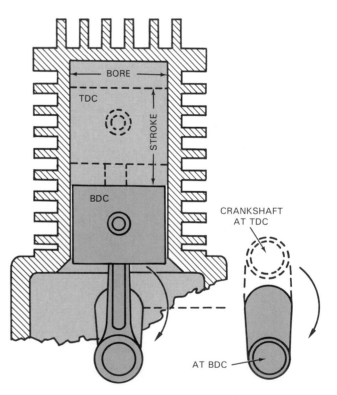

Fig. 9-3. Engine ''bore'' refers to diameter of cylinder. ''Stroke'' indicates length piston travels as it moves from TDC to BDC.

to the centerline of the crankshaft. A 2 in. offset would produce a 4 in. stroke, Fig. 9-3.

When the bore diameter is the same as the stroke, the engine is referred to as *square.* When the bore diameter is greater than the stroke, it is termed *over square.* Where bore diameter is less than the stroke, the engine is called *under square.*

ENGINE DISPLACEMENT

In a single-cylinder engine, engine size, or *displacement*, refers to the total volume of space increase in the cylinder as the piston moves from the top to the bottom of its stroke.

To work out a given engine's displacement, first determine the circular area of the cylinder (0.7854 by diameter^2). Then multiply that answer by the total length of the stroke (piston travel). The formula is:

$$\text{Engine Displacement} = 0.7854 \times D^2 \times \text{Length of Stroke}$$

If the engine has more than one cylinder, multiply the answer to the above formula by the number of cylinders. For example, say that a two-cylinder engine has a bore of 3 1/4 in. and a stroke of 3 1/4 in. Using the displacement formula, you would have:

$$0.7854 \times D^2 (10.563) \times \text{Length of Stroke (3 1/4 in.)} \times \text{Number of Cylinders (2)} = 53.9 \text{ cu. in.}$$

Fig. 9-4 illustrates piston displacement. In drawing A, the piston is at TDC (top dead center). Blue liquid has been added to fill up the space left between the piston and the cylinder head. The piston in drawing B is at the bottom of its stroke. Note the increased space above the piston now. In drawing C, red liquid is added until the cylinder is once again filled. The amount of red liquid (in cubic inches) would represent the piston displacement for this cylinder.

COMPRESSION RATIO

The *compression ratio* of an engine is a measurement of the relationship between the total cylinder volume when the piston is at the bottom

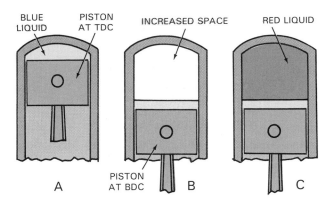

Fig. 9-4. Engine displacement is difference in volume of the cylinder and combustion chamber above piston when it is at TDC and when it is at BDC. Red area shows displacement.

of its stroke (BDC) and the volume remaining when the piston is at the top of its stroke (TDC).

For example, if cylinder volume measures 6 cu. in. when the piston is at BDC (view A in Fig. 9-5) and 1 cu. in. when at TDC, (view B in Fig. 9-5), the compression ratio of the engine is 6 to 1. Many small gasoline engines have 5 or 6 to 1 compression ratios. Certain motorcycle engines have 9 or 10 to 1 compression ratios.

FORCE

Forces are being applied all around us. *Force* is the pushing or pulling of one body on another. Usually two bodies must be in contact for force to be transmitted.

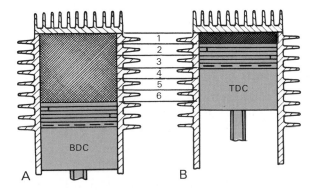

Fig. 9-5. Compression ratio is relationship between cylinder volume with piston A at BDC and piston B at TDC. Volume has been compressed to one-sixth of its original size, which indicates a 6 to 1 compression ratio. (Briggs and Stratton Corp.)

For example, as you read this you are applying a force to a chair if you are sitting or to the floor if you are standing. The force is equal to the weight of your body.

You can easily measure it with a scale, Fig. 9-6. This is known as gravitational force and it acts upon all materials on and around the earth.

Some forces are stationary (motionless); others are moving. For example, if you push against a wall, force is applied but the wall does not move. The use of force may or may not cause motion. Force itself cannot be seen, but there are many ways of using it.

Centrifugal force acts upon a body whenever it follows a circular or curved path. The body tries to move outward from the center of its path. Modern examples are the man-made satellites that orbit the earth. The circular path and speed of the satellite produces a centrifugal force outward that is equal to the earth's gravitational force inward so that each is opposed and balanced. Therefore, the satellite neither goes up nor comes down. This is one case where a force is applied without one body touching another body. A ball swung on a string applies centrifugal force as shown in Fig. 9-7.

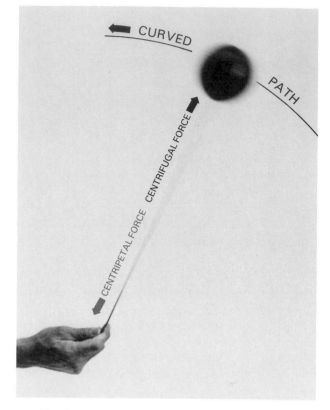

Fig. 9-7. An object in motion tends to travel in a straight line. The additional force required to cause deviation (or change) is called centripetal force. Reaction to centripetal force is called centrifugal force.

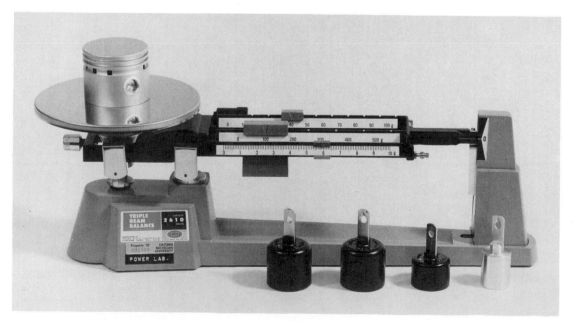

Fig. 9-6. A triple beam balance is used to measure the force of gravity applied to an engine piston. Weight is read from three scales. The top scale is graduated to ten gram increments, the center scale is graduated in 100 gram increments, and the bottom scale graduated in single grams.

Many forces interact when a gasoline engine is operating. The rotational speed of the crankshaft and flywheel create centrifugal force, which causes *tensile stress* (tension or pull) within the materials making up these parts. If the outward pulling force becomes greater than the strength of the material, the engine could fly apart. The rapid reciprocation (backward and forward or up and down motion) of the piston may put high forces on the connecting rod, crank journal, and piston pin. Fig. 9-2 shows how these parts work together.

One of the forces used efficiently in the gasoline engine is that which is applied to the top of the piston by the rapidly expanding gases. This force is produced by burning gasoline mixed with air. The greater the force applied to the piston, the greater the amount of power and work that can be done by the engine.

Force is measured in units of some standard weight such as pounds, ounces, or grams. For example, to support an engine weighing 16 lbs., a person would have to apply a lifting force of 16 lbs. Obviously, only half the lifting force would be needed to support an 8 lb. engine, Fig. 9-8.

Force and pressure are often confused. These terms should be understood in the way they are applied. *Pressure* is a force per given unit of area.

For example, a piston with a face area of 5 sq. in. may have a total force of 500 lbs. applied to it by the expanding gases. However, the pressure being applied is 500 lbs. divided by 5 sq. in., which equals 100 psi (pounds per square inch). This means that every square inch on the piston face has the equivalent of a 100 lb. weight pushing on it.

When we speak of pressures in mathematical calculations, we use letters to represent several words. For example, psi means "pounds per square inch." As an example:

$$PSI = \frac{Force}{Area}$$

or $$Force = PSI \times Area$$

or $$Area = \frac{Force}{PSI}$$

The area of a circle can be found by multiplying π ($\pi = 3.1416$) by the radius squared. (Writ-

Fig. 9-8. A small gasoline engine being lifted with a force equal to, or slightly greater than, its own weight.

ten: Area $= \pi r^2$.) Another method is to multiply the constant .7854 by the diameter squared. (Written: Area $= .7854\ D^2$.)

For example, we shall calculate a force applied to a 3 in. diameter piston, Fig. 9-9, if the cylinder pressure is 125 psi:

$$Area = \pi r^2 \text{ or } 3.14 \times 1.5 \times 1.5$$

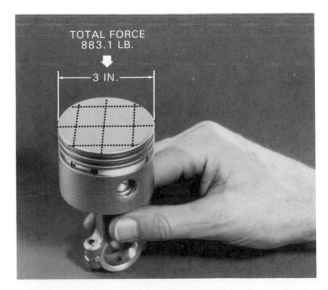

Fig. 9-9. Total force applied to piston face is equal to its area in square inches multiplied by psi (pounds per square inch).

Then the piston area equals: 7.065 sq. in.

Force = PSI × Area or 125 × 7.065

This results in a total force of 883.1 lbs.

WORK

Work is accomplished only when a force is applied through some distance. If a given weight is held so that it neither rises nor falls, no work is done even though the person holding the weight may become very tired. If the weight is raised some distance, then work is being done. The *amount* of work performed is the product or result of the *force* and the *distance* through which the weight is moved.

If a weight of 20 lbs. is lifted 3 ft., then 60 ft.-lbs. of work is accomplished. The distance must always be measured in the same direction as the applied force. This results in the formula:

$$\text{Work} = \text{Force} \times \text{Distance}$$

Because the formula calls for multiplying feet times pounds, the answer is expressed in *foot-pounds* (ft.-lbs.).

The gasoline engine utilizes the principles of a number of simple machines. These machines include the lever, the inclined plane, the pulley, the wheel and axle, the screw, and the wedge.

With each of these simple machines, you will find that to increase the output force with a given input force, the input distance will have to be increased in the same proportion or percentage. Therefore, without considering loss through friction, the foot-pounds of output are equal to the foot-pounds of input.

By using the principle of the lever, for example, a mechanical advantage is possible. Fig. 9-10 illustrates how a heavy load can be moved a short distance with a small force exerted through a relatively great distance.

The formula for computing leverage, as it applies to Fig. 9-10, is as follows:

$$\text{MA} \genfrac{}{}{0pt}{}{\text{(mechanical}}{\text{advantage)}} = \frac{\text{ED (effort distance)}}{\text{RD (resistance distance)}}$$

$$\text{MA} = \frac{\text{ED}}{\text{RD}} = \frac{6}{2} = 3$$

$$\text{E (effort)} = \frac{\text{R(resistance)}}{\text{MA}} = \frac{600}{3} = 200 \text{ lbs.}$$

$$\text{or:} \quad \text{X:6} = 2:600$$

$$\text{X} = \frac{2 \times 600}{6}$$

$$\text{X} = 200 \text{ lbs.}$$

POWER

In studying the formula for work, note that it does not consider the time required to do the work. For example, if a small gasoline engine weighing 50 lbs. is lifted 3 ft. from the floor to the workbench, 150 ft.-lbs. of work is done. The same amount of work would be performed whether it took 50 seconds to lift the engine or only 5 seconds.

Because it is important to know the rate at which work is done, the word power enters the picture. *Power* is the rate at which work is performed. The rate (amount of time) is given in seconds. Power can then be considered as foot-pounds per second.

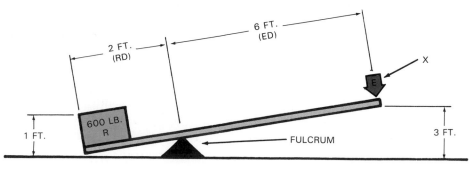

Fig. 9-10. The principle of the lever is the same as those applied to other forms of simple machines.

The formula for power is:

$$\text{Power} = \frac{\text{Work}}{\text{Time}} \quad \text{or}$$

$$\text{Power} = \frac{\text{Feet} \times \text{Pounds}}{\text{Seconds}} \quad \text{or}$$

$$\text{Power} = \text{ft.-lbs. per Second}$$

When the engine is lifted in 5 sec., 150 ft.-lbs. of work is performed. Using the power formula, it can be seen that:

$$\text{Power} = \frac{\text{Work}}{\text{Time}} \quad \text{or}$$

$$\text{Power} = \frac{150 \text{ lbs.}}{5 \text{ Seconds}} \quad \text{or}$$

$$\text{Power} = 30 \text{ ft.-lbs. per Second}$$

When lifting the engine in 50 seconds, the formula shows:

$$\text{Power} = \frac{\text{Work}}{\text{Time}} \quad \text{or}$$

$$\text{Power} = \frac{150 \text{ lbs.}}{50 \text{ Seconds}} \quad \text{or}$$

$$\text{Power} = 3 \text{ ft.-lbs. per Second}$$

Work is a force applied to an object that causes the object to move, and power is the rate at which the work is done. The standard unit of power is termed *horsepower.*

ENERGY

Energy is the capacity to perform work. It is grouped into various types, including potential energy (PE), kinetic energy (KE), mechanical energy (ME), chemical energy (CE), and thermal energy (TE). Typical sources of the various types of energy are shown in Fig. 9-11.

Many of the words just used will be unfamiliar to you. We will explain them as we go along. To begin with, energy is something we cannot define. It puts "life" into matter, giving it warmth, light, and motion. Energy cannot be seen, weighed, or measured. It does not take up space. However, we know it is there. The warmth and light of a bonfire, the electrical spark that jumps the gap of a spark plug, or the turning of a wheel are things we can sense. Thus, they must exist.

Matter and energy cannot be destroyed. Only the nature of matter and the forms of energy

	Types of Energy				
	PE	KE	ME	CE	TE
Coal, Oil, Gasoline				X	
Clock, Spring-Wound	X				
Moving Auto		X			
Water Behind A Dam	X				
Jacking Up An Auto	X				
A Cooking Fire					X
Dynamite, Stored				X	
Dynamite, Exploding			X		X
Crankshaft, Turning		X			
Man, Running		X			

Fig. 9-11. Potential, kinetic, mechanical, chemical, and thermal energy and some of their sources. (Go-Power Corp.)

change. For example, when a piece of charcoal burns and disappears, we may think it is completely gone. But the charcoal material has combined with air and formed a like quantity of ash, water, and gases. The energy that did this existed as flame (light energy) and heat (heat energy). All of this will continue until another change takes place.

Whenever a form of matter can be separated from other forms of matter so that a part or all of its energy can be released, it is said to contain potential energy. Examples of such matter include crude oil and the gasoline taken from it. We have learned different ways to release a part of the energy stored in these substances.

Engines are designed to release and change the potential energy of gasoline into mechanical power. Mechanical power does the work at hand.

HORSEPOWER

For hundreds of years, men used horses to perform work. It was only natural that when machines were invented, their ability to perform work would be compared to the horse.

In his work with early steam engines, James Watt wanted some simple way to measure their power output. In measuring the power or rate of work performed by a horse, he found that most work horses could lift 100 lbs. a distance of 330 ft. in 1 min.

If 1 lb. is lifted 1 ft. in 1 min., 1 ft.-lb. of work is done. The horse lifted 100 lbs. a distance of 330 ft. in 1 min. Using the work formula (Work = Distance × Force), Watt found that the horse performed 33,000 ft.-lbs. of work.

In determining the *rate of power* developed by the horse, we use the formula:

$$\text{Power} = \frac{\text{Work}}{\text{Time}} = \frac{33,000 \text{ ft.-lbs. (Work)}}{1 \text{ minute}}$$

$$\text{or} \quad \frac{550 \text{ ft.-lbs.}}{1 \text{ second}} = 1 \text{ Horsepower}$$

The 550 ft.-lbs. sec. (ability to lift 550 lbs. a distance of 1 ft. in 1 sec.) was then established as 1 hp. This standard is still in use today.

HORSEPOWER FORMULA

Engine horsepower can be calculated by dividing the total rate of work (ft.-lbs. per sec.) by 550 (ft.-lbs. per sec.)

If an engine lifted 330 lbs. a distance of 100 ft. in 6 sec., its total rate of work would be:

$$5500 \frac{\text{ft.-lbs.}}{\text{sec.}}$$

Dividing this by:

$$550 \frac{\text{ft.-lbs.}}{\text{sec.}}$$

(1 horsepower), you find the engine rated at 10 hp. The horsepower formula would then be:

$$1 \text{ Horsepower} = \frac{\text{Rate of Work in } \frac{\text{ft.-lbs.}}{\text{sec.}}}{550 \frac{\text{ft.-lbs.}}{\text{sec.}}}$$

This formula may be used also to determine the exact horsepower needed for other tasks.

KINDS OF HORSEPOWER

The word horsepower is used in more than one way. Some of the common terms include: brake horsepower, indicated horsepower, frictional horsepower, and rated horsepower.

BRAKE HORSEPOWER

Brake horsepower (bhp) indicates the actual usable horsepower delivered at the engine crankshaft. Brake horsepower is not always the same. It increases with engine speed. At very high and generally unusable engine speeds (depending on engine design), the horsepower output will drop off somewhat.

Fig. 9-12 shows how horsepower increases with speed for two different engine models. The top speeds in this chart do not run high enough to show a drop in horsepower.

HORSEPOWER

RPM	ENGINE MODELS	
	ACN	BKN
1600	2.5	3.5
1800	2.9	4.0
2000	3.5	4.4
2200	3.7	4.9
2400	4.2	5.4
2600	4.5	5.8
2800	4.8	6.2
3000	5.2	6.5
3200	5.6	6.7
3400	5.8	6.9
3600	6.0	7.0

Fig. 9-12. Brake horsepower increases with engine speed. Note that bhp at 3600 rpm is about twice that developed at 1600 rpm. (Wisconsin Motors Corp.)

MEASURING ENGINE BRAKE HORSEPOWER

Brake horsepower can be measured by using a Prony brake or an engine dynamometer.

The *Prony brake* is a friction device that grips an engine-driven flywheel and transfers the force to a measuring scale, Fig. 9-13. One end of the Prony brake pressure arm rests on the scale and the other wraps around a spinning flywheel driven by the engine under test. A clamp is used to change the frictional grip on the spinning flywheel.

To check brake horsepower, the engine under test is operated with the throttle wide open. Then engine speed is reduced to a specific number of

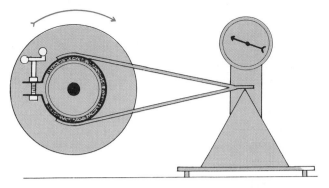

Fig. 9-13. When using a Prony brake, one end of the pressure arm surrounds a spinning flywheel driven by the engine. By tightening the friction device, torque is transmitted to and measured at the scale.

revolutions per minute by tightening the pressure arm on the flywheel. At exactly the right speed, the arm pressure on the scale is read. By using the scale reading (W) from the Prony brake, the flywheel rpm (R), and the distance in feet from the center of the flywheel to the arm support (L), brake horsepower can be computed.

The formula used to determine brake horsepower on the Prony brake is:

$$BHP = \frac{2\pi \times R \times L \times W}{33,000} \quad \text{or:}$$

$$BHP = \frac{R \times L \times W}{5252}$$

R = Engine rpm or speed.
L = Length from center of flywheel to the point where beam presses on scale in feet.
W = Weight as registered on scale in pounds

NOTE: One hp is:

$$550 \frac{\text{ft.-lbs.}}{\text{sec.}}$$

Since engine rpm is on a per-minute basis, it is necessary to multiply the 550 by 60 (60 sec. per min.), giving the figure 33,000.

ENGINE DYNAMOMETER

As with the Prony brake, the *dynamometer* loads the engine and transfers the loading to a measuring device. Instead of using a dry friction loading technique (clamping pressure arm to a spinning wheel), the dynamometer utilizes either hydraulic or electric loading. Several different types are illustrated in Fig. 9-14.

In Fig. 9-14A, the engine drives an electric generator that is attached to a spring scale. When an electrical load is placed in the circuit, the generator housing (an enclosure holding the moving parts) attempts to spin, applying a force to the scale. In Fig. 8-14B, a hydraulic water brake is attached to the scale. (Hydraulic means moved or operated by a liquid.)

The engine is loaded by admitting more and more water into the brake, causing the housing to try to rotate. This exerts a force on the scale. Fig. 8-14C shows an Eddy Current brake, and Fig. 8-14D illustrates the Prony brake.

A hydraulic load cell is used with one type of water brake dynamometer. When the housing tries to rotate, it creates a force on the hydraulic

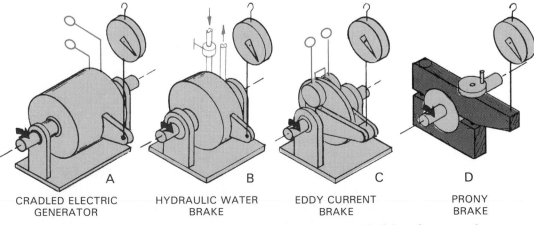

A	B	C	D
CRADLED ELECTRIC GENERATOR	HYDRAULIC WATER BRAKE	EDDY CURRENT BRAKE	PRONY BRAKE

Fig. 9-14. Variety of dynamometers shown utilize different principles of construction. (Go-Power Corp.)

Fig. 9-15. A small gas engine is readied for testing on a water brake dynamometer. (Go-Power Corp.)

load cell. This cell is a type of piston, which pushes against the water. The "push" is transmitted through tubing to the gage, where it is read in pounds. Fig. 9-15 shows a small gas engine set-up for testing on such a dynamometer. Note the load cell. See how it is attached to the housing. Fig. 9-16 pictures a dynamometer that was built to test model airplane engines.

INDICATED HORSEPOWER

Indicated horsepower (ihp) measures the power developed by the burning fuel mixture inside the cylinder. To measure ihp, you must determine the pressure inside the cylinder during the intake, compression, power, and exhaust strokes. A special measuring tool is used to provide constant checking of cylinder pressure. This pressure information is placed on an indicator graph, like the one in Fig. 9-17.

At this point, the *mean effective pressure* (mep) must be determined. To do this, we sub-

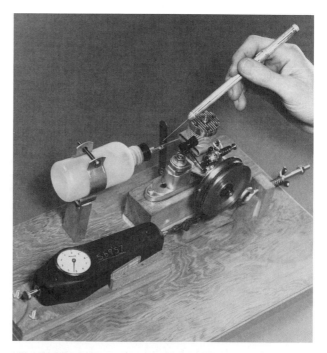

Fig. 9-16. A dynamometer built with a precision spring scale for testing a model airplane engine. Notice vibration tachometer being used to record rpm.

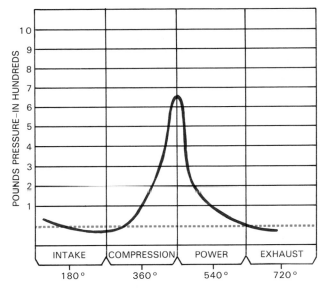

Fig. 9-17. Graph shows simulated cylinder pressure developed in cylinder of a specific four cycle engine. Atmospheric pressure is shown by dotted line. Graph makes it possible to establish mean effective pressure (mep).

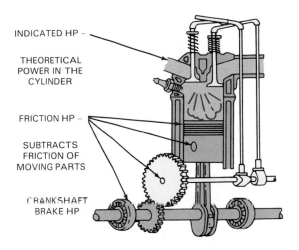

Fig. 9-18. Frictional horsepower is determined by subtracting crankshaft brake horsepower from indicated horsepower. (Deere & Co.)

tract the average pressure during the intake, compression, and exhaust strokes from the average pressure developed during the power stroke. The mean effective pressure changes according to engine type and design. After finding the mep, the following formula is used to determine indicated horsepower:

$$\text{Indicated Horsepower (IHP)} = \frac{\text{PLANK}}{33,000}$$

P = mep in pounds per square inch.

L = length of piston stroke in feet.

A = cylinder area in square inches.

N = power strokes per minute:

$\frac{\text{rpm}}{2}$ (four cycle engine).

K = number of cylinders.

FRICTIONAL HORSEPOWER

Frictional horsepower (fhp) represents that part of the potential or indicated horsepower lost because of the drag of engine parts rubbing together, Fig. 9-18.

Despite smooth contact surfaces and proper lubrication, a certain amount of friction (resistance to movement between two objects that are

rubbing together) is always present and represents a sizable horsepower loss. Actual loss will vary with engine design and use, but will generally run about 10%. Friction loss is not always the same. It increases with engine speed.

Frictional horsepower is determined by subtracting brake horsepower from indicated horsepower or:

$$\text{FHP} = \text{IHP} - \text{BHP}$$

RATED HORSEPOWER

An engine used under a brake horsepower load that is as great as the engine's highest brake horsepower rating will overheat. Excessive pressure on the bearings (loading) will seriously shorten the engine's service life. In some cases, complete engine failure can occur in a very short period of time.

As a general rule, never load an engine to more than 80% of its highest brake horsepower rating. For example, if a job requires a horsepower loading of 8 hp, you would use an engine with at least a 10 hp rating. Then, the load would be no more than 80% of the engine's maximum hp.

An engine's **rated horsepower** generally will be 80% of its maximum brake horsepower. Note in Fig. 9-19 how the rated horsepower (recommended maximum operating bhp) is less than engine maximum bhp.

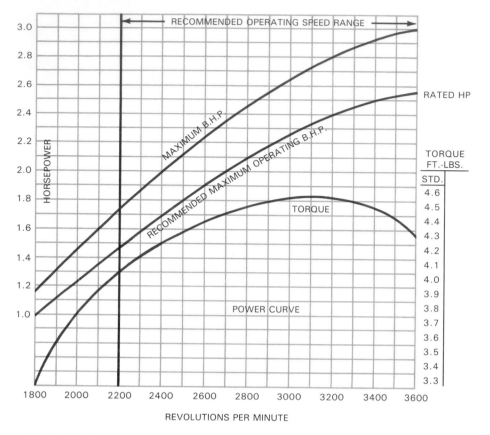

Fig. 9-19. Maximum operating brake horsepower loading is charted for a specific engine. At all speeds, "rated" hp is about 80% of maximum bhp. (Briggs and Stratton Corp.)

CORRECTED HORSEPOWER

Standard brake horsepower ratings are based on engine test conditions with the air dry, temperature at 60°F, and a barometric pressure of 29.92 in. Hg (sea level). Horsepower, however, can be greatly affected by changes in atmospheric pressure, temperature, and humidity (amount of moisture in the air).

Corrected horsepower is a "guess" at the horsepower of a given engine under specific operating conditions that are not the same as those present during actual dynamometer testing. Some facts to consider are:
- For each 1000 ft. of elevation above sea level, horsepower will drop around 3 1/2%.
- For each 1 in. drop in barometric pressure, horsepower will drop another 3 1/2%.
- Each 10°F of temperature increase results in a horsepower loss of 1%.

- New engines will develop somewhat less horsepower (due to increased friction) until they have been operated a number of hours.
- An increase of 200°F-400°F in head operating temperature can lower horsepower by 10%.
- Quality of fuel, mechanical conditions, and "state of tune" can also affect horsepower.

When horsepower tests are conducted under conditions varying from standard, corrections must be applied to establish true horsepower.

CORRECTION FACTOR

The correction factor (a factor is a condition which would change an answer) is determined by using the following formula:

Correction Factor =
Temperature Correction × Pressure Correction × Humidity Correction

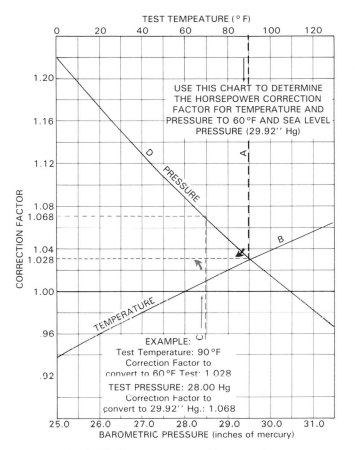

Fig. 9-20. Temperature and barometric pressure correction chart. (Go-Power Corp.)

For example, suppose that the dynamometer tests were carried out at a temperature of 90°F with an atmospheric pressure of 28 in. Hg and a wet bulb temperature (determines humidity) of 73.5°F. First, refer to the chart in Fig. 9-20. Follow dotted line A from top of chart (90°F temperature) down until it crosses the temperature line B. See that by now moving left along the chart, the dotted line shows a temperature correction factor of 1.028.

Now, follow dotted line C up from the 28 in. Hg marking at the bottom of the chart until it crosses pressure line D. By moving left at this point, a pressure correction factor of 1.068 is shown.

To find the humidity correction factor, use the chart in Fig. 9-21. Follow the dotted line up from the 90°F dry bulb temperature mark until it meets the 73.5°F dotted wet bulb temperature line. Move across to the right and note that the humidity correction factor is 1.0084.

Using the correction factor formula:

Correction Factor =
Temperature Correction × Pressure Correction × Humidity Correction   or
Correction Factor = 1.028 × 1.068 × 1.009 or
Correction Factor = 1.108

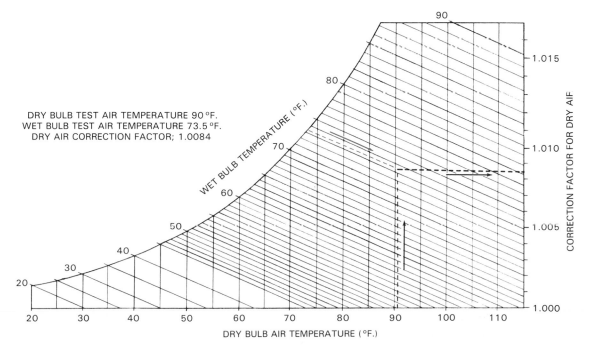

Fig. 9-21. Chart for determining humidity correction factor. (Go-Power Corp.)

If the dynamometer test had shown 3.15 horsepower, this reading could be changed to standard test conditions by applying the correction factor of 1.108 as determined above. Thus:

Corrected HP =
Correction Factor × Test HP   or
Corrected HP = 1.108 × 3.15   or
Corrected HP = 3.49 HP

TORQUE

Torque is a twisting or turning force. Therefore, any reference to engine torque means the turning force developed by the rotating crankshaft.

In order to find torque, we must know the force (in pounds) and the radius (distance, in feet, from the center of the turning shaft to the exact point at which the force is measured). The formula would read:

Torque = Force × Distance (Radius)   or
Torque = Pounds × Feet   or
Torque = lbs.-ft.

Fig. 9-22 shows a torque wrench attached to a rotating crankshaft. Imagine that it is attached with a friction device much like the Prony brake pressure arm shown in Fig. 9-13. This will allow the crankshaft to turn while still applying turning force to the stationary torque wrench. Suppose a scale placed exactly two feet from the center of

the crankshaft indicates a force of 100 lbs. By using the formula for determining torque (Torque = lbs.-ft. or Torque = 100 lbs. x 2 ft.), we find that this engine is developing 200 lbs.-ft. of torque.

If the scale is 3 ft. from the shaft center and the force is 50 lbs., the torque will be 150 lbs.-ft. When measuring torque, the reading is given in lbs.-ft. When measuring work, the reading is given in ft.-lbs.

TORQUE IS NOT CONSTANT

Engine torque, for any engine and set of test conditions, will change according to engine speed. The pressure of the burning air-fuel mixture against the piston is transferred to the crankshaft by the connecting rod. The greater the pressure, the more torque the crankshaft will develop.

The point where gas pressure will be highest is the speed at which the engine takes in the largest volume of air-fuel mixture. This point will vary according to engine design but will always be at a lower speed than that at which the greatest horsepower is reached. Horsepower generally increases to quite a high rpm before it finally begins to drop off. Torque, on the other hand, decreases at a much lower rpm.

As engine speed is increased beyond idle, its torque increases. As it continues to speed up, a

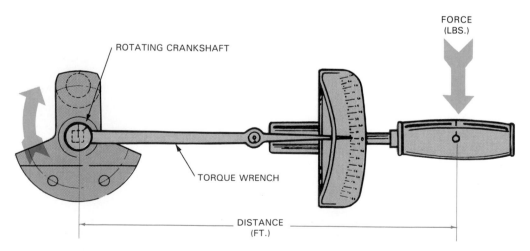

Fig. 9-22. Torque is determined by multiplying turning effort in pounds by the distance from shaft center to point at which force is read. (Dresser Industries Inc.)

point will be reached where the natural restriction to airflow through the carburetor, intake manifold, and valve ports begins to limit the speed at which the air-fuel mixture can enter the cylinder. At this point, the highest torque is developed.

When engine speed goes higher than this point, the intake valve will open but the piston moves far down on the intake stroke before the mixture can get into the cylinder. This cuts down the amount of air-fuel mixture entering the cylinder. As a result, burning pressure is lowered as well as the torque. Beyond this point, torque will decrease as speed increases.

VOLUMETRIC EFFICIENCY

How well an engine "breathes," or draws the air fuel mixture into the cylinder, is referred to as its *volumetric efficiency*. It is measured by comparing the air-fuel mixture actually drawn in to the amount that could be drawn in if the cylinder were completely filled, Fig. 9-23.

Volumetric efficiency changes with speed. At high engine speeds, it can be very low. The reason for this is simple. As engine revolutions increase beyond a certain point, the piston moves down (intake stroke) so rapidly that it travels far down the cylinder before the air-fuel mixture begins to flow into the cylinder. The intake cycle can be complete (piston moving upward on compression stroke) before the cylinder is much more than half full.

Other factors can change the volumetric efficiency, including atmospheric pressure; air temperature; air cleaner, carburetor and intake manifold design; size of intake valve; engine temperature; throttle position; valve timing; and camshaft design.

Efficiency can be increased by using a larger intake valve, altering cam profiles (shapes) or cam timing, increasing the size of the carburetor air horn, straightening and increasing the diameter of the intake manifold, improving exhaust flow, and/or adding a supercharger.

Fig. 9-24 illustrates how volumetric efficiency varies with speed. There is a gradual buildup to a certain rpm, followed by a rapid decline as speed is increased. Remember that the speed at which top volumetric efficiency is reached will vary with engine design.

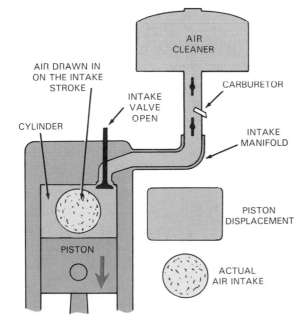

Fig. 9-23. Volumetric efficiency is the measurement of an engine's "breathing" ability. It compares intake of air-fuel mixture with piston displacement. Note that actual air intake is considerably less than piston displacement. (Deere & Co.)

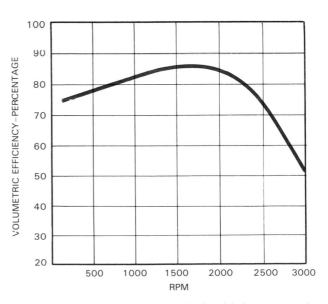

Fig. 9-24. Graph shows close relationship between volumetric efficiency and engine rpm. As engine speed reaches a certain point, efficiency declines rapidly. Torque is also greatest at the point of highest volumetric efficiency.

TORQUE AND HORSEPOWER

Unlike torque (which drops off when engine revolutions per minute exceed point of maximum volumetric efficiency), horsepower continues to increase until engine speed is very high. Beyond a certain speed, however, horsepower will actually decrease.

Keep in mind that torque measures the twisting force generated by the crankshaft while horsepower measures the engine's ability to perform work. Even though torque may decline at higher speeds, the shaft is turning much faster. Therefore, it is able to perform work at a greater rate.

Fig. 9-25 shows the relationship between torque and horsepower curves for one specific engine. Note the arrow indicating the rated horsepower. This is the horsepower at which the engine can be operated continuously without damage.

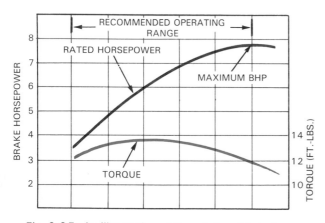

Fig. 9-25. An illustration of the relationship between torque an horsepower for one particular engine. (Clinton Engine Corp.)

PRACTICAL EFFICIENCY

In theory, each gallon of gasoline contains enough energy to do a certain amount of work. This may be thought of as potential energy. Unfortunately, engines are not efficient enough to use all the potential energy in the fuel. Practical efficiency takes into consideration power losses caused by friction, incomplete burning of the air-fuel mixture, heat loss, etc. *Practical efficiency* is simply an overall measurement of how efficiently an engine uses the fuel supply.

MECHANICAL EFFICIENCY

Mechanical efficiency is the percentage of power developed in the cylinder (indicated horsepower) compared to the power that is actually delivered at the crankshaft (brake horsepower). Brake horsepower is always less than indicated horsepower. The difference is due to friction losses within the engine. Mechanical efficiency runs about 90%, indicating an internal friction loss of about 10%.

The formula for mechanical efficiency is:

$$\text{Mechanical Efficiency} = \frac{\text{Brake Horsepower}}{\text{Indicated Horsepower}} = \frac{\text{BHP}}{\text{IHP}}$$

THERMAL EFFICIENCY

Thermal efficiency (heat efficiency) indicates how much of the power produced by the burning air-fuel mixture is actually used to drive the piston downward.

Much of the heat developed by the burning gas is lost to such areas as the cooling, exhaust, and

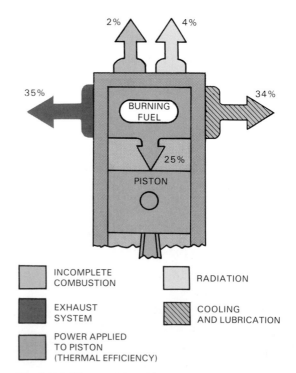

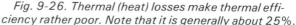

Fig. 9-26. Thermal (heat) losses make thermal efficiency rather poor. Note that it is generally about 25%.

lubricating systems. Thermal efficiency will run about 20 to 25%. Keep in mind that the percentages are only "about right" and will vary depending upon engine design and operation, Fig. 9-26. The exhaust system siphons off about 35% of the heat. The cooling and lubricating systems combine to absorb a like amount. The rest is lost through radiation and incomplete combustion.

Use this formula for computing brake thermal efficiency:

$$\text{Brake Thermal Efficiency} = \frac{\text{Brake Horsepower (BHP} \times 33,000)}{778 \text{ Fuel Heat Value} \times \text{Weight of Fuel Burned per Minute}}$$

The 778 in the above formula is Joule's equivalent (a number which is equal to another). The fuel heat (calorific) value is based on the Btu (British thermal unit) per pound.

SUMMARY

Engine bore is the diameter across the top of the cylinder. Stroke is the up or down movement of the piston. Length of stroke is determined by the distance the piston moves from its uppermost position to its lowest position.

Engine displacement refers to the total volume of space increase in the cylinder as the piston moves from the top of its stroke to the bottom of its stroke.

An engine's compression ratio is a measurement of the relationship between total cylinder volume at BDC compared to the volume remaining when the piston is at TDC.

Force is the pushing or pulling of one body on another. The greater the force applied to the piston, the greater the amount of power and work that can be done by the engine. Pressure is a force per given unit of area.

Work is accomplished only when a force is applied through some distance. Power is the rate at which work is performed. The standard unit of power is horsepower.

Energy is the capacity to perform work. Types of energy include potential energy, kinetic en-ergy, mechanical energy, chemical energy, and thermal energy.

Engine horsepower is calculated by dividing the total rate of work (ft.-lbs. per sec.) by 550 (ft.-lbs. per sec.)

Brake horsepower indicates the actual usable horsepower delivered at the crankshaft. Brake horsepower increases with engine speed.

Indicated horsepower refers to the power developed by the burning fuel mixture inside the cylinder. To measure indicated horsepower, you must determine the pressure inside the cylinder during the intake, compression, power, and exhaust strokes.

Frictional horsepower represents the part of the indicated horsepower lost because of engine parts rubbing together. Frictional horsepower is determined by subtracting brake horsepower from indicated horsepower.

An engine's rated horsepower is generally 80% of its maximum brake horsepower.

Standard brake horsepower ratings are based on ideal engine test conditions. Horsepower can be greatly affected by changes in atmospheric pressure, temperature, and humidity. Corrected horsepower is an estimation of the horsepower of a given engine under specific operating conditions.

Engine torque refers to the turning force developed by the crankshaft. Engine torque will change according to speed.

Volumetric efficiency is the measurement of how well an engine draws the air-fuel mixture into the cylinder.

Practical efficiency is an overall measurement of how efficiently an engine uses the fuel supply.

Mechanical efficiency is the percentage of power developed in the cylinder (indicated horsepower) compared to the power that is actually delivered at the crankshaft (brake horsepower).

Thermal efficiency indicates how much of the power produced by the burning air-fuel mixture is actually used to drive the piston.

KNOW THESE TERMS

Bore, Stroke, Crank offset, Displacement, Compression ratio, Force, Tensile stress, Pressure, Work, Power, Horsepower, Energy, Brake horsepower, Prony brake, Dynamometer, Indicated horsepower, Mean effective pressure, Frictional horsepower, Rated horsepower, Corrected horsepower, Torque, Volumetric efficiency, Practical efficiency, Mechanical efficiency, Thermal efficiency.

REVIEW QUESTIONS–CHAPTER 9

1. What type of small gasoline engines are discussed in this text?
2. How do you compute engine displacement?
3. What force acts opposite to the direction of centrifugal force?
4. The force applied opposite to the way a compressive force is applied is called _____ force.
5. If an object is raised 3 ft. and is then moved 5 ft. sideways, at the same level, work is being done over a distance of:
   a. 8 ft.
   b. 5 ft.
   c. 3 ft.
   d. 2 ft.
6. Give the formula for work.
7. If a force of 35 lbs. is applied to an area of 5 sq. in., what is the pressure in psi?
8. The top area of a 2 1/2 in. diameter piston would be how many sq. in.?
9. If 60 lbs. is lifted 5 ft. in 6 sec., what amount of power is exerted in ft.-lbs. per sec?
10. What is the definition of horsepower? How much is 1 hp?
11. A measure of the horsepower delivered at the engine crankshaft is called _____.
12. A Prony brake with a 12-in. arm is applied to an engine flywheel. At 1200 rpm, the scale registers 20 lbs. Calculate the horsepower to two decimal places.
13. What is indicated horsepower?
14. When testing an engine on the dynamometer, what are the standard test conditions?
15. What is the engine horsepower reduction for each 1000 ft. of elevation?
16. Explain the horsepower correction factor.
17. Volumetric efficiency is greatest at _____ speed.
18. Horsepower of an engine is greatest at maximum rpm. True or False?
19. On the average, what percentage of the energy from fuel is used to produce power?
20. An engine's rated horsepower is approximately _____ percent of its maximum brake horsepower.

SUGGESTED ACTIVITIES

1. Using the principles studied in this chapter, determine the horsepower required of several individuals to walk up one flight of stairs. The following items will be needed: a stopwatch, tape measure and bath scale.
2. Design and build a Prony brake for a small gasoline engine.
3. Use a dynamometer to develop a graph like the one in Fig. 9-25. Compare the graph with the manufacturer's graph for the same model engine.
4. On an engine with the head removed, measure the position of the top of the piston (in relation to the top of the block) at TDC and again at BDC. Determine the stroke of the engine.
5. On the same engine, measure the diameter of the piston and determine if the engine is square, oversquare or undersquare.
6. Determine the cubic inch displacement of the engine from the facts learned in activities above.

CHAPTER 10

TOOLS AND MEASURING INSTRUMENTS

After studying this chapter, you will be able to:
- ☐ Explain why quality tools and measuring instruments should be used when servicing small gas engines.
- ☐ Summarize the reasons that small engine components must be measured carefully.
- ☐ Demonstrate several common measuring techniques.
- ☐ Use common hand tools properly.

High-quality tools and measuring instruments should always be used when servicing a small gas engine. Many dimensions are critical to proper engine operation. Therefore, it is extremely important to be able to measure various engine parts and clearances accurately. When making engine repairs, measurements must be made to determine if parts are within specified limits or if replacements and adjustments must be made.

Quality tools allow you to service small engines easily and effectively. To avoid damage to engine parts, always use the tools recommended by the manufacturer. Keep tools clean and in proper working condition.

Some tools and measuring instruments are common to most engine work, while others may have only one or two specific applications. Special-purpose tools may be designed by a manufacturer for limited use on only one engine make or model. This chapter will describe how to use common tools and measuring instruments. Some special-purpose tools will also be examined, along with examples of their applications.

MICROMETERS

The *micrometer* is a precision instrument designed to accurately measure pistons, crankshafts,

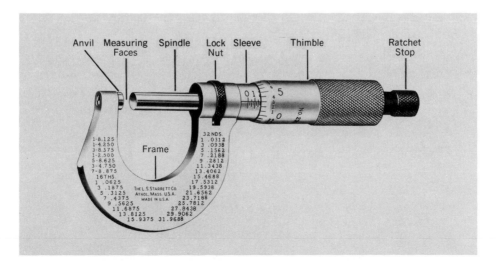

Fig. 10-1. Note the parts of an outside micrometer. This particular micrometer is graduated in thousandths of an inch (.001). (L. S. Starrett Co.)

valve stems, and other small engine components. You must be able to read a micrometer correctly to make judgments about the condition of various engine parts.

There are several varieties of micrometers available. Each type is designed for a specific purpose. An outside micrometer is used to measure thicknesses and outside diameters, Fig. 10-1. The inside micrometer, Fig.10-2, is designed for taking measurements of internal dimensions. A blade micrometer, such as the one shown in Fig. 10-3, is used to take measurements in narrow slots. Fig. 10-4 illustrates an inside micrometer being used to measure the piston pin bore of a connecting rod and an outside micrometer being used to measure the piston pin.

Because the micrometer is made of metal, it will expand when heated and contract when cooled. The micrometer should not be held in the hand for long periods of time. Body heat can

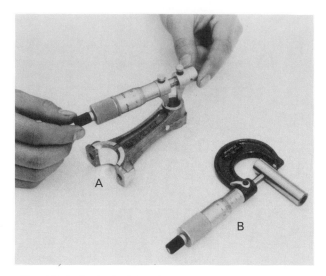

Fig. 10-4. A—An inside micrometer being used to measure the internal diameter of a piston pin hole in a connecting rod. B—An outside micrometer being used to measure a piston pin. (Kubota Tractor Corp.)

cause inaccuracy in the instrument. Always hold the micrometer properly. Fig. 10-5 shows the proper one-hand method for measuring a valve stem diameter with a micrometer. Fig. 10-6 shows a micrometer being held with two hands. Note the minimal contact between the hands and the micrometer. Measuring instruments are quite expensive and should always be handled with care.

Micrometers are available in several sizes. Most micrometers have the capability to measure

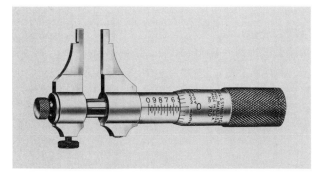

Fig. 10-2. The inside micrometer is designed to measure internal dimensions. (L. S. Starrett Co.)

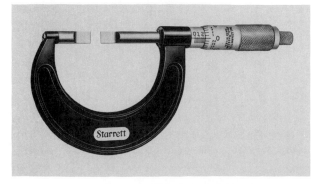

Fig. 10-3. The blade micrometer is designed to enable the spindle and anvil to measure in narrow grooves. (L. S. Starrett Co.)

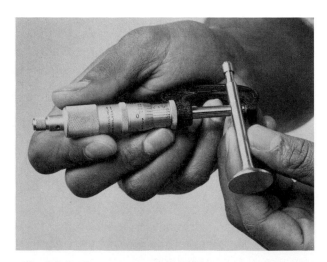

Fig. 10-5. The proper way to hold a micrometer in one hand while holding the piece to be measured in the opposite hand. Note small finger in frame of micrometer.

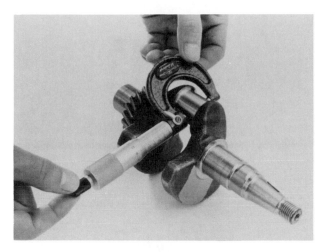

Fig. 10-6. The proper technique for measuring with a micrometer when two hands are necessary. Use very light pressure when turning the thimble of the micrometer. (Kubota Tractor Corp.)

in 1 inch increments. For example, a 0-1 in. micrometer can be used to measure objects smaller than one inch. A 2-3 in. micrometer, on the other hand, is designed to measure objects that are between 2 in. and 3 in.

CLEANING AND CALIBRATING A MICROMETER

Before attempting to measure any object, make sure that the micrometer's anvil and spindle faces are clean. If necessary, clean the faces by gently closing the micrometer on a piece of clean, white paper and drawing the paper from between the faces. To assure accuracy, the micrometer should always be checked for proper calibration before use. A 0-1 in. micrometer can be checked by simply closing it and observing the reading on the sleeve and thimble. If the reading is not zero, clean the anvil and spindle again and retest the micrometer. If it still does not read zero, check the manufacturer's instructions for adjusting the instrument. Micrometers larger than 0-1 in. require the use of a gage block, cylindrical gage, or gage pin to verify calibration. These blocks are precision ground to exact dimensions and are used to check the micrometer for accuracy. For example, a 1 in. gage block can be used to check the calibration of a 1-2 in. micrometer by simply measuring the block. The micrometer should

read exactly one inch. If it does not, it must be recalibrated.

USING A MICROMETER

To use a micrometer, simply place the object to be measured against the anvil and turn the thimble until the spindle touches the object. If the micrometer has a ratchet-stop, Fig. 10-1, click it just once after making contact with the work surface. If the micrometer does not have a ratchet-stop, the thimble must be turned very gently so the anvil and spindle faces lightly contact the outer surfaces of the part being measured. Over-tightening can permanently damage the micrometer. If it is necessary to remove the micrometer from the part to read it, the lock nut can be used to secure the spindle so that it does not turn during removal.

STANDARD AND VERNIER MICROMETERS

Some micrometers provide more accurate readings than others. The *standard micrometer* is graduated in thousandths of an inch (.001), Fig. 10-1. The *Vernier micrometer*, on the other hand, is graduated in ten-thousandths of an inch (.0001 in.). The Vernier micrometer has an additional scale on its sleeve, Fig. 10-7. Part tolerances are specified in three or four decimal place numbers. A Vernier micrometer can be used for either measurement, but the standard micrometer is only accurate to three decimal places.

READING THE STANDARD MICROMETER

The first step in reading any micrometer is to familiarize yourself with the divisions (graduations) on the sleeve and the thimble.

The micrometer's sleeve graduations reflect the fact that the spindle moves 1/40 in. (0.025 in.) for each revolution of the thimble. Therefore, the micrometer's sleeve is divided into 40 equal spaces. Each line on the sleeve represents 1/40 in. (0.025 in.). Every fourth line is numbered. These numbers represent 1/10 in. (.100 in.), Fig. 10-8. Looking at the example in Fig. 10-10A, the thimble has been rotated four turns, or one tenth of an inch (4 × .025 = .100 in.).

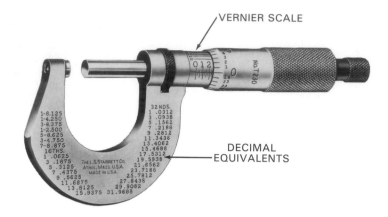

Fig. 10-7. A Vernier micrometer can measure to four decimal places with the special scale that is located on its sleeve. The fourth decimal place is determined by the two lines that coincide with each other. (L. S. Starrett Co.)

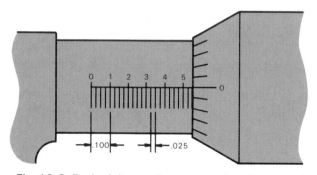

Fig. 10-8. Each of the small spaces on the micrometer's sleeve is equal to one fortieth of an inch (.025). Since there are four small spaces between each of the numbers printed on the sleeve, the distance between numbers is equal to one tenth of an inch (.100). The reading on this micrometer is five-hundred and fifty thousandths of an inch.

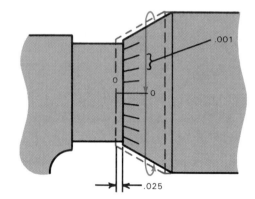

Fig. 10-9. Each small space on the thimble of the micrometer is equal to one one-thousandth of an inch (.001). One complete rotation of the thimble moves the spindle twenty-five thousandths of an inch (.025), or one small space on the sleeve.

The micrometer's thimble is divided into 25 equal parts. Each line on the thimble represents 0.001 in. Fig. 10-9 illustrates a typical micrometer thimble. Fig. 10-10B, shows a thimble that has been rotated ten full turns (.250 in.), plus .008 in. more, totaling .258 inch. Now, read the micrometer scale in Fig. 10-10C.

READING A VERNIER MICROMETER

To obtain readings to four decimal places, a Vernier micrometer must be used. The Vernier micrometer has an additional scale, called the Vernier scale, located on the top of its sleeve, Fig.

10-11. The first three decimal places on a Vernier micrometer are read in the same way as they are on the standard micrometer. The fourth decimal number, however, is obtained from the Vernier scale. Unless the zero line on the thimble is aligned with the sleeve's horizontal reference line, only one line of the Vernier scale will be perfectly aligned with one of the lines on the thimble. If the sixth line of the Vernier scale is aligned with one of the lines on the thimble, the fourth decimal place number would be six. Fig. 10-11A illustrates a 0-1 in. Vernier micrometer displaying a reading of .2586 inches. Can you read the measurements in Figs. 10-11B and 10-11C.

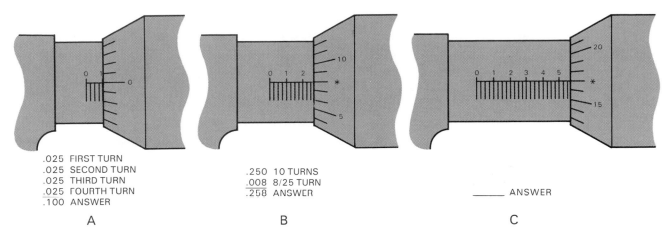

.025 FIRST TURN
.025 SECOND TURN
.025 THIRD TURN
.025 FOURTH TURN
.100 ANSWER

A

.250 10 TURNS
.008 8/25 TURN
.258 ANSWER

B

_____ ANSWER

C

Fig. 10-10. Study the micrometer readings on the 0-1 in. micrometers in A and B.
Can you read the micrometer in C? (Answer = .567)

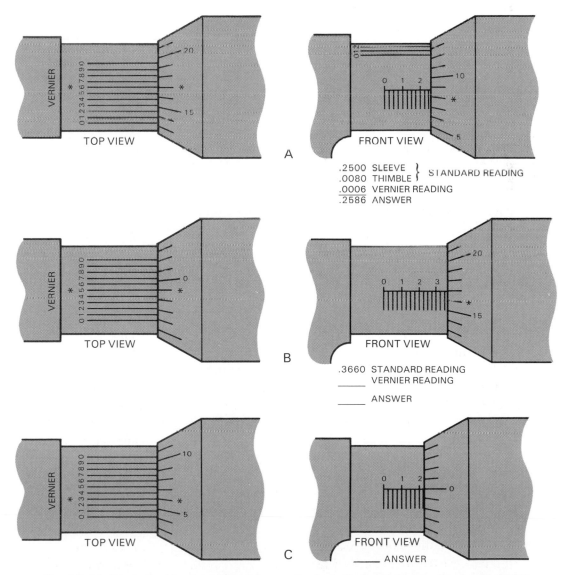

TOP VIEW

FRONT VIEW

A

.2500 SLEEVE } STANDARD READING
.0080 THIMBLE
.0006 VERNIER READING
.2586 ANSWER

TOP VIEW

FRONT VIEW

B

.3660 STANDARD READING
_____ VERNIER READING

_____ ANSWER

TOP VIEW

FRONT VIEW

C

_____ ANSWER

Fig. 10-11. Study the Vernier micrometer reading in A. Complete the reading for B (.3665).
Read C (.2253). Assume that these are 0-1 in. micrometers.

Modern digital micrometers eliminate reading anything but the numbers in a window. Remember, always handle measuring instruments carefully and keep them clean to maximize accuracy and reliability. Once you have practiced a few readings, you will see how easily and quickly micrometers can be used.

TELESCOPING GAGES

Telescoping gages are "transfer-type" measuring instruments. They do not provide a direct dimensional reading. A telescoping gage, like the one in Fig. 10-12, can be used to transfer the distance from A to B to a micrometer, Fig. 10-17.

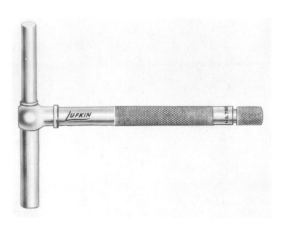

Fig. 10-13. Telescoping gages can be purchased singly. (Lufkin Rule Co.)

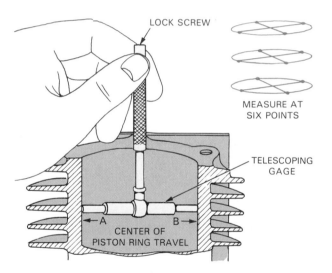

Fig. 10-12. A telescoping gage can be used to measure the inside diameter of a cylinder. 1–Depress the spindle and lock with lock screw. 2–Place gage in cylinder and release lock screw. Spindle will spring out to touch walls of cylinder. 3–Tighten lock screw. 4–Tilt gage handle and remove the gage. 5–Measure across spindle faces with an outside micrometer. (Briggs and Stratton Corp.)

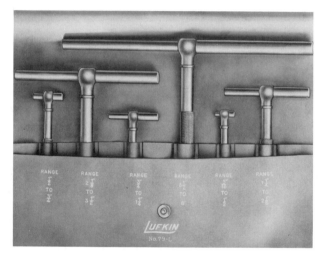

Fig. 10-14. Telescoping gages are available in sets to accommodate a variety of measurements. (Lufkin Rule Co.)

Telescoping gages can be purchased singly, Fig. 10-13, or in sets, providing a wide range of sizes to accommodate a variety of measurements, Fig. 10-14. The spindle faces are curved so that each end has only one point of contact when curved surfaces are being measured, Fig. 10-15. To use the telescoping gage to determine the diameter of a cylinder, loosen the lock screw on the end of the handle so that the telescoping parts can

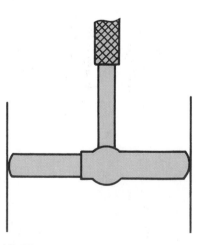

Fig. 10-15. The spindle faces of telescoping gages are curved so spindle touches cylinder at only one point.

be retracted, locked, and placed in the cylinder. Once the gage is located in the cylinder, release the lock screw to allow the telescoping ends to extend. The handle must be held perfectly in line with the centerline of the cylinder being measured. Also, it is essential that the telescoping ends be aligned exactly across the true diameter of the cylinder, Fig. 11-11. The general practice to obtain this position is to hold the handle gently with the thumb and forefinger while sliding the gage up and down the cylinder walls. Spring tension will cause the gage to seek the true diameter of the cylinder. When the true diameter has been located, use the opposite hand to gently tighten the lock screw, securing the telescoping ends. When removing the gage from the cylinder, tilt the handle so that it can be removed without changing the setting, Fig. 10-16. After the gage is removed, measure the distance between the telescoping ends with a micrometer, Fig. 10-17.

SMALL HOLE GAGE

The *small hole gage* is similar to the telescoping gage. It is intended for measuring holes that are too small for the smallest telescoping gage. Small hole gages are usually provided in sets to accommodate a variety of hole sizes, Fig. 10-18. The small hole gage is expanded by turning a knurled screw on the end of its handle until the split ball touches the inner walls of the hole, Fig. 10-19. Test for correct fit by moving the gage back and forth in the hole until you feel a slight drag.

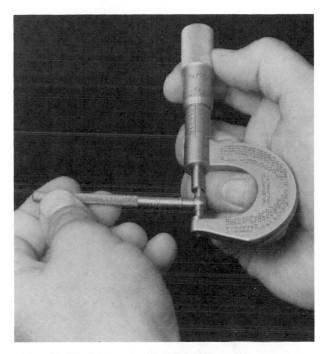

Fig. 10-17. After removing the telescoping gage from the cylinder, measure the gage with a micrometer. (L. S. Starrett Co.)

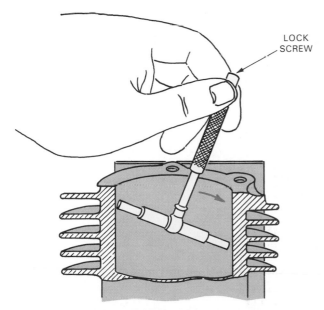

Fig. 10-16. Tilting the telescoping gage allows easy removal and will not change the position of the spindles.

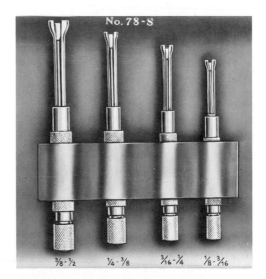

Fig. 10-18. Small hole gages are used to measure the diameter of holes that are too small for telescoping gages.

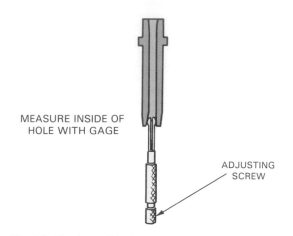

MEASURE INSIDE OF
HOLE WITH GAGE

ADJUSTING
SCREW

Fig. 10-19. A small hole gage used to measure the inside diameter of a valve guide. Insert the gage and expand the split ball end by turning the adjusting screw
until contact is felt. Move gage up and down in the
hole as adjustment is made. (Deere & Co.)

Fig. 10-21. A thickness gage is being used to measure
piston ring end gap. Note that the ring is placed in the
cylinder. (Kubota Tractor Co.)

Fig. 10-20. Measuring the small hole gage with a
micrometer. (L. S. Starrett Co.)

Fig. 10-22. Measuring valve stem clearance with a
thickness gage. This gage is often referred to as a
''feeler gage'' because the sensitivity of feel is required
to use the gage accurately. (Kubota Tractor Corp.)

Remove the gage and measure across the ball with
a micrometer, Fig. 10-20.

THICKNESS GAGES

Thickness gages are sometimes called "feeler
gages" because they rely on the user's sense of feel
for accuracy. They are used to measure small
spaces and gaps between surfaces. Thickness
gages are used to measure spark plug gap, breaker
point gap, and piston ring gap, Fig. 10-21. Fig.
10-22 illustrates a thickness gage being used to
check valve stem clearance.

Thickness gages consist of a set of metal leaves
that vary in thickness. Each leaf has its decimal
thickness etched onto its surface, Fig. 10-23.

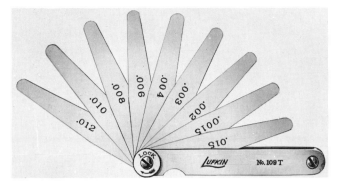

Fig. 10-23. The leaves of a thickness gage are marked with their exact thickness. When several leaves are used to obtain a desired thickness, each must be flat and clean. Dirty or bent leaves will cause inaccurate readings. (Lufkin Rule Co.)

Fig. 10-24. A thickness gage being used to measure crankshaft end-play. Begin with a thin leaf and progress to thicker leaves until one will not enter the space. Then, return to the thickest leaf that will fit the space. (Deere & Co.)

Fig. 10-25. Valve spring tension tester. A conversion plate located on the tester converts pounds-inch and pounds-foot of torque to pounds of applied force. If specifications are in metric newton meters, one newton meter is equal to 1.36 pounds-foot.

Some leaves are as thin as .0005 in. and may be damaged if not handled carefully. It is sometimes necessary to select several leaves to equal a desired thickness. When using a thickness gage, the leaves must be perfectly clean. Avoid bends and distortions. These conditions will increase the total thickness of the leaves and cause measurement errors.

Fig. 10-21 shows a thickness gage being used to check for proper ring gap. To use the gage, select a leaf that is thinner than the gap to be measured and place it in the gap. Progressively use thicker leaves until you can feel a slight contact or drag from the gap's edge surfaces. At this point, compare the leaf thickness with the amount of gap specified. If the gap is too small, it may be corrected by carefully filing the ring ends. See Fig. 12-19. If the gap is too wide, a new ring will be required. Do not use leaf-type thickness gages for measuring gaps on used spark plugs or pitted breaker points. Fig. 10-24 shows a thickness gage being used to measure crankshaft end play.

VALVE SPRING TENSION TESTER

The function of the valve springs is to close the valves quickly and securely. If spring tension is inadequate, valve leakage and poor valve timing can occur. The *valve spring tension tester*, Fig. 10-25, is designed to test valve spring tension with the aid of a torque wrench. The information plate on the base of the tester converts pounds-inch and pounds-foot of torque to pounds of applied force. Some torque specifications are provided in metric newton meters (N.m).

COMBINATION SQUARE

The *combination square* has many uses. It is often used to measure the length of valve springs and check them for straightness, Fig. 10-26. For accuracy, always place spring and square on a flat, machined surface.

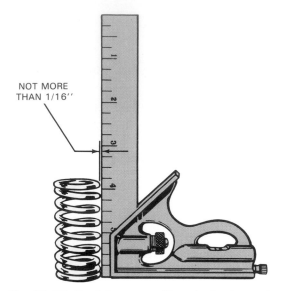

NOT MORE
THAN 1/16"

Fig. 10-26. Straightness and length of valve springs
can be measured on a flat surface with a combination
square. (Deere & Co.)

DIAL INDICATOR

Dial indicators are precision instruments that are very useful for measuring the movement of various parts, Fig. 10-27. They are also used to check for surface irregularities and run-out. The dial indicator is equipped with a spring-loaded spindle, which is placed against the part to be measured. A needle on the instrument's dial is used to indicate the amount of movement made by the part being tested. The needle rotates over the dial indicator's face, which is calibrated in thousandths of an inch, ten-thousandths of an inch, or hundredths of a millimeter. The indicator's bezel can be revolved to locate the zero line on the dial face under the needle. The zero line coincides with the initial position of the needle and serves as a reference point when checking for movement. The dial indicator can be mounted on a variety of devices that are designed to hold it in the desired location. One convenient holding device has a magnetic base that can be

Fig. 10-28. A dial indicator is often used to measure
crankshaft end-play. (Kubota Tractor Co.)

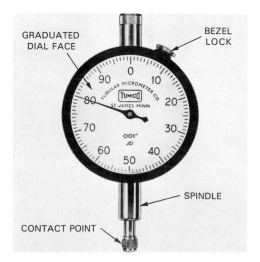

Fig. 10-27. The dial indicator can be used to measure
linear movement. Each space on this dial indicator represents one one-thousandth of an inch (.001).

Fig. 10-29. A dial indicator being used to measure side
movement of a valve stem in its valve guide.
(Kubota Tractor Co.)

turned on and off. When the mounting surface is nonmagnetic, however, a clamp-type holding device must be used. Fig. 10-28 shows a dial indicator being used to measure crankshaft end-play. The side movement of a valve in its guide is being checked in Fig. 10-29.

SCREW PITCH GAGE

The *screw pitch gage*, Fig. 10-30, is used to determine the number of threads per inch on bolts, screws, nuts, and in threaded holes. Each leaf of the gage is marked with the number of threads per inch it will match when the leaf is placed onto the screw, bolt, etc., Fig. 10-31. Keep trying leaves until one fits exactly into the threads. Read the number on the leaf to determine the number of threads per inch on the item being checked. Screw pitch gages are available for both standard and metric threads.

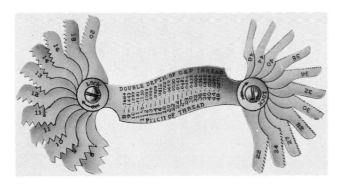

Fig. 10-30. The screw pitch gage has many blades. Each blade is marked with the number of threads per inch or threads per millimeter that it will match. The use of this gage is a trial and error attempt to match the teeth of a gage blade to the threads of a screw.

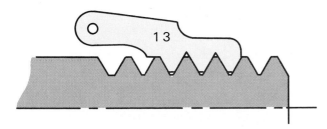

Fig. 10-31. A screw pitch blade matched to the threads of a screw. Threaded bolts must match the threads of a threaded hole to prevent damage. A bolt should never be forced to enter a threaded hole if it will not turn freely.

WRENCHES

There are many types of wrenches available to suit practically every situation encountered when servicing small engines. These include box-end wrenches, open-end wrenches, adjustable wrenches, Allen wrenches, socket wrenches, and torque wrenches. The type of wrench used may depend on the kind of fastener to be installed or removed. Box-end, open-end, adjustable, and socket wrenches are used on hexagon head bolts and nuts.

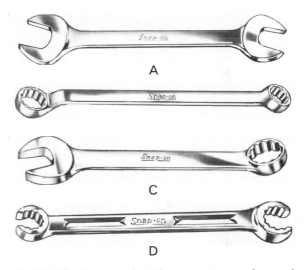

Fig. 10-32. Always select the correct type of wrench for the job at hand. A—Open-end wrench. B—Box-end wrench. C—Combination box/open-end wrench. D—Tubing wrench. (Snap-on Tools Corp.)

Open-end wrenches, Fig. 10-32A, should only be used when it is not possible to encompass the nut or bolt head with a box-end wrench or a socket. *Box-end wrenches,* Fig. 10-32B, can be used where partial or full-turn clearance is available. Box-end wrenches may be six or twelve sided and are less likely to slip around the bolt head corners than open-end wrenches or adjustable wrenches. *Combination wrenches* have a box on one end and are open on the other, Fig. 10-32C. A *tubing wrench,* Fig. 10-32D, is used on metal tubing connection fittings. When applying force to a wrench, always pull the tool instead of pushing on it, Fig. 10-33. This will prevent hand and knuckle injury if the wrench should accidentally slip from the bolt head.

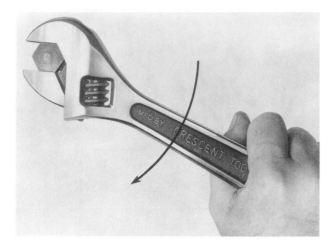

Fig. 10-33. A wrench should always be pulled in the proper direction to prevent it from slipping off a bolt or nut and rounding the corners.

Adjustable wrenches should only be used as a last resort when other wrenches are not available. Adjustable wrenches are used similarly to open-end wrenches and should also be pulled in the correct direction, Fig. 10-34. Due to the movable jaw in adjustable wrenches, they are prone to loosening and slipping around the corners of bolts and nuts.

Many varieties of *socket sets* are available. For small engine work, sets with standard and deep-length sockets, one ratchet wrench, one spark

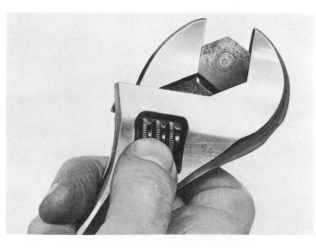

Fig. 10-34. Adjustable wrenches should always be pulled in the correct direction when tightening or loosening bolts and nuts. The movable jaw should always face the turning direction. The jaws should be adjusted to fit the bolt or nut as tightly as possible.

plug socket, and several extensions will meet most needs, Fig. 10-35. Sockets are extremely useful when bolts or nuts are recessed in counterbored holes. There are many occasions when, due to obstructions, other wrenches cannot be applied or turned very far. In these cases, a socket and ratchet wrench is the only way a bolt or nut can be removed or tightened. At other times, sockets can simply save a lot of time compared with other wrenches.

Allen wrenches are used to remove or install hex socket-head screws. Allen wrenches may be the conventional right angle style, Fig. 10-36, or

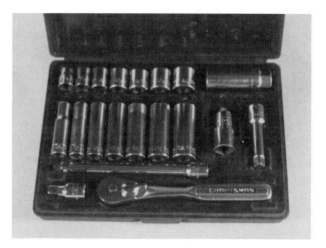

Fig. 10-35. A socket set is almost a necessity for small engine work. It permits installation and removal of bolts in hard-to-reach places.

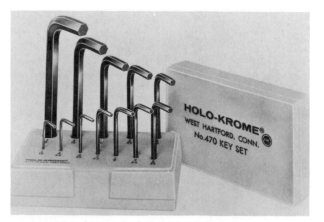

Fig. 10-36. Right angle, hexagon Allen wrenches are used on hex socket-head screws. The correct size should be used carefully to avoid stripping the socket, making the screw difficult to remove. (Holo-Krome)

straight with tee handles, Fig. 10-37. The correct size Allen wrench should be used to avoid slipping in the hex socket recess and damaging the socket. When this happens, screw removal may be difficult, if not impossible.

Torque wrenches are used to tighten threaded fasteners to a specific torque setting. *Torque* is the turning force applied to the fastener. Fig. 10-38 shows how to use a "preset" torque wrench. After the desired amount of torque is set, the wrench socket is placed on the bolt head and the handle is drawn until a click can be felt and heard. The socket should be held down firmly while pulling on the handle.

The torque wrench shown in Fig. 10-39 uses a pointer that moves up the scale as torque is applied. With this wrench, the mechanic pulls the wrench handle until the pointer reaches the correct torque reading. Torque data charts, like the one in Fig. 10-40, supply the necessary data to correctly tighten critical parts.

Most small engine torque charts specify both inch pounds (in.-lbs.) and foot-pounds (ft.-lbs.). Torque wrenches may be calibrated in either of these increments (units). The scale on the wrench is clearly marked whether the reading is given in in.-lbs. or ft.-lbs.

Fig. 10-37. Tee-handle Allen wrenches.

Fig. 10-38. Proper procedure for using a torque wrench is to support socket with one hand and apply turning effort at right angles to the handle.

Fig. 10-39. A variety of torque wrenches are available for small engine use. This particular wrench uses a pointer and a flexible arm.

FOUR CYCLE TORQUE SPECIFICATIONS

		In. Pounds	Ft. Pounds
Cylinder Head Bolts		140 - 200	12 - 16
Connecting Rod Lock Nuts	1.5 - 3.5 H.P.	65 75	5.5 - 6
	4 - 6 H.P.	86 - 100	7 - 9
Cylinder Cover or Flange to Cylinder		65 - 110	5.5 - 9
Flywheel Nut		360 - 400	30 - 33

Fig. 10-40. Torque specifications are provided in engine manuals for all critical bolts and nuts. (Tecumseh Products Co.)

The torque reading is the product of the length of the wrench handle (in feet or inches) and the applied force. For example, applying one pound of force through a handle one foot long would produce 1 ft.-lb. or 12 in.-lbs. of torque. In order to convert ft.-lbs. to in.-lbs., multiply the ft.-lb. reading by 12.

PLIERS

Pliers are extremely useful tools for gripping, bending, pulling, and, in some cases, cutting wires. They should not be used to replace wrenches. Because most pliers are not designed for tightening purposes, they usually damage surfaces when used in this way. Fig. 10-41 shows a variety of plier types that are helpful in small engine repair work. "Vise-Grip" pliers, Fig. 10-41A, are designed to apply great clamping pressure and have large gripping teeth. They are sometimes used as a last resort to loosen something that is rusted or frozen in place. Once the part is removed, however, it is usually damaged and needs to be replaced. A good penetrating oil should be used to assist in removing rusted parts. Needle nose pliers, Fig. 10-41B, are very useful for bending small wires and gripping items that have fallen into small recesses. Diagonal side-cutting pliers are used for cutting various types of electrical wire, Fig. 10-41C. Combination slip-joint pliers, Fig. 10-41D, are general purpose pliers and are less expensive than other types of pliers. They can also be used for cutting soft, solid wire. Pliers should not be allowed to get hot since they will become distorted and lose the hardness of their cutting jaws.

RETAINING RING PLIERS

Inside and outside retaining rings are used to keep a part from moving axially on a shaft (in a direction parallel to the centerline of a shaft). For example, a sliding gear may be limited from moving too far by a retaining ring set in a groove machined into a shaft. Retaining rings are made from spring steel and must be installed and removed with special *retaining ring pliers*, such as those shown in Fig. 10-42. To install an outside retaining ring, the nibs of the pliers are inserted into the small holes in the ring and the ring is expanded by squeezing the plier handles. The ring is then slid over the shaft to the machined groove, where it is released. For internal retaining rings, the ring is compressed to fit inside a cylindrical hole when the plier handles are squeezed. The ring is then inserted into the hole and released into

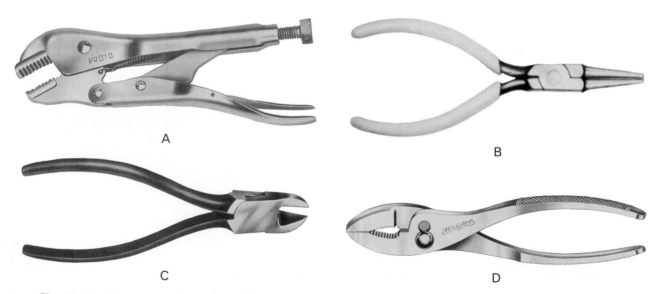

A

B

C

D

Fig. 10-41. Pliers are used for gripping, bending, pulling, and cutting wires. They should not be used to replace wrenches on nuts and bolts. A–''Vise-Grip'' pliers use mechanical advantage to increase grip. B–Needle nose pliers. C– Diagonal, side-cutting pliers. D– Combination slip-joint pliers. (Duro Metal Products Inc., Stanley-Proto)

Fig. 10-42. Retaining ring pliers are used to remove and install retaining rings on shafts or in cylindrical holes. Retaining rings are made of spring steel and can fly off if they slip from the plier nibs. Always wear safety glasses when using these pliers.

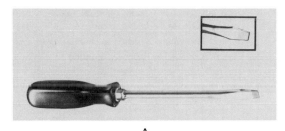

A

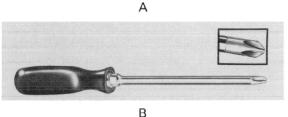

B

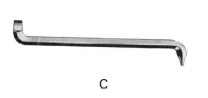

C

Fig. 10-43. Many types of screwdrivers are used for engine work. A–Standard screwdriver. B–Phillips screwdriver. C–Offset screwdriver. (Snap-on Tools Corp.)

a machined groove. Some pliers are only for inside or outside rings, while others are designed to accommodate both types.

Always wear safety glasses when using these pliers because retaining rings can slip off and fly with considerable velocity.

SCREWDRIVERS

Several types of screwdrivers are frequently used when servicing small engines. *Standard screwdrivers* are available in a variety of shapes and sizes, Fig. 10-43A. The proper size blade should be used to match the length and thickness of the screw head slot, Fig. 10-44. In addition to installing and removing screws, there are times when screwdrivers can be used to apply leverage to move or separate parts. Quality screwdrivers are very strong and will not be harmed by these applications.

Phillips screwdrivers of various sizes are useful for Phillips head screws and bolts, which have a cross-shaped recess, Fig. 10-43B. These screwdrivers are available in a variety of sizes to accommodate various sizes of screws and bolts.

In tight situations, the length of a regular screwdriver may prohibit its use. When this occurs, an *offset screwdriver* can be very useful, Fig. 10-43C. Offset screwdrivers are available with standard and Phillips-type heads.

HAMMERS

Hammers are extremely useful for small engine work. *Ball peen hammers* are used for tapping things into place. They are often used in

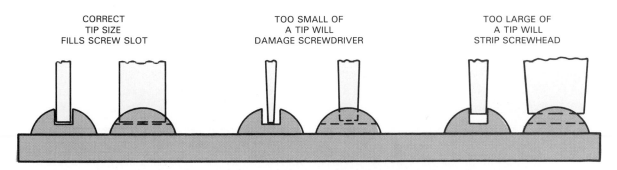

CORRECT TIP SIZE FILLS SCREW SLOT

TOO SMALL OF A TIP WILL DAMAGE SCREWDRIVER

TOO LARGE OF A TIP WILL STRIP SCREWHEAD

Fig. 10-44. To avoid damaging screw head, screwdriver blade must fit the slot or recess in the screw properly.

conjunction with other tools, such as punches and chisels. The ball peen hammer is considered to be a hard-faced hammer because its head is made of steel, Fig. 10-45. When using a ball peen hammer for small engine work, care must be taken so that parts are not dented or deformed by the hard faces. The ball peen hammer can be used to tap on a wrench to loosen a stubborn bolt or nut. The ball peen hammer is also used with pin punches to install locking pins in holes or with cold chisels to shear bolts, pins, or sheet metal. Ball peen hammers are available in a variety of sizes and are rated according to weight. A hammer's weight is usually stamped on the side of its head.

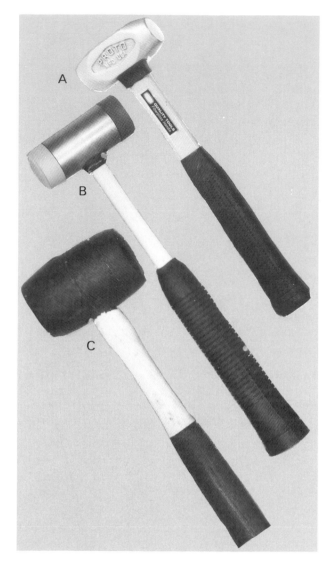

Fig. 10-46. Soft-faced hammers have faces that are softer than the objects they are used on. A–Brass or lead hammer. B–Plastic-faced hammer. C–Rubber mallet. (Duro Metal Products Inc., Stanley-Proto)

Fig. 10-45. The ball peen hammer has a cylindrical flat face on one end of the head and a ball shaped end on the other. Hammers are classified by head weight (in ounces), which is usually stamped on the head. (Duro Metal Products Inc.)

Soft-faced hammers are used to tap on parts that are easily damaged by hard-faced hammers. See Fig. 11-7. Soft-faced hammers are made from a variety of materials that are softer than steel. Lead, copper, brass, leather (rawhide), wood, rubber, and plastic are commonly used to make soft-faced hammers. Fig. 10-46 shows several soft-faced hammers.

PUNCHES

Many types of punches are used for small engine work. *Center punches* have hardened steel points and are used to make depressions in metal surfaces before drilling. The depression helps to start the drill and keep it in the desired location. Fig. 10-47A illustrates a center punch with a 90° point angle. Prick punches are similar to center punches. However, the point angle of a prick punch is 60°. The *prick punch* is used to make a very small depression prior to using the center

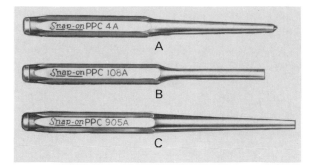

Fig. 10-47. There are several types of punches available for small engine use. A—Center punch. B—Pin punch. C—Drift punch. (Snap-on Tools Corp.)

punch to enlarge the depression for drilling. Center punches and prick punches are driven with a ball peen hammer.

Pin punches of various diameters and lengths are available for driving straight pins, tapered pins, and roll pins in and out of holes, Fig. 10-47B. A ball peen hammer is used to apply the driving force to the punch. Sometimes, a rusty bolt can be driven out of its hole with a pin punch.

Drift punches are tapered and are used to align holes in mating parts, allowing a bolt or pin to be passed through the parts, Fig. 10-47C.

COLD CHISELS

The **cold chisel** is a cutting tool that can shear bolts, pins, rivets, sheet metal, rods, and other materials, Fig. 10-48.

The cold chisel is made of a special tool steel, which is hardened and tempered by heat-treatment, Fig. 10-49. The cutting edge is very hard and sharpened to an angle between 60° and 90°. A 60° angle is used for shearing sheet metal. A 90° angle is for shearing bolts and rivets. The cutting edge can be sharpened when it becomes dull and may need heat treatment after several grindings. It should have a slightly curved edge for shearing and a straight edge for cutting flat surfaces, Fig. 10-50. Although the portion immediately behind the cutting edge is tempered for toughness, the chisel's shank portion is annealed (made soft) to prevent it from shattering. When the shank end of the chisel becomes flared, it should be reground to remove the flared portion,

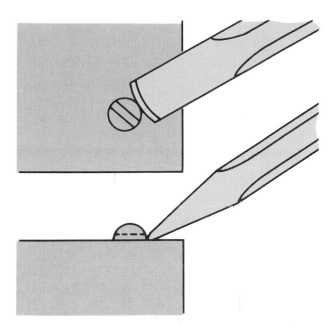

Fig. 10-48. Cold chisels are used for shearing bolts, screws, rivets, sheet metal, rods, or other materials.

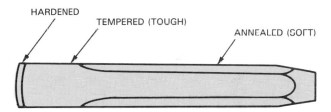

HARDENED TEMPERED (TOUGH) ANNEALED (SOFT)

Fig. 10-49. The cold chisel is a heat treated tool with a hardened and tempered cutting edge and an annealed shank.

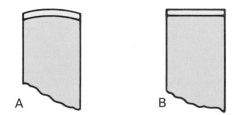

A   B

Fig. 10-50. A—The cutting edge of a cold chisel should be slightly curved for shearing. B—The cold chisel's edge should be straight for cutting flat materials.

Fig. 10-51. The flared part becomes work hardened and highly stressed from being hammered.

⚠ Safety glasses should always be worn when using punches or chisels. Pieces from the flared end of a chisel can fracture and fly at high velocity.

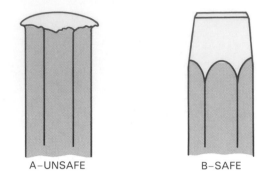

A–UNSAFE   B–SAFE

Fig. 10-51. A–The hammered end of a cold chisel becomes work hardened and flared after extended use. B–The flared end is very dangerous and should be ground to a slight taper.

GEAR PULLERS

Gear pullers are used for pulling gears and bearings from shafts, Fig. 10-52. Gears and bearings are often press fit onto shafts and have to be pulled with considerable force to be removed. The gear puller is designed to provide this pulling force. Gear pullers can be adapted for many applications and are often used to remove flywheels from small gas engines, Fig. 10-16. Also, timing

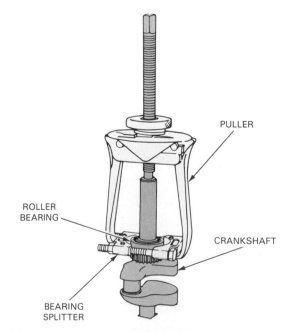

PULLER

ROLLER
BEARING

CRANKSHAFT

BEARING
SPLITTER

Fig. 10-52. A gear puller is used to remove gears and bearings from shafts. When pulling bearings, a tool called a bearing splitter should be used to avoid damaging the bearing while force is being applied. (Tecumseh Products Co.)

gears can be removed from crankshafts with a gear puller. When roller bearings are pulled, a tool called a bearing splitter must be used to avoid damage to the bearings and the outer race, Fig. 10-52.

PROBE AND PICKUP TOOLS

The *probe and pickup tools* shown in Fig. 10-53 assist when bolts, screws, washers, or other small items are dropped into crevices where they cannot reached with the hands or fingers. The mirror probe can help locate items that cannot be seen. Ferrous items (iron bearing) can be removed with the magnetic probe. Nonferrous (nonmagnetic) items can be gripped and removed with the finger pickup tool.

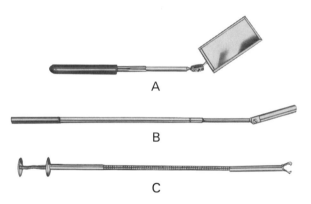

A

B

C

Fig. 10-53. Probe and pickup tools are helpful in locating small parts that may drop into crevices or hard-to-reach places. A–Mirror probe. B–Magnetic pickup Tool. C–Finger pickup tool.

VISE

The *machinists vise*, Fig. 10-54, is extremely useful for holding parts while they are being worked on. Some vises can be swiveled for convenient positioning of the work piece. The jaws are hardened steel and have a rough gripping surface. When parts that must not be scratched are clamped in a vise, soft jaw covers should be placed over the steel jaws. Sheet copper, soft aluminum, or wood can be used to pad the jaws. The anvil portion of the vise can be used to flatten sheet metal or to straighten bent parts. Always be careful not to distort or damage parts by over-

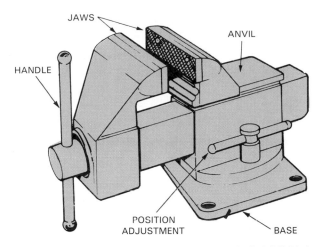

Fig. 10-54. The machinists vise can be a helpful "third hand" for holding parts while they are being worked on. Soft copper or aluminum jaw covers can be placed over the steel jaws to protect delicate parts. (Snap-on Tools Corp.)

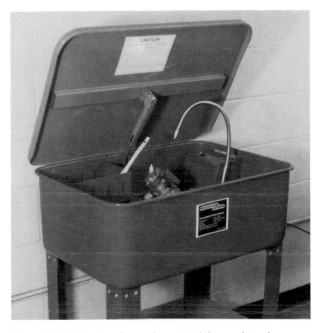

Fig. 10-55. A cleaning solvent tank is used to clean engine parts. This tank is fitted with a fusible link, which will automatically close the lid in case of fire. It also has a built-in pump and filter system for cleaning the solvent.

tightening the vise. Never clamp critical engine components, such as pistons, in a vise. Many engine parts are extremely delicate and the slightest distortion will render them useless.

The machinists vise can also be used for holding parts or materials that need to be drilled, filed, formed, or sawed. When filing or sawing an object in a vise, adjust the piece so that the work is done close to, but not in contact with, the vise jaws. This tends to minimize vibration.

CLEANING TANK

A *cleaning tank* is used to safely clean small engine parts, Fig. 10-55. A nontoxic, nonflammable solvent should be used to scrub engine parts after old gaskets and excess grease have been removed. The tank has a pump to recirculate and filter the solvent and direct it through a flexible tube. The tube can be directed at the parts for flushing. After the parts are clean, a low-pressure safety blow gun can be used to remove excess solvent.

SUMMARY

Quality tools and measuring instruments should be used when servicing small engines.

A micrometer is a precision instrument used to measure crankshafts, pistons, and other compo-

nents. A standard micrometer is used to measure to three decimal places. A Vernier micrometer can be used to obtain readings to four decimal places.

Telescoping gages transfer dimensions to an outside micrometer. A small hole gage is used to measure holes that are too small for the telescoping gage. The small hole gage also transfers dimensions to the outside micrometer. Thickness gages are used to measure small spaces and gaps between surfaces. They are often called "feeler" gages.

A valve spring tester is used to check valve spring tension. It is usually used in combination with a torque wrench. A combination square is often used to measure the length of valve springs and to check them for straightness.

The dial indicator is used to measure the movement of various parts. Many holding devices are available to hold the indicator in the desired location.

A screw pitch gage is used to determine the number of threads on a screw, bolt, or nut. They are available for both standard and metric threads.

Many types of wrenches are necessary for small engine work. These include the box end wrench, open-end wrench, adjustable wrench, Allen wrench, socket wrench, and torque wrench. Pliers are used for gripping, bending, pulling, or cutting. They should never be used for tightening purposes. Retaining ring pliers are used to remove inside and outside retaining rings. Always wear safety glasses when working with retaining ring pliers.

Several types and sizes of screwdrivers are used for small engine work, including the straight blade screwdriver, the Phillips screwdriver, and the offset screwdriver.

Hammers are used for tapping things into place. Hard-faced hammers have a head that is made of steel. The head on soft-faced hammers is made of lead, copper, brass, leather, wood, or plastic. Soft-faced hammers are used on parts that would be easily damaged by a hard-faced hammer.

Punches are used to make depressions in metal surfaces before drilling, drive pins from holes, and align holes in mating parts. The cold chisel is used to shear bolts, pins, rivets, sheet metal, and other materials.

Gear pullers are used to remove gears and bearings from shafts.

Probe and pickup tools assist when small items are dropped into areas that cannot be reached with hands and fingers.

A machinist's vise is useful for holding parts while they are being worked on. Sheet copper, aluminum, or wood can be used to pad the vise jaws when working on parts that must not be scratched. A cleaning tank is often used to clean small engine parts.

KNOW THESE TERMS

Micrometer, Vernier micrometer, Telescoping gage, Small hole gage, Thickness gage, Valve spring tester, Combination square, Dial indicator, Screw pitch gage, Open-end wrench, Box-end wrench, Allen wrench, Torque wrench, Torque, Pliers, Retaining ring pliers, Standard screwdriver, Phillips screwdriver, Offset screwdriver, Ball peen hammer, Center punch, Prick punch, Pin punch, Drift punch, Cold chisel, Gear puller, Vise.

REVIEW QUESTIONS–CHAPTER 10

1. Name several types of measuring instruments used to determine if engine parts are within manufacturer's tolerance.
2. Each space on the thimble of a micrometer represents:
   a. .0001 in.
   b. .001 in.
   c. .01 in.
   d. .1 in.
3. Each space on the sleeve of a micrometer represents what part of an inch?
   a. .0025 in.
   b. .100 in.
   c. .050 in.
   d. .025 in.
4. The discrimination of a Vernier micrometer is:
   a. .001 in.
   b. .01 in.
   c. .0001 in.
   d. .000001 in.
5. The telescoping gage is used for measuring:
   a. Inside diameters.
   b. Outside diameters.
   c. Depths of cylinders.
   d. Valve angles.
6. Which measuring instrument is similar to the telescoping gage?
7. An instrument called a _____ is used to determine the number of threads per inch on bolts or nuts.
8. Telescoping gages and small hole gages transfer inside dimensions to an outside _____.
9. Dial indicators are used to check for _____, _____, or _____.
10. What kind of wrench is open on one end and boxed on the other?
11. Which kind of wrench should be used only as a last resort if others are not available?
12. To remove a spark plug, what kind of wrench would be best to use to avoid damaging the plug during removal?

13. A torque wrench is used to measure the turning force that is applied to a threaded fastener. True or False?

14. Why must safety glasses be worn when using retaining ring pliers?

15. To locate a position to be drilled, a hammer and a _____ _____ should be used.

16. When using punches and cold chisels, it is very important to make sure the hammered end of the punch or chisel does not have a _____ end.

SUGGESTED ACTIVITIES

1. Clean the anvil and spindle of a micrometer and check its calibration accuracy.
2. Practice measuring engine parts with a standard outside micrometer.
3. Practice measuring engine parts with a Vernier micrometer.
4. Measure a cylinder bore with a telescoping gage and outside micrometer.
5. Measure a valve guide bore with a small hole gage and outside micrometer.
6. Check the straightness and length of a small engine valve spring. Does it meet the manufacturer's specifications?
7. Check ring groove and piston ring clearance with a thickness gage.
8. Check ring end clearance in a cylinder with a thickness gage.
9. Check the thread pitch of several different size bolts or screws with a screw pitch gage.
10. Measure the end play of a crankshaft with a dial indicator.
11. Identify the various types of wrenches, pliers, punches, chisels, screwdrivers, and other tools in your shop.
12. Demonstrate the proper way to use wrenches, chisels, screwdrivers, pliers, punches, and other tools in your shop.

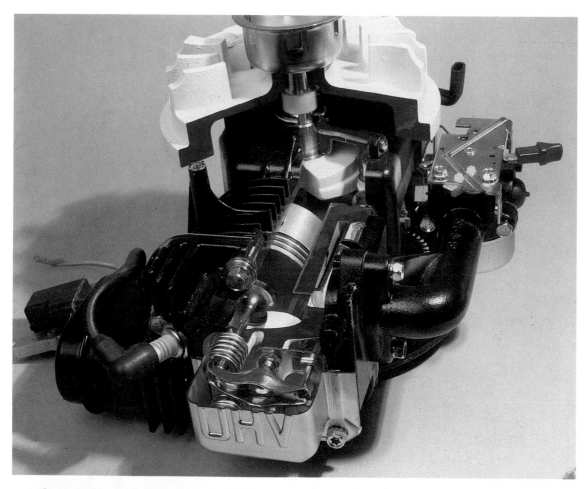

Cutaway view of a 5 hp, vertical shaft, four cycle engine. This engine features overhead valves, electronic ignition, an oil pump, and a float-type carburetor. (Tecumseh Products Co.)

CHAPTER 11

TROUBLESHOOTING, SERVICE, AND MAINTENANCE

After studying this chapter, you will be able to:
- [ ] Describe systematic troubleshooting.
- [ ] Use manufacturer's service manuals to determine engine specifications and explain why this information is necessary when servicing a small engine.
- [ ] Change the oil in a 4 cycle engine.
- [ ] Mix fuel and oil correctly for a 2 cycle engine.
- [ ] Perform preventive maintenance on various engine systems, including the crankcase breather, air cleaner, and muffler.
- [ ] Prepare a water cooling system for storage.

Most small engine service and repair jobs can be done without taking the whole engine apart. If the engine will not start, is hard to start, runs rough, or lacks power, troubleshooting may be necessary. Troubleshooting is simply a number of tests and steps you go through to find the problem.

Sometimes the cause of an engine problem is easy to find. At other times, checking probable causes requires a certain amount of reasoning and the use of the process of elimination. Also, more than one fault can exist at the same time, making it harder to locate the trouble.

SYSTEMATIC TROUBLESHOOTING

Always take a systematic approach when troubleshooting small engines. *Systematic troubleshooting* involves checking and/or testing one component after another until the problem is located and corrected. There are two basic principles to keep in mind when trying to pinpoint small engine problems:
1. Look for the easiest things first.

2. Verify the five fundamental operating requirements.

EASIEST THINGS FIRST

Always start troubleshooting with the simplest, most probable possibilities first. If an engine will not start, the problem could be something as simple as an empty fuel tank or a disconnected spark plug wire. Do not start working on the carburetor or the ignition system until you have made a few basic checks to determine that a simple remedy will not cure the problem.

FUNDAMENTAL OPERATING REQUIREMENTS

In order to start and run properly, an engine must meet five fundamental operating requirements. These requirements include:
1. Proper carburetion—clean, fresh fuel must be delivered in the correct proportion with combustion air.
2. Correct ignition system operation—strong ignition spark must be precisely timed for best performance and efficiency.
3. Adequate lubrication—the proper amount of high-quality lubricating oil must reach critical engine components.
4. Sufficient cooling—ample supply of cooling air (no more than 20°F hotter than outside ambient air) must reach engine.
5. Proper compression—30-45 psi minimum for starting and 90 psi minimum for efficient operation and sufficient power.

Keep these operating requirements in mind when troubleshooting small gas engines. Through the process of elimination, you can easily isolate problems. For example, if an engine will not start but will spin normally, you can eliminate lubrication system problems because the engine is not locked-up. By spinning the engine, you can also determine whether or not it has sufficient compression. If the engine will not start after it has cooled down, the cooling system can be eliminated as a potential problem. In a matter of seconds, you have determined that your troubleshooting efforts should be concentrated in the areas of carburetion and ignition.

The engine's owner can also provide assistance with your troubleshooting efforts. Ask a few questions about the engine's performance before it stopped. Relate the answers to the operating requirements. For example, if an engine runs for 30-45 minutes and then stalls, you should ask if it restarts immediately after it stops. If the answer is yes, the problem is probably an ignition component that is intermittently experiencing heat-related breakdown. If the engine must cool before it will restart, vapor lock or sticking valves are possible problems. If a metallic snap is evident in the engine during the cooling period, the valves are likely to be the problem. The time you spend to ask a few pertinent questions can save a lot of time in the long run by eliminating additional problem possibilities.

SERVICE INFORMATION

Before starting engine work, look at the manufacturer's *service manual* or troubleshooting chart. Manuals generally include *service procedures.* These procedures list the steps to take in order to accomplish a task effectively. Manuals also contain *exploded views* of assemblies and systems. These detailed drawings can help you to disassemble and reassemble parts in the right order. *Troubleshooting charts* list the most common engine troubles along with possible causes and suggested remedies. See Fig. 11-1. Experienced mechanics have learned what most symptoms mean and go about solving problems without using troubleshooting charts.

ENGINE TROUBLESHOOTING CHART

CAUSE	REMEDY
Engine Fails to Start or Starts with Difficulty	
No fuel in tank	Fill tank with clean, fresh fuel.
Shut-off valve closed	Open valve.
Obstructed fuel line	Clean fuel screen and line. If necessary, remove and clean carburetor.
Tank cap vent obstructed	Open vent in fuel tank cap.
Water in fuel	Drain tank. Clean carburetor and fuel lines. Dry spark plug and points. Fill tank with clean, fresh fuel.
Engine overchoked	Close fuel shut-off and pull starter until engine starts. Reopen fuel shut-off for normal fuel flow.
Improper carburetor adjustment	Adjust carburetor.
Loose or defective magneto wiring	Check magneto wiring for shorts or grounds; repair if necessary.
Faulty magneto	Check timing, point gap, and if necessary, overhaul magneto.
Spark plug fouled	Clean and regap spark plug.
Spark plug porcelain cracked	Replace spark plug.
Poor compression	Overhaul engine.
No spark at plug	Disconnect ignition cut-off wire at the engine. Crank engine. If spark at spark plug, ignition switch, safety switch, or interlock switch is inoperative. If no spark, check magneto.
Crankcase seals and/or gaskets leaking (2 cycle only)	Replace seals and/or gaskets.
Exhaust ports plugged (2 cycle only)	Clean exhaust ports.

Fig. 11-1. Typical troubleshooting chart for small gasoline engines.

(Continued)

ENGINE TROUBLESHOOTING CHART

CAUSE	REMEDY
Engine Knocks	
Carbon in combustion chamber	Remove cylinder head and clean carbon from head and piston.
Loose or worn connecting rod	Replace connecting rod.
Loose flywheel	Check flywheel key and keyway; replace parts if necessary. Tighten flywheel nut to proper torque.
Worn cylinder	Replace cylinder.
Improper magneto timing	Time magneto.
Engine Misses Under Load	
Spark plug fouled	Clean and regap spark plug.
Spark plug porcelain cracked	Replace spark plug.
Improper spark plug gap	Regap spark plug.
Pitted magneto breaker points	Replace pitted breaker points.
Magneto breaker arm sluggish	Clean and lubricate breaker point arm.
Faulty condenser	Check condenser on a tester; replace if defective.
Improper carburetor adjustment	Adjust carburetor.
Improper valve clearance	Adjust valve clearance to recommended specifications.
Weak valve spring	Replace valve spring.
Reed fouled or sluggish (2 cycle only)	Clean or replace reed.
Crankcase seals leak (2 cycle only)	Replace worn crankcase seals.
Engine Lacks Power	
Choke partially closed	Open choke.
Improper carburetor adjustment	Adjust carburetor.
Magneto improperly timed	Time magneto.
Worn rings or piston	Replace rings or piston.
Air cleaner fouled	Fill crankcase to the proper level.
Lack of lubrication (4 cycle only)	Clean air cleaner.
Valves leaking (4 cycle only)	Grind valves and set to recommended specifications.
Reed fouled or sluggish (2 cycle)	Clean or replace reed.
Improper amount of oil in fuel mixture (2 cycle only)	Drain tank; fill with correct mixture.
Crankcase seals leak (2 cycle only)	Replace worn crankcase seals.
Engine Overheats	
Engine improperly timed	Time engine.
Carburetor improperly adjusted	Adjust carburetor.
Air flow obstructed	Remove any obstructions from air passages in shrouds.
Cooling fins clogged	Clean cooling fins.
Excessive load on engine	Check operation of associated equipment. Reduce excessive load.
Carbon in combustion chamber	Remove cylinder head and clean carbon from head and piston.
Lack of lubrication (4 cycle only)	Fill crankcase to proper level.
Improper amount of oil in fuel mixture (2 cycle only)	Drain tank; fill with correct mixture.
Engine Surges or Runs Unevenly	
Fuel tank cap vent hole clogged	Open vent hole.
Governor parts sticking or binding	Clean and, if necessary, repair governor parts.

(Continued)

Fig. 11-1. Continued.

ENGINE TROUBLESHOOTING CHART

CAUSE	REMEDY
Engine Surges or Runs Unevenly	
Carburetor throttle linkage or throttle shaft and/or butterfly binding or sticking	Clean, lubricate, or adjust linkage and deburr throttle shaft or butterfly.
Intermittent spark at spark plug	Disconnect ignition cut-off wire at the engine. Crank engine. If spark, check ignition switch, safety switch and interlock switch. If no spark, check magneto. Check wires for poor connections, cuts or breaks.
Improper carburetor adjustment	Adjust carburetor.
Dirty carburetor	Clean carburetor.
Engine Vibrates Excessively	
Engine not securely mounted	Tighten loose mounting bolts.
Bent crankshaft	Replace crankshaft.
Associated equipment out of balance	Check associated equipment.
Engine Uses Excessive Amount of Oil (4 Cycle Only)	
Engine speed too fast	Using tachometer adjust engine RPM to specifications.
Oil level too high	To check level, turn dipstick cap tightly into receptacle for accurate level reading.
Oil filler cap loose or gasket damaged causing spillage out of breather	Replace ring gasket under cap and tighten cap securely.
Breather mechanism damaged or dirty causing leakage	Replace breather assembly.
Drain hole in breather box clogged causing oil to spill out of breather	Clean hole with wire to allow oil to return to crankcase.
Gaskets damaged or gasket surfaces nicked, causing oil to leak out	Clean and smooth gasket surfaces. Always use new gaskets.
Valve guides worn excessively thus passing oil into combustion chamber	Ream valve guide oversize and install 1/32'' oversize valve.
Cylinder wall worn or glazed, allowing oil to bypass rings into combustion chamber. Piston rings and grooves worn excessively	Bore hole or deglaze cylinder as necessary. Reinstall new rings and check land clearance and correct as necessary.
Piston fit undersized	Measure and replace as necessary.
Piston oil control ring return holes clogged	Remove oil control ring and clean return holes.
Oil passages obstructed	Clean out all oil passages.

Fig. 11-1. Continued.

TOLERANCES AND CLEARANCES

Engine *tolerances* and *clearances* are given in chart form in service manuals and/or bulletins. See Fig. 11-2. When checking and adjusting spark plug gap, breaker point gap, ignition timing, etc., this chart gives the correct dimensions and piston locations.

Note that in Fig. 11-2, the spark plug gap setting is .030 in. and breaker point gap (magneto point gap) is .020 in. Ignition timing shows the crankshaft throw at 22° before top dead center (BTDC), which is the fully retarded position.

Some engine charts show piston height in thousandths of an inch BTDC rather than in degrees.

When tolerance specifications show two values, the actual dimension must be within that range. In Fig. 11-2, for example, the cylinder bore (diameter) must measure somewhere between 2.1260 in. and 2.1265 in., a range of .0005 in.

TORQUE SPECIFICATIONS

Before attempting troubleshooting or maintenance of any kind, you must be familiar with *torque specifications* for fasteners used in as-

TABLE 1. TOLERANCES AND CLEARANCES FOR THE J-321 ENGINE

Cylinder Bore	2.1265 2.1260	Spark Plug Gap	.030
Piston Skirt Diameter	2.1227 2.1220	Magneto Point Gap	.020
Piston Ring Width	.0925 .0935	Ignition Timing in Degrees B.T.D.C.	22° (Fully Retarded)
Piston Pin Diameter	.5001 .4999	Piston Skirt to Cylinder Clearance	.0033 .0045

Fig. 11-2. Typical tolerance and clearance chart furnished by an engine manufacturer. (Jacobsen Mfg. Co.)

sembling different parts of the engine. Torque refers to the effort extended to turn something. Bolts and nuts must be "torqued" to a given tightness to hold mating parts together under specified tension.

Obviously, if a bolt or nut is too loose, vibration may loosen it more. If the same bolt is too tight, the threads may be stripped or the bolt may be broken off in the hole.

If the installed bolt does not break or strip the threads, expansion from heat will weaken the metal in the bolt. This expansion may go beyond the elastic limit of the bolt (a point where the bolt stretches but does not return to its original length upon cooling). When this happens, the bolt or part may fail.

There is still another reason why bolts and nuts should not be overtightened. Excessive internal stresses within a part may cause warpage or failure of the part.

When a number of bolts are required to fasten mating parts together (such as cylinder heads to cylinder block), it is necessary to tighten them evenly and according to a certain sequence (order) to prevent warping and leaks. When gaskets are used, overtightening of fasteners will crush the gasket under the bolt heads and distort the metal between the bolts.

Fig. 11-3 shows the correct cylinder head bolt tightening sequence in a typical small gasoline engine. To illustrate the procedure, the bolts are numbered in the particular order or sequence in which they are to be tightened. After the first bolt is tightened to 100 in.-lbs., the one directly oppo-

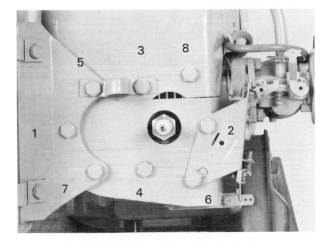

Fig. 11-3. Head bolts should be tightened evenly and in the sequence recommended by the manufacturer. (Deere & Co.)

site is tightened to the same torque reading. Next, the bolts that lie perpendicular to a line between the first two are similarly torqued. Then, the bolts 90° to the left and right are tightened, etc.

Generally, bolts are tightened in a sequence from one side to another in a kind of extending rotation, first tightening each bolt to 100 in.-lbs., then going back to tighten each an additional 20 in.-lbs. until specified torque is reached.

CHECKING RPM

When servicing small engines, it is often necessary to test or set maximum idle rpm or governor rpm. One way of doing this is by using a device that converts engine vibration from power pulses to rpm. See Fig. 11-4.

Troubleshooting, Service, and Maintenance 181

Fig. 11-4. Testing speed of engine through vibrations caused by power pulses of the piston. Note the adjustable slide. (Vibra-Tach)

Fig. 11-5. Flashing light of stroboscope unit can be adjusted to match revolutions of crankshaft, which gives engine rpm.

Place the nose of the instrument against the running engine. A thin wire is moved in or out of the instrument barrel until it vibrates into a fan pattern. When the fan shape is at its widest point, take the rpm reading from the scale on the barrel of the instrument.

Many mechanics use an electronic unit called a *stroboscope* to check rpm, Fig. 11-5. This method is very accurate. To prepare for the test, make a chalk mark on the crankshaft of the engine.

The stroboscope produces a high intensity light, which flashes on and off at a controlled rate. Aim the light at the chalk mark on the rotating shaft and adjust the flashing frequency until the chalk mark appears to stand still. At this point, the time interval between each crankshaft rotation is equal to the time interval between each flash of the light. The scale on the stroboscope shows the number of light flashes per minute. This, in turn, is equal to the rpm of the crankshaft.

⚠️ When a stroboscope is used, parts that are actually moving at a high rate of speed appear to be stopped. Never attempt to touch the parts or the chalk mark.

TESTING COMPRESSION

A cylinder *compression test* can be a first step toward determining the condition of the upper major mechanical parts of the engine. This test is especially valuable if an engine lacks power, runs poorly, and shows little or no improvement after fuel system and ignition adjustments. Use the following procedure for a compression test:

1. Run the engine until it is warm.
2. Disconnect all drives to the engine.
3. Open choke and throttle valves wide.
4. Remove the air cleaner.
5. Remove the spark plug and insert compression gage, Fig. 11-6.
6. Crank the engine as fast as possible to obtain an accurate test. Repeat to ensure accuracy.

Engines equipped with compression release camshafts may have to be cranked in reverse rotation to obtain an accurate test. Most can be cranked forward.

An engine producing a compression less than the minimum suggested by the manufacturer usually has one or more of the following problems:

1. Leaking cylinder head gasket.
2. Warped cylinder head.
3. Worn piston rings.
4. Worn cylinder bore.
5. Damaged piston.

Fig. 11-6. Compression test can indicate condition of various mechanical components of engine.

6. Burned or warped valves.
7. Improper valve clearance.
8. Broken valve springs.

To determine whether the valves or rings are at fault, pour a tablespoonful of SAE 30 oil into the spark plug hole. Crank the engine several times to spread the oil and repeat the compression test. The heavy oil will temporarily seal leakage at the rings. If the compression does not improve, the rings are satisfactory and leakage is due to valves, cylinder head, or a damaged piston. If the compression is much higher than the original test, the leakage is due to defective piston rings.

GENERAL PREVENTIVE MAINTENANCE

There are certain maintenance tasks that must be performed regularly to keep an engine working properly. These tasks come under the heading of *preventive maintenance* because they help prevent premature engine wear.

KEEPING ENGINE CLEAN

Cleaning a small air-cooled engine periodically can prevent overheating. For proper cooling action, air must pass across the extended metal surfaces (cooling fins) of the cylinder block and cylinder head. If the *cooling fins* are insulated by dirt, leaves, and/or grass clippings, engine parts will retain most of the combustion heat. Parts will expand, probably distort, and possibly seize. Therefore, all finned surfaces should be cleaned to the bare metal.

Methods for cleaning small air-cooled engines vary. You can scrape the dirty areas with a piece of wood and wipe them with a clean cloth. Or, you can blow debris from the fins with compressed air and then use a cleaning solvent. Various aerosol spray cleaners are suitable for use on small engines.

⚠ When using compressed air, be extremely careful where you direct the blast of air. Wear safety goggles. Never direct the air blast toward skin or clothing.

CHECKING OIL LEVEL AND CONDITION

Crankcase oil in four cycle engines should be checked periodically. Preferably, it should be checked each time fuel is added. The engine manufacturer provides a means of visually inspecting the level and condition of the oil. Use the type and viscosity grade of oil recommended by the manufacturer and maintain it at the proper operating level.

To check the oil level, withdraw the *dipstick* and wipe it dry. Re-insert the dipstick as far as it will go. Withdraw it a second time and observe the oil level, Fig. 11-7. Add oil if the level is below the ADD or LOW mark.

Do not run the engine with oil showing above the FULL mark on the dipstick. If the crankcase oil level is high, drain some oil. Overfilling can foul plugs and cause the engine to use too much oil.

Some small gasoline engines do not have dipsticks. Instead, they have a *filler plug* that seals out dirt and seals in the oil. Fig. 11-8 shows the proper method of loosening one type of filler plug. When the plug is removed, the oil level should be at the top of the filler hole or to a mark just inside of the filler hole.

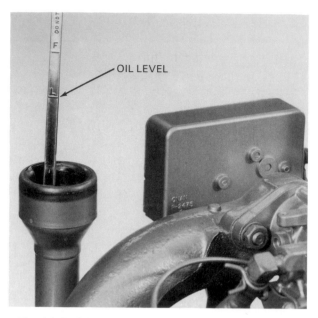

Fig. 11-7. Oil level should be maintained between the full and low marks. Never overfill. (Onan Corp.)

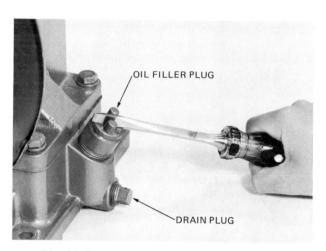

Fig. 11-8. Removing oil filler plug from engine crankcase. (Briggs and Stratton Corp.)

If the engine oil level drops at an excessive rate (requires addition of oil frequently), look for the cause. Refer to the troubleshooting chart for the particular engine at hand. Typical causes are: external leaks, worn oil seals around the crankshaft, worn valve guides, worn piston rings, or a hot running engine.

The color of used oil is not always an accurate indication of its condition. Additives in the oil may cause it to change color, while not decreasing its lubricating qualities.

WHEN TO CHANGE OIL

The small engine manufacturer will recommend oil changes at intervals based on hours of running time. A new engine should have the first oil drained after only a few hours of running to remove any metallic particles from the crankcase. After that, the time specified may vary from 10 hours to 50 hours.

Engine oil does not "wear out." It always remains slippery. However, oil used for many hours of engine operation becomes contaminated with dirt particles, soot, sludge, varnish-forming materials, metal particles, water, corrosive acids, and gasoline. These contaminants finally render the oil "useless." The harm they cause outweighs the lubricating quality of the oil.

The time interval for oil changes is selected so that the oil never reaches a "loaded" level of contamination. *Loaded oil* cannot absorb any more contaminants and still be an effective lubricant. When oil reaches a loaded condition, varnish deposits begin to form on the piston and rings and sludge collects in the crankcase.

CHANGING OIL

Draining engine oil is not difficult. First, run the engine until it is thoroughly warmed up. Warm oil will drain more completely, and more contaminants will be removed if the oil is agitated.

Stop the engine and disconnect the spark plug. The oil drain plug is located at a low point on the crankcase, usually along the outside edge of the base, Fig. 11-9. Some engines are drained through the filler cap and have no drain plug.

Clean the dirt from the drain plug area, then remove the plug with a proper wrench. Drain the oil for approximately five minutes to remove as much contaminated oil as possible. If possible, tilt the engine toward the drain hole if it is located on the side or top. When draining is complete, replace the drain plug.

If the engine is equipped with a disposable oil filter, it should be replaced each time the oil is changed. See Fig. 11-10. To prevent oil leaks, always coat the O-ring seal with a light coat of clean oil before installing the filter.

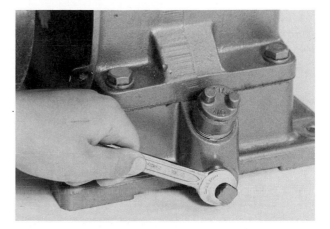

Fig. 11-9. Oil drain plug is located at a low point in the crankcase to permit complete drainage of used oil.

Fig. 11-10. Typical disposable oil filter. To prevent leaks, apply a light film of oil around O-ring seal before installation.

Before putting fresh oil in the engine, clean the filler opening, funnel (or other container), and the top of the can of oil. Be sure to use the type, viscosity grade, and quantity of oil recommended by the manufacturer. Pour it in the engine's crankcase and check the level. Then replace the filler cap and connect the spark plug lead to the spark plug.

Start and run the engine for a few minutes. Stop the engine and recheck oil level. Check for oil leaks.

⚠️ WARNING: After changing the oil, wash or destroy oily rags. Storing them may cause spontaneous combustion. Spontaneous combustion occurs when combustible material self-ignites.

If any oil is left in the can, keep it clean during storage by installing a plastic top from a vacuum-packed food container. If oil is kept in screw top cans, keep the cap tightly sealed to prevent condensation.

MIXING OIL AND FUEL

A two cycle engine can be seriously damaged by improperly mixing gasoline and oil, even though the recommended proportions are used. The proper way to mix gasoline and oil is to pour some of the gasoline into a clean metal container first. Add the oil to the gasoline and agitate (shake) the partial mixture. Add the rest of the gasoline and agitate again thoroughly.

Many service station attendants will pour the oil into the container first, then allow pump pressure to mix the gasoline and oil. However, this method does not always thoroughly mix the two elements.

Once the gasoline and oil are thoroughly mixed, the oil will remain in suspension indefinitely. If fuel is to be stored for several weeks (or more), add a gasoline stabilizer to prevent the formation of oxidation, varnish and corrosive acids, which can ruin an engine.

For safety, always store fuel in containers clearly marked for "gasoline." If possible, keep the storage container full to prevent moisture condensation.

Condensation forms in partially filled containers because of temperature changes. During the day, the air above the fuel may become very warm and hold a considerable amount of moisture. At night, the warm air cools, contracts, and loses the water vapor.

As water droplets gather on the inside walls of the container, they run together and flow into the fuel. Each night, more water is added to fuel in this manner. Partially filled fuel tanks react to temperature changes in this way. For this reason, fuel tanks should be kept full of gasoline or gasoline/oil fuel mixture.

AIR CLEANER SERVICE

The carburetor *air cleaner* should be cleaned before each season of operation and at regular

Fig. 11-11. Carburetor air filters must be kept clean. Their purpose is to trap dirt, but, eventually, airflow is restricted. (Jacobsen Mfg. Co.)

intervals thereafter. A plugged air filter, like the one in Fig. 11-11, can cause hard starting, loss of power, and spark plug fouling. Remove old oil from oil bath cleaners and wash them in solvent. Refill with engine oil to the correct level.

Polyurethane foam air filters should be washed and rinsed in kerosene or a similar solvent. About two tablespoons of clean SAE 30 oil should be evenly distributed in the filter material by compressing it in the hand.

Under severe dust conditions, air filters should be cleaned more often. If a dry canister filter is used, replace it with a new one.

CRANKCASE BREATHER SERVICE

If the small gasoline engine has a *crankcase breather*, it should be removed and cleaned periodically. The breather assembly is located over the valve stem chamber. It is held in place with two or more screws.

To service the breather, remove the screws and the cover. Under the cover is a filter element and a reed valve unit similar to the one in Fig. 11-12. The breather allows outward airflow only. Inspect the reed valve to make sure it is not damaged or distorted. Wash the parts in a cleaning solvent and replace damaged gaskets. See that the drain hole in the reed valve body is open. It permits accumulated oil to return to the engine. After all

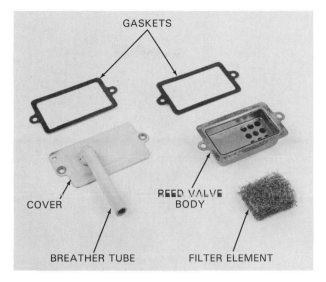

Fig. 11-12. Crankcase breather elements need periodic cleaning and inspection. Reed valve function is to only permit air to leave the crankcase. (Deere & Co.)

components have been cleaned and inspected, replace the assembly and tighten the screws.

MUFFLER SERVICE

An engine takes in large quantities of air mixed with fuel, and then burns the mixture. Unless the engine readily rids itself of the by-products of combustion, its efficiency will be greatly reduced. This is the task of the exhaust system, which, in small gasoline engines, mainly consists of exhaust port(s) and a small muffler.

The *muffler* is designed to reduce noise and allow gases to escape. When it becomes clogged with carbon soot, gases cannot get out quickly enough to allow fresh air and fuel to enter. Then, a power loss occurs along with a tendency to overheat.

If a muffler is designed to be taken apart, as shown in Fig. 11-13, it should be disassembled and cleaned in a solvent. If it is a sealed muffler and clogging is suspected, replace it with a new one and check for improved engine efficiency.

MAINTAINING WATER COOLING SYSTEMS

The water cooling systems used in small engines require maintenance similar to that employed in the automobile engine. Because the combination of water and metal sometimes produces harmful chemical reactions that attack the water jacket, a chemical *rust inhibitor* should be added to the cooling system.

Whenever a system is drained and refilled, it is advisable to add an inhibitor. If rust and scale are allowed to form and accumulate, the walls of the water jacket will become insulated. This will cause engine heat to be retained rather than removed.

Scale settling to the bottom may plug water passages in the cylinder block and clog the water tubes in the radiator. Without free circulation of water, the engine will run hot even when the thermostat is open. Local "hot spots" can occur in the engine when the passages in the block are obstructed.

In severe cases, the water may boil inside the block, and steam will prevent water from contacting and cooling the inner walls. Then, serious overheating and damage to parts of the engine are bound to occur.

The cooling fins that surround the tubes of the radiator should be kept clean for efficient heat transfer. Compressed air or pressurized water will remove any accumulations that might prevent air from passing through the fins and across the tubes. To remove the debris, direct the flow

Fig. 11-13. Accumulations of carbon in the exhaust muffler can seriously retard the scavenging of gases from the cylinder. This muffler can be disassembled for cleaning. (McCulloch)

of air or water in the opposite direction of normal airflow.

Engine blocks and radiators may be cleaned periodically by *reverse flushing* the system with pressurized water. Disconnect the hoses from the radiator and the block. Force clean water in the opposite direction of normal circulation. This will push loose sediment out. Continue flushing until the water runs clear. Flushing should be done with the engine stopped and cool.

To remove additional rust clinging to inner surfaces, use a commercial cooling system rust remover. Follow the instructions given by the manufacturer.

SYSTEMS COOLED WITH SALT WATER

Outboard engines that are operated in salt water are exposed to extremely corrosive conditions. Exposed engine parts require very careful maintenance.

Outboards used in salt water should be removed from the water immediately after operation. If the engine cannot be removed, tilt the gearcase out of the water and rinse it with fresh water. (The gearcase must be removed from the water when not in use.)

Flushing the internal cooling system of an outboard engine is extremely important. Flushing is done by attaching a fresh water hose to the water scoop or by operating the engine in a barrel of fresh water for several minutes.

Rinse the engine with fresh water and wipe all lower unit parts with a clean, oily cloth. Ignition leads and spark plug insulators should be wiped frequently to prevent an accumulation of salt residue.

STORING WATER-COOLED ENGINES

Storing water-cooled engines for lengthy periods, particularly during winter, calls for special maintenance procedures. If the engine is "radiator cooled," *antifreeze* must be added to the water to protect against freezing at the lowest possible temperatures. If rust inhibitor is not supplied in the antifreeze, it should be added.

If the engine will not be started at any time during storage, drain the cooling system completely. Then, tag the engine to indicate its drained condition.

When storing outboard engines, remove all plugs from the gearcase and drive shaft housing. This allows accumulated water in the gearcase and cooling system to drain off.

Failure to take this precaution when winterizing may result in a cracked cylinder block and/or gearcase, plus possible damage to water channels and tubes.

Rock the engine from side to side to make certain all water has drained. Refill the gearcase with the type of grease specified by the engine manufacturer. Attend to all other lubrication recommendations made by the manufacturer for care of engines being stored.

SUMMARY

When troubleshooting a small gas engine, always take a systematic approach. Check the simplest, most probable possibilities first. Make sure the engine meets the five fundamental operating requirements including: proper carburetion, correct ignition system operation, adequate lubrication, sufficient cooling, and proper compression.

A manufacturer's service manual should be used when servicing small gas engines. These manuals contain assembly drawings, troubleshooting charts, and information on engine tolerances and clearances.

Fasteners should always be tightened to proper torque specifications. Refer to service manuals for manufacturer's recommendations.

Engine rpm can be checked with a wire-type tester or a stroboscope. Moving parts appear to be stopped when using a stroboscope. Do not touch moving parts.

A compression test can help to determine the internal condition of an engine. This information is valuable when engine lacks power, runs poorly, or does not respond to ignition and fuel system adjustments.

Preventive maintenance will prevent premature wear of engine parts. Cleaning a small engine regularly will help prevent overheating. Keep the cooling fins free from dirt, leaves, grass clippings, and other obstructions.

Check crankcase oil (four cycle engines) each time you add gas to the engine. Use only the type and viscosity grade oil recommended by the manufacturer. Oil change intervals are based on hours of running time. Time may vary from 10 to 50 hours. If necessary, replace oil filter when changing oil.

Although proper proportions are used, two cycle engines can be damaged if the fuel and the oil is not mixed correctly.

If engine is equipped with a crankcase breather, it should be removed and cleaned periodically. The carburetor air cleaner should be cleaned at regular intervals. If necessary, the muffler may have to be cleaned or replaced to prevent power loss and overheating.

Water cooling systems are prone to the formation of rust and scale. Rust inhibitor should be added to these systems. Engine blocks and radiators can be cleaned periodically by reverse flushing the systems.

Outboard engines operated in salt water are exposed to extremely corrosive conditions. These engines should be rinsed and flushed with fresh water. All lower unit parts should be wiped with a clean, oily rag.

When storing water cooled engines, antifreeze should be added to protect against the lowest possible freezing temperatures. If engine will not be started when in storage, drain the cooling system. Remove all plugs from gearcase and drive shaft housing on outboard engines.

KNOW THESE TERMS

Systematic troubleshooting, Service manual, Service procedures, Exploded view, Troubleshooting chart, Tolerances, Clearances, Torque specifications, Stroboscope, Compression test, Preventive maintenance, Cooling fins, Dipstick, Filler plug, Loaded oil, Condensation, Air cleaner, Crankcase breather, Muffler, Rust inhibitor, Reverse flushing, Antifreeze.

REVIEW QUESTIONS–CHAPTER 11

1. Systematic troubleshooting involves looking for the _____ problem possibilities first.
2. In order to run properly, an engine must have proper carburetion, correct ignition system operation, adequate lubrication, sufficient cooling, and _____.
3. Manufacturer's service manuals contain:
   a. Service procedures.
   b. Troubleshooting charts.
   c. Tolerance and clearance specifications.
   d. All of the above.
4. Engine rpm can be checked with a _____, which flashes a light on and off at a controlled rate.
5. A compression test can be used to determine the condition of the spark plug. True or False?
6. Preventive maintenance helps protect against premature engine wear. True or False?
7. Keeping an engine clean can help prevent _____.
8. Never run engine if the oil level is above full mark. True or False?
9. Although oil does not wear out, excess _____ make the oil useless.
10. When changing oil, the engine should be:
    a. Cold
    b. Warm
    c. Running
    d. None of the above.
11. Fuel tanks should be kept full to prevent the formation of _____.
12. Clogged mufflers can cause power loss and overheating. True or False?
13. All mufflers can be cleaned by soaking them in solvent. True or False?
14. To prevent the formation of rust and scale, a rust _____ should be added to the water (antifreeze) in the cooling system.
15. Outboard engines operated in salt water are exposed to _____ conditions.
16. When storing outboard engines, all plugs should be removed from the _____ and _____ _____ housing.

SUGGESTED ACTIVITIES

1. Locate several malfunctioning engines. Try to determine the cause of the problems using the systematic troubleshooting method. Remember the five fundamental operating requirements.
2. Review several manufacturers' service manuals. Study trouble code charts and exploded-view assembly drawings. Check maintenance procedures for engines in your shop.
3. Perform preventive maintenance procedures on several engines in your shop. Check the oil level and condition. If necessary, change the oil and the filter. Make sure the air filter element, crankcase breather (if applicable), and cooling fins are clean. If engine is equipped with a water cooling system, check engine block and radiator for signs of rust and corrosion.

CHAPTER 12

FUEL SYSTEM SERVICE

After studying this chapter, you will be able to:
☐ Test a fuel pump for proper operation.
☐ Summarize basic carburetor adjustments.
☐ Test two cycle engine reeds for leakage.
☐ Explain basic procedures for inspecting, overhauling, and adjusting diaphragm and float-type carburetors.
☐ Troubleshoot diaphragm and float-type carburetors.

TROUBLESHOOTING THE FUEL SYSTEM

If the symptoms of an engine's malfunction point to the fuel system, the problem could be located in several areas. It could involve the fuel pump, carburetor, reed valves (in two cycle engines), fuel lines, filters, or air cleaner. Troubleshooting will involve checking and/or testing one part after another until the trouble is located and corrected.

FUEL PUMP

If fuel is not being delivered to the carburetor in a *gravity-fed system*, check the following:
• Is there fuel in the gas tank?
• Is fuel flow blocked by a clogged filter or obstructions in the line?
If engine is equipped with a *fuel pump* and fuel is not being delivered to carburetor, check to see that:
• There is fuel in the gas tank.
• Fittings connecting the fuel line to the tank and the pump are tight; otherwise, the pump will draw air.
• Pump filter is clean and gasket on the filter bowl is in good condition. The method of in-

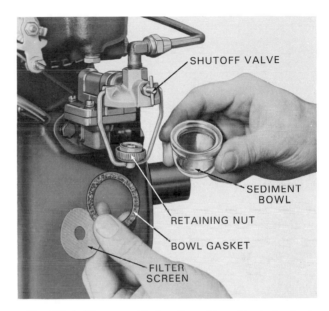

Fig. 12-1. Disassembled sediment bowl includes fuel bowl, strainer, and gasket. If water was present, it would settle to the bottom of the bowl.

specting the filter and gasket is shown in Fig. 12-1. Water and contaminants in the fuel are easy to see. Water, being heavier than fuel, separates and collects at the bottom of the bowl, along with other contaminants such as dirt and rust. Generally, visual inspection can be made without dismantling the pump. If disassembly is required, consult an exploded view of the pump, Fig. 12-2.
• Pump is actually working. Disconnect the fuel line between pump and carburetor. Turn the engine with the starter. There should be a well-defined spurt of fuel at every stroke of the pump (every two revolutions of the engine).
When fuel pump problems are suspected, make sure that the fuel flow from the supply tank to the

Fuel System Service 191

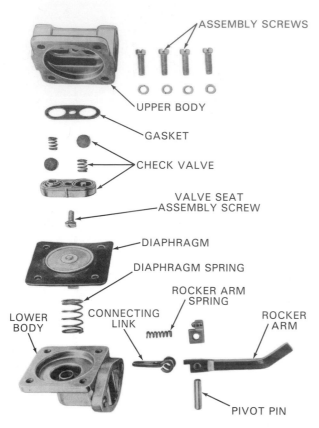

Fig. 12-2. Exploded view of fuel pump is typical of those found in engine service manuals. These are a useful aid to repair and reassembly. (Wisconsin Motors Corp.)

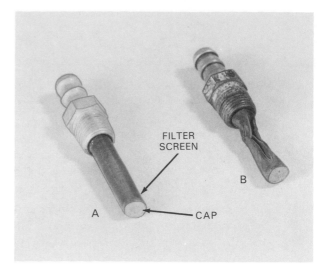

Fig. 12-3. Damaged gas tank filter can block fuel flow. A–New filter. B–Filter damaged when gasoline can nozzle was inserted too far into fuel tank.

HIGH SPEED AND IDLE MIXTURE ADJUSTMENT

Each engine manufacturer will give "rough" settings for the high speed adjustment and idle mixture adjustment. This will permit the engine to be started. It should then be warmed up before further adjustments are made.

pump is not interrupted. Fig. 12-3 shows a filter that allowed the engine to start and idle, but caused the engine to stall whenever the throttle was opened. A replacement filter (A) solved the problem.

CARBURETOR ADJUSTMENTS

If the fuel pump is working properly, but the engine still surges or lacks power, the cause may be poor carburetor adjustment or carburetor defects. Most carburetors have two needle valve adjustments: the *high speed adjustment* and the *idle mixture adjustment.* A third adjustment that is found on carburetors is the *idle speed stop screw.* Needle valves are not always found in the same location on the carburetor body. It is always a good practice to refer to the manufacturer's manual for needle positions and instructions on proper settings.

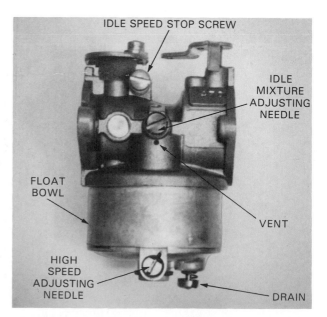

Fig. 12-4. External carburetor adjustments are the high speed adjusting needle, the idle mixture needle, and the idle speed stop screw. (Jacobsen Mfg. Co.)

The general adjustment procedure is to open the throttle wide and turn the high speed adjusting needle forward and backward slowly until maximum speed is reached. See the carburetor in Fig. 12-4 for the general location of external parts. After reaching the maximum speed, turn the needle counterclockwise very slightly so the engine is running a little rich.

To adjust the idle mixture needle, move the throttle to the slow running position. Turn the idle mixture needle slowly, first in one direction and then in the other. Continue turning the needle until idle is smooth. If necessary, adjust the idle speed stop screw to obtain the idle speed recommended in the manual. Check the manufacturer's recommendations carefully. Each make and model may be adjusted differently.

Unlike Carburetor A in Fig. 12-5, not all carburetors have two needle settings. Carburetor B has only a high speed needle adjustment and carburetor C does not have fuel adjustment needles. Each carburetor has an idle speed adjustment screw. Some of the needle adjustments are omitted because the manufacturer has preset the adjustments and sealed them.

TESTING TWO CYCLE ENGINE REEDS

To determine if the *reeds* are leaking on a two cycle engine, remove the air cleaner from the carburetor intake. Run the engine while holding a clean strip of paper about one inch from the carburetor throat, Fig. 12-6. If the paper becomes spotted with fuel, the reeds are not seating properly.

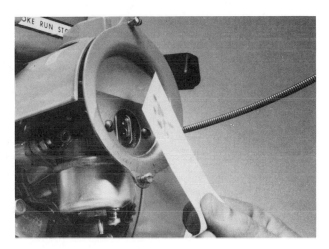

Fig. 12-6. Reeds can be tested for leakage by placing a clean strip of paper one inch from the carburetor intake. If spots of fuel show on the paper, the reeds are leaking.

Fig. 12-5. Depending on the make and model, carburetors may have one or several adjustments. A–Carburetor with high speed adjustment, idle mixture adjustment, and idle speed stop screw. B–Carburetor with two points of adjustment: high speed and idle speed. C–Carburetor with idle speed stop screw only.

Operating an engine with leaking reeds will result in fuel starvation, poor lubrication, and overheating. The reeds should be replaced or repaired.

FLOAT-TYPE CARBURETOR REPAIR

Carburetion problems that cannot be corrected by adjusting mixture needles are usually the result of gummed-up fuel passages or worn internal parts. The most effective solution to these problems is to overhaul the carburetor.

When making float-type carburetor repairs, the following rules should be observed:

- Follow all instructions carefully.
- Never use drill bits or wires to clean passages.
- Never enlarge passages.
- Always remove all welch plugs before soaking carburetor in cleaner.
- Never soak carburetor in cleaner for more than 30 minutes.
- Never reuse gaskets.
- Always use new service replacement screws for choke and throttle valve plates (screws are treated with a dry adhesive to secure them).
- Always set carburetor float to manufacturer's specified height.

It should be noted that many carburetors are designed for specific engine models. Similar carburetors may contain hundreds of significant variations in detail. The procedures presented in this chapter are general and may not pertain directly to the carburetor that you are attempting to service. However, if care is taken during the disassembly phase, no difficulty will be encountered during reassembly.

CARBURETOR OVERHAUL

A carburetor *overhaul* generally consists of disassembling, cleaning, and replacing parts as recommended by the manufacturer. *Carburetor kits* are generally available from small engine repair shops or manufacturer's distribution centers. These kits contain all the parts needed for a typical carburetor overhaul. A typical carburetor diagram showing part locations is illustrated in Fig. 12-7.

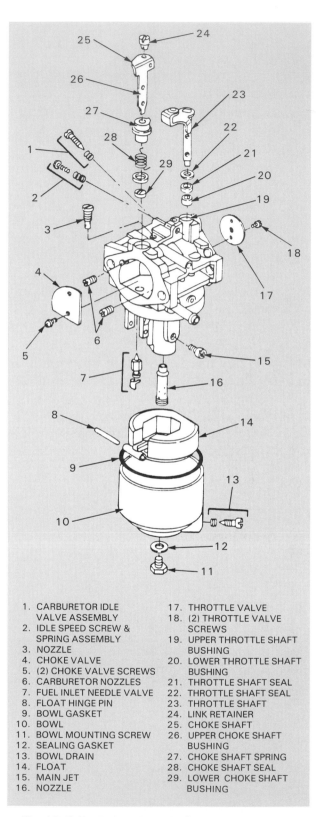

1. CARBURETOR IDLE VALVE ASSEMBLY
2. IDLE SPEED SCREW & SPRING ASSEMBLY
3. NOZZLE
4. CHOKE VALVE
5. (2) CHOKE VALVE SCREWS
6. CARBURETOR NOZZLES
7. FUEL INLET NEEDLE VALVE
8. FLOAT HINGE PIN
9. BOWL GASKET
10. BOWL
11. BOWL MOUNTING SCREW
12. SEALING GASKET
13. BOWL DRAIN
14. FLOAT
15. MAIN JET
16. NOZZLE
17. THROTTLE VALVE
18. (2) THROTTLE VALVE SCREWS
19. UPPER THROTTLE SHAFT BUSHING
20. LOWER THROTTLE SHAFT BUSHING
21. THROTTLE SHAFT SEAL
22. THROTTLE SHAFT SEAL
23. THROTTLE SHAFT
24. LINK RETAINER
25. CHOKE SHAFT
26. UPPER CHOKE SHAFT BUSHING
27. CHOKE SHAFT SPRING
28. CHOKE SHAFT SEAL
29. LOWER CHOKE SHAFT BUSHING

Fig. 12-7. Exploded view of a float-type carburetor, showing individual part locations. (Briggs and Stratton Corp.)

WARNING: Do not allow flames, sparks, pilot lights, or arching equipment near the fuel system. Ignition of fuel can result in severe personal injury or death.

CARBURETOR REMOVAL

The following removal procedures are typical to most carburetor configurations:

1. Remove air cleaner and set aside for later use.
2. Turn off fuel shut-off valve (located between fuel tank and carburetor), Fig. 12-8.
3. Drain carburetor bowl by removing drain screw.

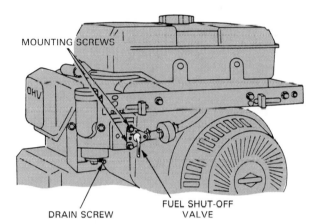

Fig. 12-8. The fuel shut-off valve is located between the fuel tank and the carburetor. (Briggs and Stratton Corp.)

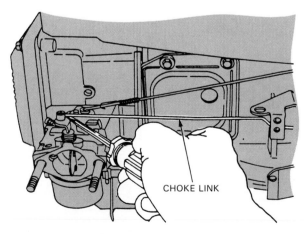

Fig. 12-9. Choke linkage can be removed from the choke lever by prying it with a screwdriver. (Briggs and Stratton Corp.)

4. Release fuel line hose clamp and disconnect hose.
5. Disconnect governor linkage and spring. (Do not deform spring or linkage.)
6. Remove throttle linkage from throttle lever.
7. Remove carburetor mounting screws. (Hold body to prevent it from falling from engine.)
8. Remove choke linkage from choke lever, Fig. 12-9.
9. Lift carburetor from engine (should be free from engine and linkage).

CARBURETOR DISASSEMBLY

Carefully disassemble the carburetor, removing all nonmetallic parts (gaskets, O-rings, etc.). Clean all metallic parts with solvent or commercial carburetor cleaner. Never soak parts in solvent for longer than 30 minutes. Nonmetallic parts should never be exposed to solvents.

WARNING: Safety goggles and rubber gloves should always be worn when working with solvents and commercial carburetor cleaners. Many commercial carburetor cleaners are extremely caustic and can cause serious burns to skin and eyes.

On some carburetor models, *welch plugs* must be removed to expose drilled passages. To remove welch plugs, sharpen a small chisel to a wedge point and drive the chisel into the plug, Fig. 12-10. Push down on the chisel and pry the plug out of position. To install a new plug after cleaning the carburetor, place the plug into the receptacle

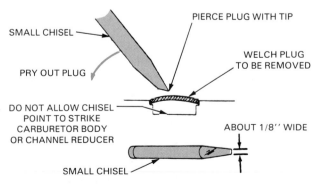

Fig. 12-10. Before soaking carburetor in solvent, welch plug is removed from body to expose drilled passages. (Tecumseh Products Co.)

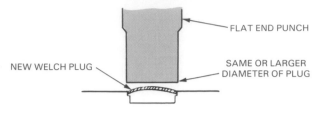

Fig. 12-11. Use a flat punch to install new welch plugs. Only flatten plug. Do not dent it. (Tecumseh Products Co.)

with the raised portion up and flatten it with a flat punch that is slightly larger in diameter than the plug itself. Do not dent the plug or drive the center of the plug below the top surface of the carburetor, Fig. 12-11.

Clean all carbon from the carburetor bore, especially where the throttle and choke plates seat. Be careful not to plug the idle or main fuel ports. Dry all passages with low-pressure air (35 psi). Avoid using wire or other objects for cleaning ports. These items frequently enlarge the diameter of important passages.

INSPECTION OF PARTS

When all parts are clean and dry, organize them neatly on a clean cloth. Carefully inspect

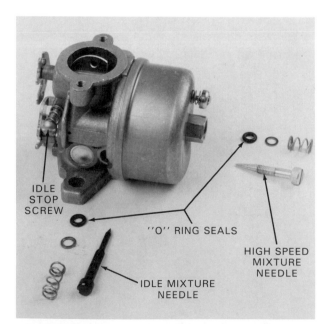

Fig. 12-12. Examine adjustment needles and replace O-rings if they are deformed. (Tecumseh Products Co.)

each part for damage and wear. Examine the idle and high-speed adjusting screw needles. The points should be straight and smooth. O-ring seals should be replaced if they are damaged, Fig. 12-12. Examine the choke and throttle plates for coded markings, which identify the way that they must be installed, Figs. 12-13 and 12-14. Study the choke shaft, throttle shaft, and bearing holes for wear.

The fuel bowl must be free of dirt and corrosion. Replace the bowl gasket or O-ring seal before installing the bowl on the carburetor body,

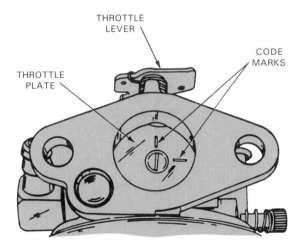

Fig. 12-13. This throttle plate must be installed with the code marks in the location shown. (Tecumseh Products Co.)

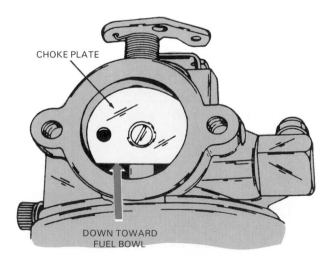

Fig. 12-14. This choke plate must be installed with the flat side down toward the float bowl. (Tecumseh Products Co.)

Fig. 12-15. Some fuel bowls will interfere with the float if not positioned correctly, Fig. 12-16.

Examine the float for damage. Hollow, brass-type floats must be free from pinholes and dents. Check the float hinge bearing surfaces for wear. The tab that contacts the inlet needle should also be inspected for wear, Fig. 12-17.

The inlet needle may seat on a synthetic rubber seat in the carburetor body. To remove this seat, pull it out with a short piece of hooked wire or force it out with a short blast of compressed air.

When installing a new seat, moisten it with oil and insert it into the carburetor body (smooth side toward the inlet needle). Press the seat into the cavity using a flat punch that is the same diameter as the seat. Make sure it is firmly seated, Fig. 12-18. Check the condition of the spring clip that is used to attach the inlet needle to the float tab, Fig. 12-19.

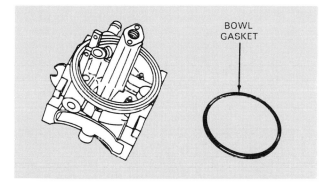

Fig. 12-15. When replacing float bowl, always install a new gasket or O-ring. (Briggs and Stratton Corp.)

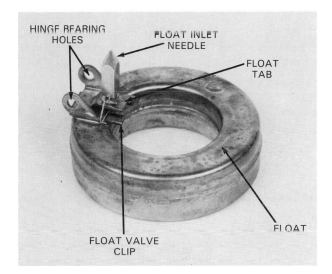

Fig. 12-17. Float hinge bearing holes should be examined for wear. Float tab is used to adjust float height per specifications. (Tecumseh Products Co.)

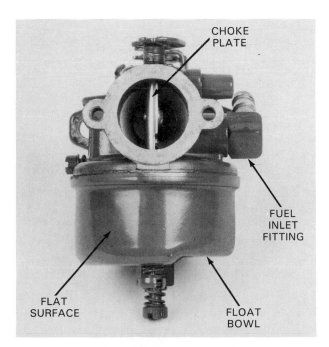

Fig. 12-16. Some float bowls must be installed in a particular way to avoid interference with the internal float. (Briggs and Stratton Corp.)

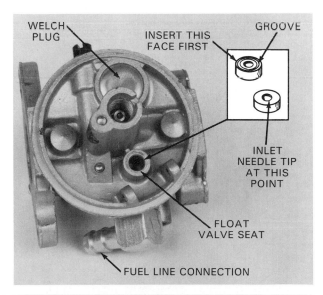

Fig. 12-18. Synthetic rubber needle seat must be installed with grooved face toward bottom of hole. Rubber seat is oiled before being pressed into hole with a flat punch. (Tecumseh Products Co.)

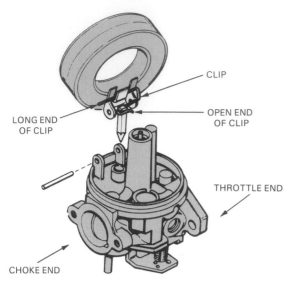

Fig. 12-19. The inlet needle is attached to the float with a specially shaped wire clip. (Tecumseh Products Co.)

CARBURETOR ASSEMBLY

After all carburetor parts have been cleaned, inspected, and replaced (if necessary), they should be reassembled in the following order:

1. Install new welch plugs (if applicable).
2. Install throttle shaft, bushings, seals, and washers.
3. Attach throttle valve plate with new screws.
4. Install idle-speed stop screw (if applicable).
5. Install choke shaft, spring, seal, washer, and bushing.
6. Attach choke valve plate with new screws.
7. Install metering nozzles and needle seats.
8. Install idle needle assembly.
9. Install high-speed needle valve assembly.
10. Attach inlet needle to float with wire clip.
11. Install float on carburetor with hinge pin.
12. Adjust float height per manufacturer's specifications, Fig. 12-20.
13. Install float-bowl over float assembly with mounting screw. Use a new gasket or O-ring to seal bowl, Fig. 12-21.
14. Install carburetor on engine. If a gasket is required, always use a new one.
15. Connect throttle, choke, and governor linkage. On some engines, linkage is connected before bolting carburetor to engine.

Fig. 12-20. Float must rest on needle at a specified height. If float is high, too much fuel will be used. If float level is low, lean mixture may cause overheating. (Tecumseh Products Co.)

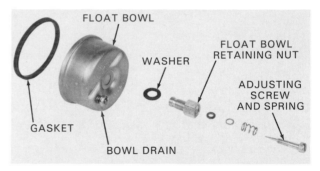

Fig. 12-21. Float bowl is assembled in the order shown. A new gasket should always be used when reinstalling the float bowl. (Tecumseh Products Co.)

16. Connect fuel line.
17. Install air cleaner.
18. Adjust idle mixture and high-speed needles to initial starting settings per manufacturer's instructions.

ENGINE PRIMING

A manually operated, plunger-type *primer* may be found on some carburetors. The main purpose of the primer pump is to force fuel through the main carburetor nozzle, producing a rich mixture for starting. Fig. 12-22 shows a bulb-type primer pump. When the rubber bulb is pressed, air pressure is increased in the float bowl, forcing fuel through the main nozzle and into the carburetor throat, Fig. 12-23.

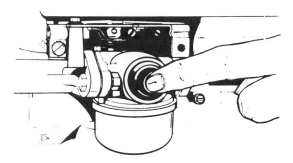

Fig. 12-22. Manually operated, bulb-type primer.
(Tecumseh Products Co.)

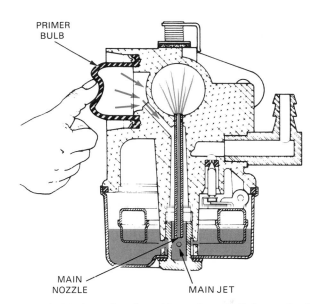

Fig. 12-23. Pressing the rubber primer bulb forces fuel
through the main nozzle, providing a rich air/fuel
mixture to help start a cold engine.
(Tecumseh Products Co.)

TROUBLESHOOTING FLOAT-TYPE CARBURETORS

The chart in Fig. 12-24 points out areas of possible float-type carburetor troubles and lists symptoms and suggested repairs. This figure is shown for illustrative purposes and is not representative of all float-type carburetors. Always consult an appropriate service manual when servicing a carburetor.

DIAPHRAGM CARBURETOR REPAIR

Except for the float and the float chamber, the procedures for servicing *diaphragm-type carbu-* *retors* are similar to those used for float-type carburetors. After removing and disassembling the carburetor, wash each part in an appropriate solvent. Lay the clean parts out on a clean, white cloth so that they are not lost or damaged.

If the carburetor is very dirty, a commercial carburetor cleaner solution may be used. Be

TROUBLESHOOTING FLOAT-TYPE CARBURETORS

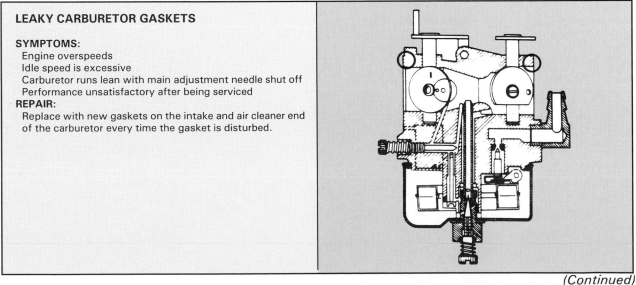

LEAKY CARBURETOR GASKETS

SYMPTOMS:
Engine overspeeds
Idle speed is excessive
Carburetor runs lean with main adjustment needle shut off
Performance unsatisfactory after being serviced
REPAIR:
Replace with new gaskets on the intake and air cleaner end
of the carburetor every time the gasket is disturbed.

(Continued)

Fig. 12-24. Troubleshooting chart for float-type carburetors. (Tecumseh Products Co.)

THROTTLE AND/OR CHOKE SHAFT WORN, THROTTLE AND/OR CHOKE SPRINGS NOT FUNCTIONING

SYMPTOMS:
Engine will not start
Engine hunts (at idle or high speed)
Engine will not idle
Engine lacks power at high speed
Idle speed is excessive
Choke does not open fully
Performance unsatisfactory after being serviced

REPAIR:
Replace all worn parts, springs, dust seals (when so equipped). If carburetor body is worn out of round, causing the leak, a new service carburetor should be used.

FUEL INLET IS PLUGGED OR LOOSE

SYMPTOMS:
Engine will not start
Engine hunts (at idle or high speed)
Engine will not idle
Engine lacks power at high speed
Carburetor leaks
Engine starves for fuel at high speed (leans out)

REPAIR:
Clean fuel system completely. Refill with clean fresh fuel, as recommended. Replace loose clamps and fittings.

DIRTY, STUCK, OR DAMAGED NEEDLE AND SEAT

SYMPTOMS:
Engine will not start
Carburetor floods
Carburetor leaks
Poor engine performance

REPAIR:
Remove old needle and seat. Install a new needle and seat according to manual instructions.

Fig. 12-24. Continued.

DAMAGED OR LEAKY O-RINGS

SYMPTOMS:
Engine hunts (at idle or high speed)
Carburetor leaks
Engine overspeeds
Idle speed is excessive
Carburetor runs with main adjustment needle shut off
REPAIR:
All rubber O-rings should be removed before cleaning and should be replaced with new ones when rebuilding the carburetor.

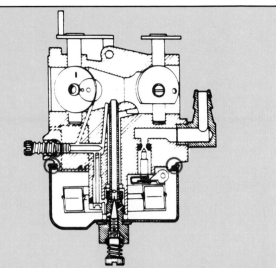

MAIN NOZZLE RESTRICTED OR PLUGGED

SYMPTOMS:
Engine will not start
Engine hunts at high speed
Engine starves for fuel at high speed (leans out)
REPAIR:
Soak carburetor in cleaner for no more than 30 minutes.
Use compressed air to clean passages.

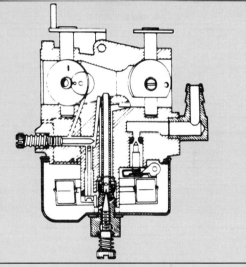

DAMAGED AND/OR WORN HINGE PIN OR FLOAT, IMPROPER FLOAT HEIGHT

SYMPTOMS:
Engine hunts (at idle or high speed)
Engine will not idle
Engine lacks power at high speed
Carburetor floods
Engine starves for fuel at high speed (leans out)
Carburetor runs with main adjustment needle shut off
Poor starting
REPAIR:
Replace float and axle. If hinge pin area of casting is worn, the carburetor body must be replaced. The float height is set using specified tool.

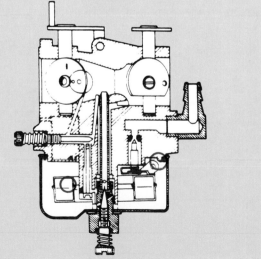

Fig. 12-24. Continued.

(Continued)

FUEL PICK-UP RESTRICTED OR PLUGGED

SYMPTOMS:
Engine will not start
Engine hunts (at idle or high speed)
Engine will not idle
Engine starves for fuel at high speed (leans out)
REPAIR:
After soaking carburetor in a commercial cleaner (no longer than 30 minutes), use compressed air to clean passages.

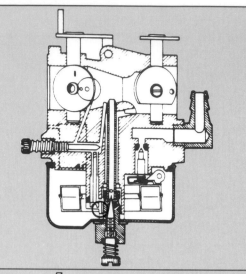

DAMAGED OR INCORRECT FUEL ADJUSTMENT NEEDLES

SYMPTOMS:
Engine will not start
Engine will not accelerate
Engine hunts (at idle or high speed)
Engine will not idle
Engine lacks power at high speed
Engine overspeeds
Engine starves for fuel at high speed (leans out)
Carburetor runs with main adjustment needle shut off
Performance unsatisfactory after being serviced
REPAIR:
Replace damaged needles with correct fuel adjustment needles. CAUTION: Do not over-seat needles.

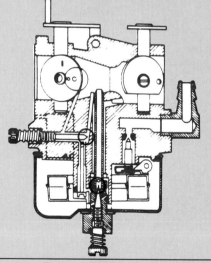

RESTRICTED OR PLUGGED AIR BLEED OR IDLE SYSTEM

SYMPTOMS:
Engine runs rich
Engine hard to start
Engine will not accelerate
Engine hunts
Engine will not idle
REPAIR:
After soaking carburetor in a commercial cleaner (no longer than 30 minutes), use compressed air to clean passages. The metering rod on Series I carburetors is not a serviceable part. If metering rod is not free, carburetor body must be replaced.

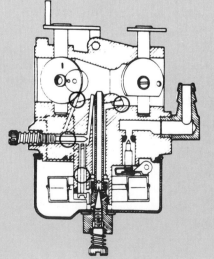

Fig. 12-24. Continued.

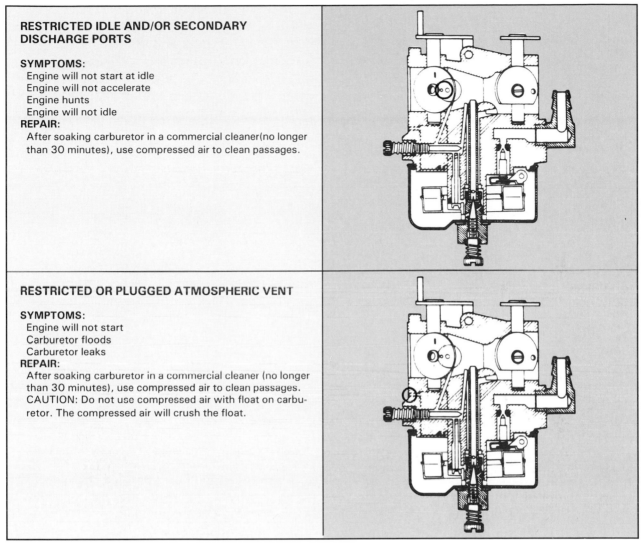

RESTRICTED IDLE AND/OR SECONDARY DISCHARGE PORTS

SYMPTOMS:
Engine will not start at idle
Engine will not accelerate
Engine hunts
Engine will not idle
REPAIR:
After soaking carburetor in a commercial cleaner(no longer than 30 minutes), use compressed air to clean passages.

RESTRICTED OR PLUGGED ATMOSPHERIC VENT

SYMPTOMS:
Engine will not start
Carburetor floods
Carburetor leaks
REPAIR:
After soaking carburetor in a commercial cleaner (no longer than 30 minutes), use compressed air to clean passages.
CAUTION: Do not use compressed air with float on carburetor. The compressed air will crush the float.

Fig. 12-24. Continued.

careful not to get any solution on hands or clothing. Wear safety glasses. Put only the metallic carburetor parts in the solution and let them soak. Nonmetallic parts can be damaged by harsh commercial cleaners.

After the parts have soaked for 30 minutes, rinse them with a milder cleaning solvent and dry with compressed air. Do not dry the parts with a rag or paper towel. Lint from the rags may get into the passages. Never clean holes or passages with wires or similar objects. These will distort the openings and may prevent the engine from running properly.

After cleaning, inspect the parts for wear, material failure, or other damage. Check the carburetor body and crankcase for cracks or worn mating surfaces. See Fig. 12-25. If necessary, replace defective parts.

Check throttle and choke shafts. They must fit closely but turn easily in their bearing holes, Fig. 12-26. If loose, they will cause poor engine performance.

The diaphragm should be checked for defects that would cause leakage. The diaphragm needle valve must be straight and fit the seat so that it seals when closed.

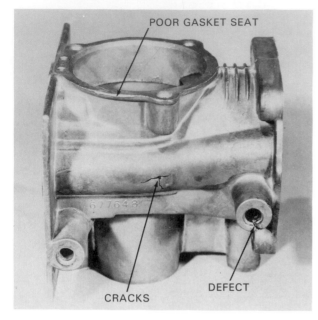

Fig. 12-25. This carburetor has several defects. If repair is not possible, carburetor should be replaced. (Lawn-Boy Power Equipment, Gale Products)

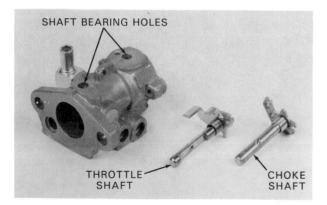

Fig. 12-26. Throttle and choke shafts must fit bores closely, but turn freely. (Tecumseh Products Co.)

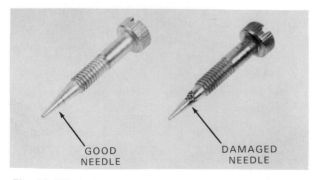

Fig. 12-27. A—Needle valve has straight, smooth taper. B—Needle valve is damaged and should be replaced.

The high speed and idle mixture needles should have straight, smooth tapers, Fig. 12-27. The O-rings or seals around the needles should be replaced if they are cut or deformed.

Some manufacturers supply diaphragm carburetor repair kits, Fig. 12-28, which include those items that would most likely need replacing. Other parts can be purchased as they are needed.

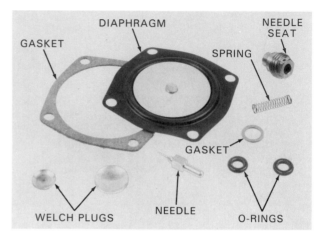

Fig. 12-28. Carburetor repair kit shown is for a diaphragm-type carburetor.

ASSEMBLING DIAPHRAGM CARBURETOR

Because of the many different carburetor designs, it is highly recommended that the manual be followed when assembling a carburetor. After all carburetor parts have been cleaned, inspected, and replaced (if necessary), the carburetor should be reassembled in the following order:

1. Place the throttle shaft in the bearing hole with the return spring on the shaft as illustrated in Fig. 12-29. The throttle plate must be placed properly in the carburetor bore. Usually, identifying marks of some sort are put on the plate to assist the assembler. On the carburetor shown in Fig. 12-29, the marks have to be placed facing outward.
2. Assemble the choke plate in the same way as the throttle. It is *vented*, which means that it has openings to allow some air to enter even when it is closed. Fig. 12-30 shows the flat side facing the mixture needles and the indentations facing outward.

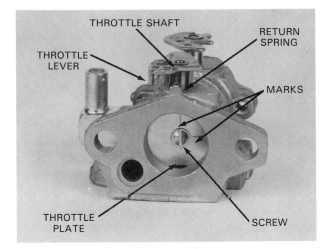

Fig. 12-29. Throttle shaft with return spring attached is placed in holes provided in the throttle body. Throttle valve must be fastened to shaft in the proper position. (Tecumseh Products Co.)

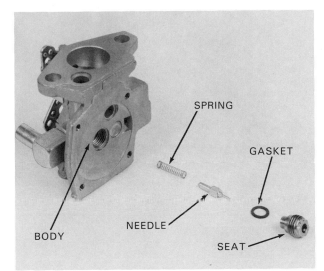

Fig. 12-31. Diaphragm needle valve is installed in carburetor body in sequence shown.

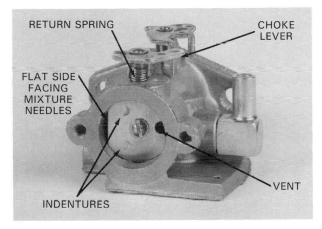

Fig. 12-30. Choke valve assembly is shown in place in carburetor air horn.

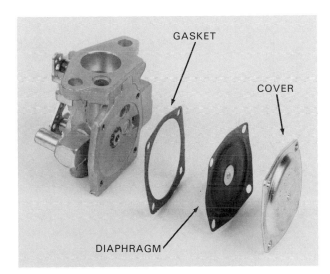

Fig. 12-32. In this application, gasket, diaphragm, and cover are fastened to the carburetor body with four screws. (Tecumseh Products Co.)

3. Install the spring and needle valve in the order shown in Fig. 12-31. The spring and needle valve are inserted into the body. The gasket and seat are then screwed into the threaded hole in the carburetor body.

4. Assemble the diaphragm as shown in Fig. 12-32. Four screws are placed through holes in diaphragm cover to tightly draw the assembly to the body. The rivet head in the center of the diaphragm is turned toward the needle valve.

5. Install the high speed and idle mixture needles in the order shown in Fig. 12-33.

6. Adjust the high speed and idle mixture needles as prescribed by the manual. This will get the engine started. Finer adjustments can be made when the engine is warmed up. The idle stop screw in Fig. 12-33 is turned one full turn after it first makes contact with the throttle lever.

7. Attach the carburetor to the engine.

8. Install the linkage.

9. Attach the air cleaner with new gaskets.

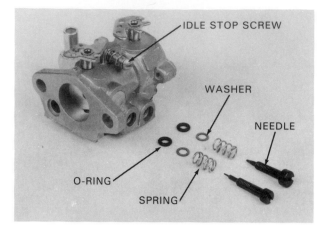

Fig. 12-33. High speed needle, idle mixture needle, and idle stop screw are installed and preset.

TROUBLESHOOTING DIAPHRAGM CARBURETORS

The troubleshooting chart in Fig. 12-34 points out areas of possible diaphragm carburetor trouble and lists symptoms and suggested repairs. This chart is presented for illustrative purposes and is not representative of all diaphragm carburetors. Always consult the manufacturer's service manual when servicing a diaphragm carburetor.

ENGINE GOVERNOR ADJUSTMENTS

Only two types of *governors* are used on small engines: the air-vane governor and the centrifugal flyweight governor. The centrifugal flyweight governor is by far the most popular. Often, the method of adjusting the governor can be determined by good judgment and reasoning if it is kept in mind that centrifugal force and spring pressure are opposed to each other and work to open or close the throttle.

Fig. 12-35 shows a governor that requires moving the spring bracket to increase or decrease governor spring tension. Loosen the bracket nut to move the bracket and then tighten nut.

TROUBLESHOOTING DIAPHRAGM CARBURETORS

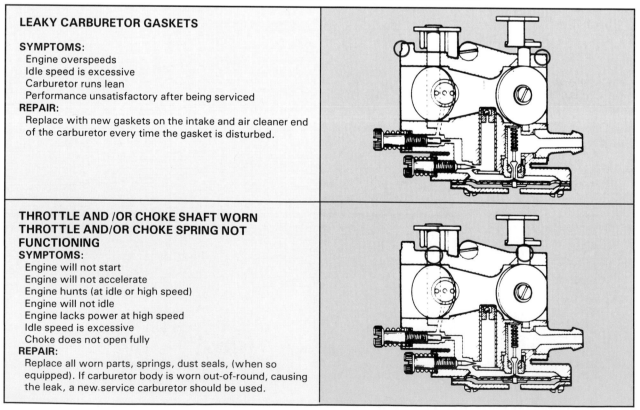

LEAKY CARBURETOR GASKETS

SYMPTOMS:
Engine overspeeds
Idle speed is excessive
Carburetor runs lean
Performance unsatisfactory after being serviced
REPAIR:
Replace with new gaskets on the intake and air cleaner end of the carburetor every time the gasket is disturbed.

THROTTLE AND /OR CHOKE SHAFT WORN THROTTLE AND/OR CHOKE SPRING NOT FUNCTIONING
SYMPTOMS:
Engine will not start
Engine will not accelerate
Engine hunts (at idle or high speed)
Engine will not idle
Engine lacks power at high speed
Idle speed is excessive
Choke does not open fully
REPAIR:
Replace all worn parts, springs, dust seals, (when so equipped). If carburetor body is worn out-of-round, causing the leak, a new service carburetor should be used.

Fig. 12-34. Troubleshooting chart for diaphragm-type carburetors. (Tecumseh Products Co.)

TROUBLESHOOTING DIAPHRAGM CARBURETORS

WHEN FUEL INLET IS PLUGGED

SYMPTOMS:
Engine will not start
Engine hunts (at idle or high speed)
Engine will not idle
Engine lacks power at high speed
Carburetor leaks
Engine starves for fuel at high speed (leans out)

REPAIR:
Clean fuel system completely. Refill with clean, fresh fuel as recommended. Replace loose clamps and fittings.

DIRTY, STUCK, OR DAMAGED NEEDLE AND SEAT

SYMPTOMS:
Carburetor floods
Carburetor leaks
Carburetor runs rich with main adjustment needle shut off

REPAIR:
Remove old needle and seat assembly. Install a new needle and seat assembly according to manual instructions.

CRACKED OR BRITTLE DIAPHRAGM OR IMPROPER POSITIONING OF DIAPHRAGM

SYMPTOMS:
Engine will not start
Engine will not idle
Engine lacks power at high speed
Carburetor floods
Idle speed is excessive
Engine starves for fuel at high speed (leans out)
Carburetor runs rich with main adjustment needle shut off
Carburetor leaks

REPAIR:
Replace diaphragm with a new one. Install with a new gasket according to the style of carburetor body that you have. Check mechanic's manual for proper sequence of diaphragm and gasket.

FUEL PICK-UP RESTRICTED OR PLUGGED

SYMPTOMS:
Engine will not start
Engine will not idle
Engine hunts (at idle or high speed)
Engine starves for fuel at high speed (leans out)

REPAIR:
Soak in a commercial carburetor cleaner (no longer than 30 minutes). Use compressed air to clean all passages.

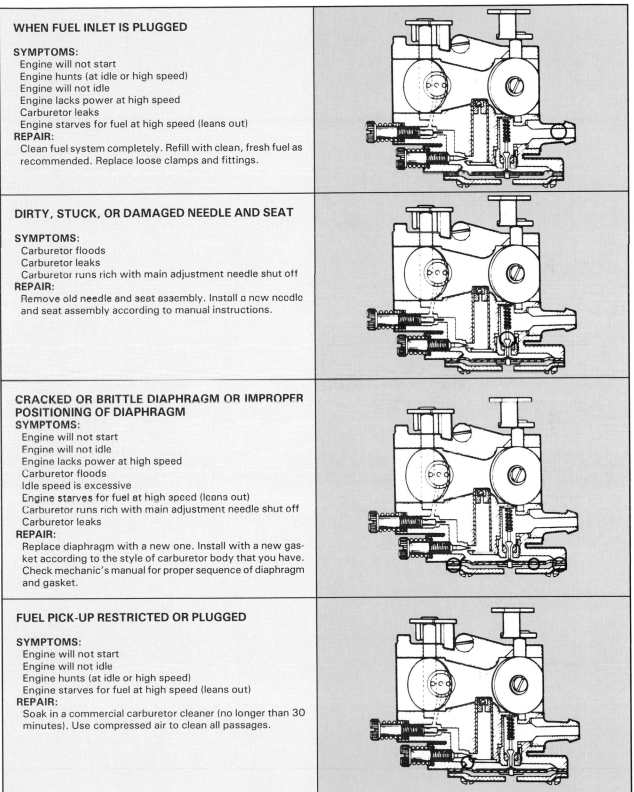

Fig. 12-34. Continued.

(Continued)

DAMAGED OR INCORRECT IDLE MIXTURE SCREW
SYMPTOMS:
Engine will not start
Engine will not accelerate
Engine hunts (at idle or high speed)
Engine will not idle
Engine lacks power at high speed
Engine overspeeds
Engine starves for fuel at high speed (leans out)
Carburetor runs rich with idle adjustment needle shut off
Performance unsatisfactory after being serviced

REPAIR:
Replace idle mixture screw and O-ring with new, clean passages. Do not overseat needle. If tip breaks off in carburetor, the body must be replaced.

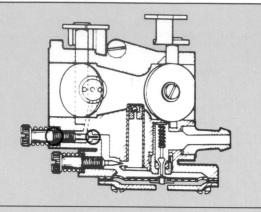

DAMAGED OR INCORRECT MAIN MIXTURE SCREW
SYMPTOMS:
Carburetor out of adjustment
Engine will not start
Engine will not accelerate
Engine hunts (at idle or high speed)
Engine will not idle
Engine lacks power at high speed
Engine overspeeds
Engine starves for fuel at high speed (leans out)
Carburetor runs rich with main adjustment needle shut off
Performance unsatisfactory after being serviced

REPAIR:
Replace main mixture screw and O-ring with a new one. Clean passages. Do not overseat mixture screw.

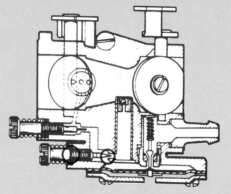

RESTRICTED OR PLUGGED AIR BLEED

SYMPTOMS:
Engine will not accelerate
Engine hunts (at idle or high speed)
Engine will not idle

REPAIR:
Soak carburetor body in a commercial cleaner (no longer than 30 minutes). Use compressed air to clean all passages.

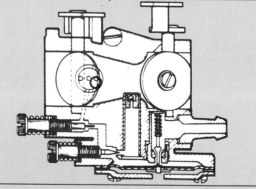

RESTRICTED IDLE AND/OR SECONDARY DISCHARGE PORTS

SYMPTOMS:
Engine will not start at idle
Engine will not accelerate
Engine hunts
Engine will not idle

REPAIR:
Soak carburetor body in a commercial cleaner (no longer than 30 minutes). Use compressed air to clean all passages.

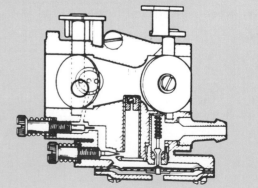

Fig. 12-34. Continued.

TROUBLESHOOTING DIAPHRAGM CARBURETORS

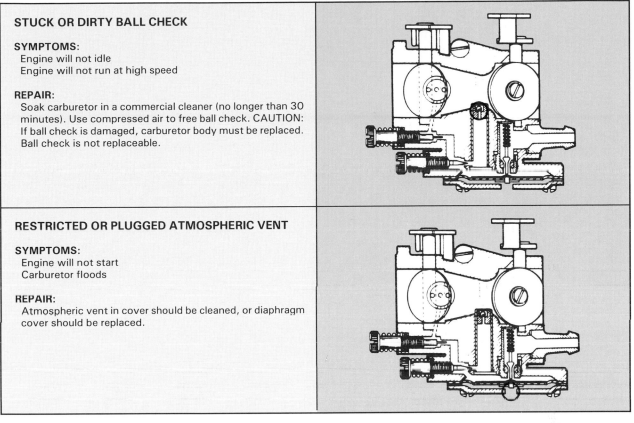

STUCK OR DIRTY BALL CHECK

SYMPTOMS:
Engine will not idle
Engine will not run at high speed

REPAIR:
Soak carburetor in a commercial cleaner (no longer than 30 minutes). Use compressed air to free ball check. CAUTION: If ball check is damaged, carburetor body must be replaced. Ball check is not replaceable.

RESTRICTED OR PLUGGED ATMOSPHERIC VENT

SYMPTOMS:
Engine will not start
Carburetor floods

REPAIR:
Atmospheric vent in cover should be cleaned, or diaphragm cover should be replaced.

Fig. 12-34. Continued.

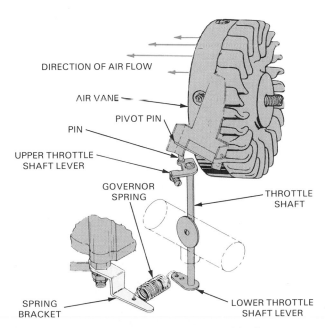

Fig. 12-35. The spring bracket on this air vane governor is moved to adjust spring tension. (Tecumseh Products Co.)

Fig. 12-36 shows a knurled thumbscrew that is used to adjust spring tension, thus changing rpm. The governor spring should be in hole number three for standard operation. In general, the closer the spring is to the pivot end of the lever, the smaller the difference between load and no-load engine speed.

If the spring is brought too close to the pivot point, the engine will begin to *hunt* (engine speed increase and decrease). The farther the spring is from the pivot end, the less the tendency to hunt. However, there will be a greater speed drop under load. If the governed speed is lowered, the spring can usually be moved closer to the pivot.

If the governor shaft has been removed or loosened, the proper adjustment is:

1. Loosen the screw holding the governor lever.
2. Hold the throttle in the high speed position.
3. With a screwdriver, turn the governor shaft *counterclockwise* as far as it will go.

4. Tighten the screw holding the governor lever to the governor shaft per specifications (typically 35-45 in./lbs. or 4.0-5.0 N·m).

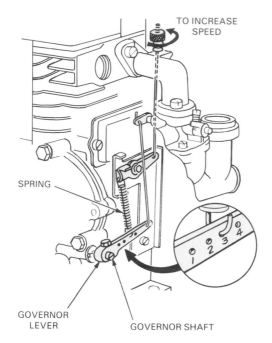

Fig. 12-36. The knurled thumbscrew on this governor is turned to adjust spring tension. Note that the spring is in hole number three of the governor lever. (Briggs and Stratton Corp.)

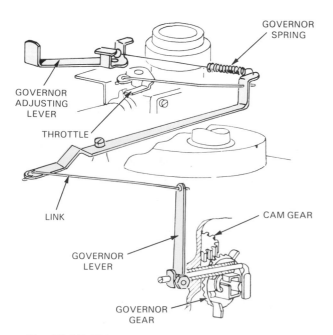

Fig. 12-37. This governor is used on a vertical shaft engine. Note the location of all parts.

5. Before starting the engine, move the linkage manually to check for binding.

Fig. 12-37 illustrates a governor used on a vertical shaft engine. It has a manually operated governor adjusting lever. If the governor lever has been removed or loosened, the proper adjustment is:

1. Loosen the nut holding the governor lever to the governor shaft.
2. Hold the throttle in the high speed position.
3. With a screwdriver, turn the governor shaft *clockwise* as far as it will go, Fig. 12-38.

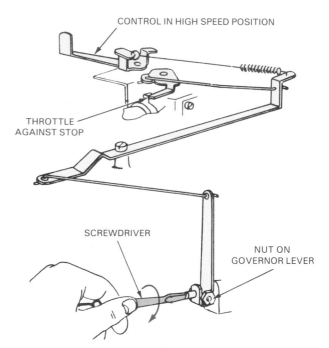

Fig. 12-38. To adjust governor, turn governor shaft clockwise as far as it will go. Note that the throttle is in the high speed position when making governor adjustments.

4. Tighten the screw holding the governor lever to the governor shaft per specifications (about 35-45 in./lbs. or 4.0-5.0 N·m).
5. Check the governor linkage for freedom of movement.

Fig. 12-39 illustrates a governor used on a horizontal shaft engine. Adjustment of the governor lever is the same as the one in Fig. 12-38.

Some governors require bending the spring attach arm, as shown in Fig. 12-40.

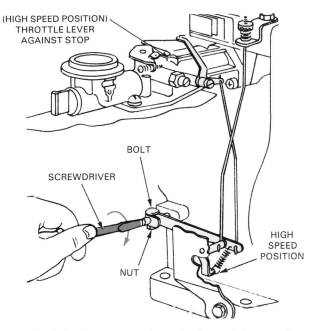

Fig. 12-39. Governor used on a horizontal shaft engine.

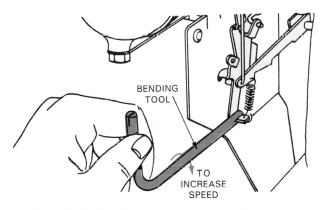

Fig. 12-40. Bending governor spring attach arm to increase spring tension and engine speed.

SUMMARY

Fuel system problems can involve the fuel pump, carburetor, reed valves, fuel lines, or fuel filter. Troubleshooting the fuel system involves testing each component in the system until the problem area is located.

When a faulty fuel pump is suspected, make sure that fuel flow is not interrupted before it gets to the pump. If the pump is in proper working condition, poor engine performance may be caused by improper carburetor adjustments or a defective carburetor. Make high speed adjust-

ment and idle adjustment according to manufacturer's specifications.

Two cycle reeds should be checked for leaks by holding a strip of paper in front of the carburetor throat when the engine is running. If the paper becomes spotted, the reeds are not seating properly. Defective reeds should be replaced or repaired.

If inspection and adjustment indicate the need for repair, the carburetor can be overhauled. Carburetor repair kits are usually available from the manufacturer. Troubleshooting charts can be used to locate specific problems. Because of the many variations on the market, it is very important to use a service manual when disassembling a carburetor. Follow all instructions carefully. Never soak carburetor parts in solvent for longer than 30 minutes. Wires or similar objects should never be used for cleaning carburetor ports.

The main purpose of an engine primer is to force fuel into the main carburetor nozzle, producing a rich mixture for starting.

There are several methods for adjusting the small engine governor. Refer to the service manual for specific information on the model being serviced. Often, the method of adjusting the governor can be determined by good judgment and reasoning.

KNOW THESE TERMS

Gravity-fed system, Fuel pump, Carburetor, Needle valve adjustments, Idle speed stop screw, High speed needle adjustment, Idle mixture adjustment, Reeds, Float-type carburetor, Overhaul, Carburetor kit, Welch plugs, Primer, Diaphragm-type carburetor, Vented, Governor, Hunt.

REVIEW QUESTIONS–CHAPTER 12

1. If the gasoline supply and fuel pump operation are satisfactory but an engine still idles and accelerates poorly, where would you check for the cause of the trouble?
2. Name the three basic carburetor adjustments in the order in which they are performed.

3. Which of the following procedures would not be recommended in carburetor maintenance?
   a. Clean nonmetallic parts in a commercial carburetor cleaner.
   b. Wash metallic parts in commercial carburetor cleaner.
   c. Dry parts with compressed air.
4. Make sure all carburetor passages and holes are open by pushing a stiff wire of the proper size through them. True or False?
5. Paper toweling is used to dry carburetor parts. True or False?
6. Never soak carburetor parts in solvent for more than _____ minutes.
7. Float bowl gaskets and O-rings are commonly reused when overhauling a float-type carburetor. True or False?
8. How can you check the condition of a reed valve without removing the carburetor and the valve?
9. The main purpose of the engine primer is to produce a rich mixture for _____.
10. Small engine governor adjustments are the same for all engines. True or False?

SUGGESTED ACTIVITIES

1. Make a complete carburetor adjustment following the procedure described in the text. Check the service manual for specifications.
2. Rebuild a float-type carburetor.
3. Rebuild a diaphragm carburetor.
4. Look up governor adjustments in a service manual for a specific engine. Explain how the governor works and demonstrate the correct adjustment procedure.

CHAPTER 13

IGNITION SYSTEM SERVICE

After studying this chapter, you will be able to:

☐ Examine spark plug deposits for signs of abnormal combustion.

☐ Clean, gap, and install spark plugs correctly.

☐ Explain the basic inspections and tests used to verify proper ignition system operation.

☐ Adjust breaker points, piston height, and ignition spark timing.

☐ Explain basic tests for breaker point and solid state ignition systems.

☐ Explain typical service procedures for battery ignition systems.

Although small engine ignition systems are durable, they do require periodic inspection and maintenance. When these services are performed, it is called an *ignition system tune-up*.

Ignition system service involves the entire ignition system, from the coil windings to the spark plugs. The small engine technician will check or test part after part until the entire system is working well. In the process, worn or defective parts must be replaced.

IGNITION SYSTEM AND SPARK PLUGS

When a small engine is difficult to start, a new *spark plug* may seem to solve the problem. However, the technician cannot assume that this is the only fault. Often, a less obvious problem has caused the plug to fail.

Although a *magneto system* may be able to supply 30,000 volts, it produces only enough voltage to jump the spark plug *gap.* Therefore, the condition of the spark plug determines the amount of voltage that other ignition parts must produce.

Spark plugs used in normal operation will wear out from erosion caused by combustion. A new plug may need only 5000 volts to fire. After many hours of operation, however, the same plug may require 10,000 volts to fire. If a pull on the starter cord produces less than 10,000 volts, an engine equipped with such a plug will not start. Certainly not all cases of hard starting are caused by a bad spark plug. Therefore, replacing the plug without further checking is not a good practice. The actual problem may lie in the malfunction of other ignition system components.

Changing the plug simply means that less voltage is needed to fire it. Carbon deposits will again build up in the cylinder and exhaust ports due to poor combustion. More carbon will form on the spark plug electrodes and cause further hard starting.

You can analyze the quality of combustion that has been taking place in a cylinder by examining the carbon deposits on the spark plug. Deposits having a beige to gray-tan color indicate normal combustion of the air-fuel mixture at the

Fig. 13-1. A normal spark plug will be clean and dry, showing a beige to gray-tan color on the porcelain shell insulator. (Champion Spark Plug Co.)

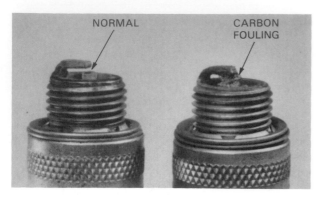

Fig. 13-2. Normal spark plug on left is compared to a carbon-fouled plug on right. (Jacobsen Mfg. Co.)

Fig. 13-3. An oil-fouled spark plug can be an indication of a mechanical malfunction. (Champion Spark Plug Co.)

proper operating temperature. Fig. 13-1 shows how a "normal," used spark plug looks. Fig. 13-2 compares a normal plug with one that is extremely carbon fouled.

An oil-fouled plug, like the one shown in Fig. 13-3, is saturated with wet oil. The eroded plug pictured in Fig.13-4 resulted from many hours of use.

Study the situation carefully before changing to a spark plug that is *hotter* than the one specified. The hot plug may stop carbon buildup, but other, more serious problems can develop. Spark plug deposits are usually caused by weak magneto voltage, incorrect carburetor adjustments, poor air cleaner maintenance, incorrect gasoline or oil, or incorrectly mixed gasoline and oil.

SPARK PLUG REMOVAL

The following four steps for removing spark plugs are simple and should become a habit. They can prevent troublesome problems from occurring later.

1. Gently rotate and pull the spark plug boot from the spark plug. *Do not grab or pull on the spark plug cable. Pull on the boot only.*
2. Use a correct-size, deep spark plug socket with a rubber sleeve installed to protect the plug insulator from breaking. NOTE: Before removing spark plugs from engines with aluminum heads, allow the engine to cool. The heat of the engine, in combination with a spark plug that has run for many hours, may cause the spark plug to seize.

Fig. 13-4. After many hours of use, spark plug electrodes tend to erode from normal combustion. (Jacobsen Mfg. Co.)

3. Before removing the loosened spark plug, blast dirt away from the area around the plug with compressed air.
4. Remove spark plug and check its appearance. Refer to the Spark Plug Analysis Table in the Useful Information section of this text.

CLEANING SPARK PLUGS

The following procedure should be observed when cleaning spark plugs:

1. Wipe all spark plug surfaces clean. Remove oil, water, dirt, and moist residues.
2. Check the firing end or tip of the spark plug for oily or wet deposits. If the spark plug has deposits, brush the plug with a nonflammable, nontoxic solvent. Then, dry the plug with compressed air to prevent caking of the cleaning compound deep within the spark plug shell.

3. Clean the spark plug in a spark plug cleaning machine. (See section on using the spark plug cleaning machine.)
4. File spark plug electrodes after cleaning to square them and to remove any oxide or scale from the surfaces. Open the gap so that a spark plug file can be put between the electrodes, Fig. 13-5.
5. Clean spark plug threads with a hand or powered wire brush, Fig. 13-6. Take care not to damage the electrodes or insulator. If threads are nicked or damaged, discard the plug.

Fig. 13-7. This is a special spark plug cleaning unit that removes carbon from the electrodes and the insulator.

Fig. 13-5. Oxides should be removed from electrodes with a spark plug file.

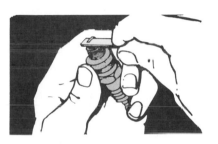

Fig. 13-6. Threads can be cleaned with a power wire brush.

USING THE SPARK PLUG CLEANING MACHINE

There are many types of *spark plug cleaning machines* available, which are designed to remove deposits from a plug. The machine in Fig. 13-7 should be used in the following manner:

1. Select the proper size adaptor and position it in the plug cleaning recess. Lock the adaptor to the cleaning recess with the retaining ring.
2. Press the firing end of the dirty spark plug firmly into the hole in the adaptor, Fig. 13-7.

3. With one hand, wobble or swivel the terminal end of the spark plug in a circular motion (one inch circle). With the other hand, depress the "Abrasive Blast" lever for three or four seconds.
4. Release the "Abrasive Blast" lever. Depress the "Air Blast" lever and wobble the plug for three or four more seconds to remove loose particles of compound that may be lodged in the spark plug.
5. Inspect the spark plug with an inspection light. If the insulator is not white, clean the plug again.

CAUTION: Prolonged application of the cleaning blast can damage the spark plug by wearing down the insulator and the electrodes.

The spark plug shown in Fig. 13-8 is being *compression tested*. A spark coil provides the voltage, while a valve wheel increases compression. The spark intensity is observed through a window on the tester. If the spark is intermittent or stops before the compression reading reaches an acceptable level, the plug should be discarded.

When installing plugs, clean the spark plug seat on the cylinder head. Also, clean the threads and make sure the gasket is in good condition.

Not all plugs require gaskets. Fig. 13-9 shows one type that does not.

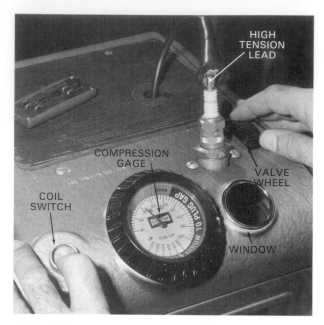

Fig. 13-8. After cleaning and gapping the plug, it can be spark tested under compression and compared to a new one. (Champion Spark Plug Co.)

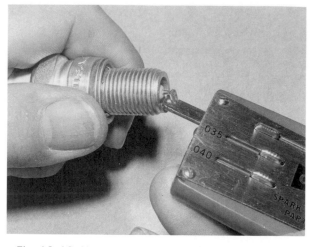

Fig. 13-10. Use a spark plug gapping tool to bend the outer electrode toward or away from the center electrode. (Champion Spark Plug Co.)

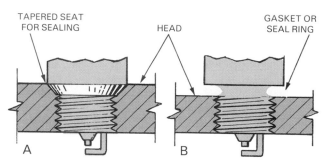

Fig. 13-9. Two types of plugs. A—Plug with tapered seat does not require a gasket. Clean the area around the plug hole for a good seal. B—Spark plug with gasket. Inspect plug gaskets carefully. A damaged gasket will not seal properly.

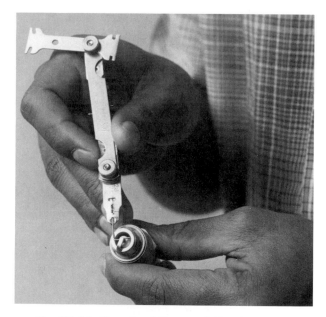

Fig. 13-11. Gap should be carefully adjusted and measured with wire gages.

GAPPING SPARK PLUGS

When *gapping* a spark plug, bend the outer electrode toward or away from the center electrode. For best results, use a gapping tool, Fig. 13-10. Note the built-in wire gages (.035, .040).

Standard *leaf-type feeler gages* may be used if the plug is new or if the electrodes are in good condition. Otherwise, wire-type thickness gages should be used, Fig. 13-11.

The reason that *wire-type thickness gages* are recommended for use on worn spark plugs is shown in Fig. 13-12. Note that the flat, leaf-type thickness gage would leave an additional gap between the worn electrodes.

ANALYSIS OF USED SPARK PLUGS

Spark plugs that are cleaned and gapped at regular intervals can provide many hours of useful life in an engine using leaded gasoline. Spark plugs in high energy ignition systems (TCI or

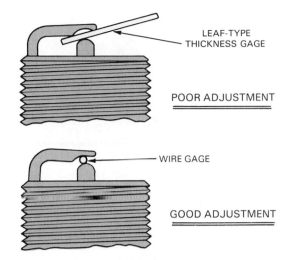

Fig. 13-12. Leaf-type thickness gages may not give an accurate measurement of gap if electrodes are not flat and parallel.

CDI) that burn unleaded fuel can provide more than twice as many hours of useful service.

Spark plugs in two different engines of the same make and model may show a wide variation in appearance. Engine condition, carburetor settings, and operating conditions, such as sustained high speeds or continual low-speed, stop-and-start operation, are all variables that affect spark plug life.

Spark plugs are sometimes incorrectly blamed for poor engine performance. Replacing an old plug may temporarily improve engine performance because of the lessened demand a new spark plug makes on the ignition system. However, it is not a cure-all for poor performance caused by worn rings or cylinders, improper carburetion, worn ignition system parts, or other engine problems.

SPARK PLUG INSTALLATION

Spark plugs must be installed properly. The heat dispersing properties of the spark plug depend on correct plug seating. If the spark plug is tightened excessively, the gasket will be crushed. Internal leakage may result. Attempting to remove an overtightened spark plug can strip cylinder head threads. Seating the spark plug too loosely can result in preignition and possible engine damage caused by spark plug overheating.

To install spark plugs:
1. Make sure cylinder head and spark plug threads are clean. If necessary, use a thread chaser and a seat cleaning tool.
2. Make sure that the spark plug gasket seat is clean. Thread the gasket to fit flush against the gasket seat on the spark plug.
3. Make sure the spark plug has the correct gap.
4. Screw the spark plug finger-tight into the cylinder head. Use a torque wrench to tighten the plug to 13-15 ft.-lbs. *Do not overtighten.*

IGNITION TESTING PROCEDURE

Certain basic inspections and tests are necessary to determine whether an ignition system is working properly. The following procedure can be used to verify ignition system operation:
1. Make certain that the thin ignition ground wire is not grounding out the entire system. Examine the high tension spark plug lead for voltage leaks. Look for worn insulation or cuts where metallic contact is made.
2. Remove the spark plug and examine the electrodes and the porcelain insulator. If the plug is carbon grounded, replace it. Adjust the plug electrodes to the specified gap. Then, install the plug and tighten it to the specified torque. If torque specifications are not available, turn the plug finger-tight and tighten it a half turn with a wrench. This will allow for proper heat transfer from plug to cylinder head.
3. Hold the plug wire by the insulation (well away from metal connector) so that the connector is 3/16 in. from the tip of the plug. See Fig. 13-13.
4. Pull the starter cord. An orange-blue spark should jump the gap between the connector and the plug. If it does, the ignition system is good.
5. If no spark occurs, hold the wire 3/16 in. from the base of the plug, Fig. 13-14. Pull the starter cord once more. If a spark occurs at base of the plug and not at the tip, the plug is failing under compression and should be replaced.

Fig. 13-13. If spark does not jump a 3/16 in. gap at the plug tip but jumps the same gap at the base of the plug, the plug is defective.

Fig. 13-14. If no spark occurs at the plug tip or the plug base, the problem is in the magneto system.

6. If a spark does not occur in either position, the problem is in the magneto system.

A spark tester can also be used to test the ignition circuit for proper operation. Most spark testers are simply attached to the secondary wire as shown in Fig. 13-15. If current is flowing in the spark plug lead, it will induce a voltage in the tester. This voltage will energize a light in the window of the tester. The greater the voltage in the secondary wire, the brighter the light in the tester's window.

MAGNETO SERVICE

In some *magneto systems,* the coil is mounted outside of the flywheel and the gap between the laminated core and the flywheel magnets is ad-

Fig. 13-15. A spark tester can be used to determine the intensity of secondary current. A light will flash in the window each time ignition spark occurs.

justable. This adjustment must be made carefully or the magnetic strength and voltage produced will be reduced. Loosen the adjustment screws and place a feeler gage or nonmagnetic shim stock of the correct thickness in the gap, Fig. 13-16. After the correct gap has been obtained, tighten the adjustment screws.

Fig. 13-16. On some engines, magneto air gap is checked with a nonmagnetic thickness gage. (Lawn Boy Power Equipment, Gale Products)

In some *solid state systems,* you can adjust the air gap between the trigger module and the flywheel trigger projection. Rotate the flywheel until the projection is next to the trigger module. Loosen the trigger retaining screw and move the trigger until the air gap is correct. Use a feeler gage and make certain that the flat surfaces on the trigger and the projection are parallel before tightening the retaining screws.

Many magneto systems are completely contained under the flywheel. The flywheel must be removed from the crankshaft to service the ignition components in these systems. The flywheel is mounted on the tapered end of the crankshaft and keyed for alignment, Fig. 13-17.

Fig. 13-18. A knock-off tool is often used to remove a flywheel. The tool must be tight on the shaft before being hit. Pry bar provides a valuable assist. (Jacobsen Mfg. Co.)

Fig. 13-17. Crankshaft end is tapered and keyed to hold the flywheel in exact position. The flywheel often needs a special tool for removal from the taper. (Deere & Co.)

REMOVING THE FLYWHEEL

Since the *flywheel* is fastened to the crankshaft with a nut, it may be hard to remove. One way is to remove the flywheel nut and other accessory parts and place a *knock-off tool* on the crankshaft. Thread the tool all the way onto the crankshaft and tap it with a hammer. See Fig. 13-18.

The sudden jolt will loosen the flywheel. Never use the flywheel nut as a knock-off tool or the crankshaft threads will be damaged.

Another way to remove the flywheel is to use a *wheel puller,* like the one shown in Fig. 13-19. After installing the puller, tighten the center bolt snugly. If the flywheel does not break loose from

Fig. 13-19. A flywheel puller is often used to prevent the flywheel from distorting or cracking during removal. (Kubota Tractor Corp.)

the crankshaft when the center bolt is tightened, tap the bolt with a hammer. Use extreme care when removing aluminum flywheels.

INSPECTING THE FLYWHEEL

When the flywheel is removed, inspect it. Make sure that it is not cracked and that the mounting hole and keyway are not damaged. Inspect the key. If it has begun to shear, as in Fig. 13-20A, or if the key is too narrow, as in Fig. 13-20B, the engine will be out of time. In either case, the key must be replaced.

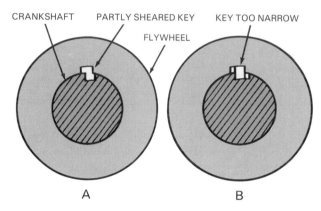

CRANKSHAFT   PARTLY SHEARED KEY   KEY TOO NARROW

FLYWHEEL

A   B

Fig. 13-20. Partly sheared flywheel keys or keys that do not fit well can put the ignition system out of time.

Fig. 13-21. Magnetism is the heart of the magneto system. Test magnets by placing a socket on them and shaking the flywheel. Socket should remain in place.

To check the strength of the flywheel magnet, place a 1/2 in. socket on the magnet and shake the flywheel. The socket should remain in place on the magnet, Fig. 13-21. The magnet can lose its magnetism.

TESTING IGNITION SYSTEM COMPONENTS

After removing the flywheel, inspect the magneto system components. Check all parts for dents, cracks, and gouges. If system is equipped with breaker points, check for improper alignment and pitted contact surfaces. Make sure the insulation on all wire leads is in good condition.

An ohmmeter can be used to check the condition of coils and condensers. To check a coil's primary windings for *continuity,* set the ohmmeter to its lowest resistance range. Connect one meter probe to the coil ground connection and the

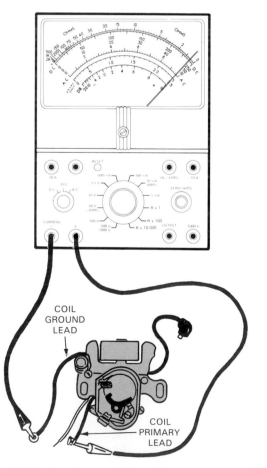

COIL GROUND LEAD

COIL PRIMARY LEAD

Fig. 13-22. Testing a primary coil for continuity with an ohmmeter.

other to the coil primary lead, Fig. 13-22. The meter should indicate almost no resistance between these connections. To check the secondary coil for continuity, set the ohmmeter to the R x 100 or R x 1000 resistance range. Connect one

meter probe to the ground connection and the other to the high tension output lead, Fig. 13-23. The resistance between these connections may range from several hundred ohms to several thousand ohms. Check manufacturer's specifications, Fig. 13-24. If the meter registers infinite resistance in either test, the coil is faulty and it must be replaced.

To test a condenser for continuity, set the ohmmeter to the highest resistance scale. Connect the test leads as illustrated in Fig. 13-25. The meter

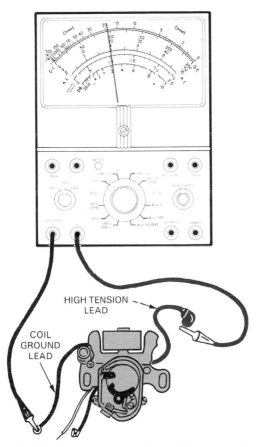

Fig. 13-23. Testing a secondary coil for continuity. Meter readings will vary, depending on coil type. Check manufacturer's specifications.

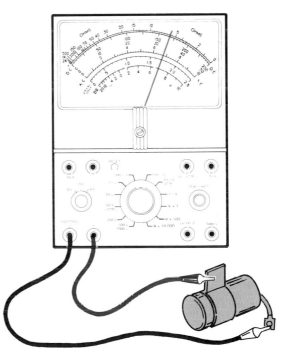

Fig. 13-25. An ohmmeter can be used to test a condenser for continuity.

SMALL ENGINE COIL SPECIFICATIONS

Coil No.	Mfg. Model No.	Mfg. No.	Operating Amperage	Primary Resistance Min.-Max.	Secondary Continuity Min.-Max.
26787	Fairbanks-Morse	HX-2477	1.90		50–60
28259	Wico	X-11654	2.1		40–55
29176	Phelon	FG-4081	2.8		40–60
29632	Lauson	5022 (Syncro)	2.3	.5–1.5	40–55
30560	Lauson VH60 John Deere (Replaces 30546)	5160 (Syncro)	2.9	.35–.45	40–55
32014	Lauson HH80	8	1.4	.37–.45	55–65

Fig. 13-24. Coil specifications are often provided by test equipment manufacturer. (Merc-O-Tronic Instrument Corp.)

should initially indicate a low resistance, but the reading should rise rapidly as the condenser takes on a charge from the meter's battery. If the reading remains low, the condenser is faulty and it must be replaced.

Solid state ignition systems are offered in many small engines. They offer dependable service, with few moving parts to oil or adjust. Certain solid state components are manufactured as assemblies and cannot be serviced.

If a solid state system fails to produce a spark between the plug cable terminal and the cylinder head, check the condition of the cable. If it has cracked insulation or shows the aftereffects of arcing (black powdery discoloration or deposits), replace the pulse transformer and the high tension lead assembly.

Examine the low-tension lead. If the insulation is cracked or in poor condition, replace it. The ignition charging coil, electronic triggering system, and mounting plate are available as an assembly only. If you must replace this assembly, carefully follow the manufacturer's instructions.

Before replacing the pulse transformer, attach the leads from a new transformer, ground the unit, and test for spark. If you have spark where none existed before, replace the old unit with the new one.

In many cases, the condition of ignition components can be checked with a special ignition tester. The ignition tester checks components under conditions similar to operating conditions. These testers can be used to verify the operation of condensers, coils, and solid state ignition components, Fig. 13-26. Recommended testing procedures vary among manufacturers. Always consult an appropriate shop manual before performing tests on ignition components. Follow the manufacturer's recommendations carefully. Typical testing procedures for three common ignition testers are illustrated on pages 332-335 in the Useful Information section of this text. When testing ignition system components, perform tests on a wooden table or an insulated tabletop. This will prevent current leakage, which can cause "shorting" or shock hazards.

ADJUSTING BREAKER POINT IGNITION SYSTEMS

The first step in adjusting a breaker point ignition system is to set the breaker points to the correct gap. Look in the engine manual for the proper setting. Remove the dust cover from the stator plate to expose the breaker points. Turn the crankshaft slowly until the high point of the cam

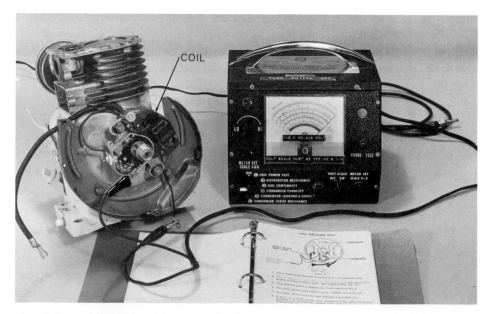

Fig. 13-26. A special ignition tester is often used to check individual ignition components for proper operation. This tester is being used to check coil ground connections.

lobe is directly under the wear block or cam follower. Then, slightly loosen the point adjustment screw, as demonstrated in Fig. 13-27. NOTE: On the magneto system shown in these illustrations, a screwdriver can be used to move the stationary part of the breaker points. A fulcrum point of leverage is provided, Fig. 13-28.

To adjust the breaker point gap, use a feeler gage of the specified thickness. First, move the

Fig. 13-29. A feeler gage is used to set breaker point gap. When gap is correct, set screw is tightened to lock in adjustment.

Fig. 13-27. To adjust breaker point gap, loosen point adjustment set screw first.

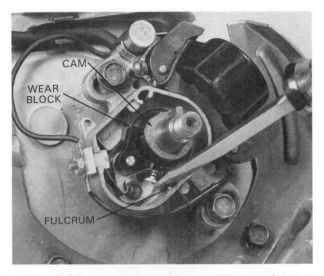

Fig. 13-28. Using a screwdriver as a lever resting on the fulcrum, move stationary breaker point to vary the size of the gap. Wear block should be on the high point of the cam.

stationary point until it lightly touches the feeler gage, Fig. 13-29. Then, tighten the adjusting screw that holds down the stationary breaker point. Finally, recheck the point gap. The stationary point may move when the adjusting screw is retightened. Repeat the adjusting procedure until the gap setting is right.

ADJUSTING PISTON HEIGHT

Ignition spark occurs at the instant the breaker points open. When this happens, the piston must be at the proper position (near top dead center) in the cylinder. Fig. 13-30 shows piston heights for a full line of Tecumseh engines. Model A, for example, requires the piston to be between .060 in. and .070 in. BTDC (below top dead center).

One method of setting piston height is to use a **timing tool,** Fig. 13-31. Install the cylinder head over the cylinder and fasten it with one or two head bolts. Do not use the gasket. Locate the head so that the spark plug hole is over the piston and both valves can move freely.

To use the timing tool, screw it into the spark plug hole with its graduated rod riding on the piston top. Note that each marking on the rod is 1/32 in. (.031 in.). Place the nut on the crankshaft

FOUR CYCLE ENGINE TABLE OF SPECIFICATIONS

Model	A		C	D	E V & H40	F V & H60	G V & H50	H	I
Displacement	6.207	7.35	7.61	8.9	11.04	13.53	12.176	7.75	9.06
Stroke	1 3/4''	1 3/4''	1 13/16''	1 13/16''	2 1/4''	2 1/2''	2 1/4''	1 27/32''	1 27/32''
Bore	$\frac{2.125}{2.127}$	$\frac{2.3125}{2.3135}$	$\frac{2.3125}{2.3135}$	$\frac{2.5000}{2.5010}$	$\frac{2.5000}{2.5010}$	$\frac{2.6250}{2.6260}$	$\frac{2.6250}{2.6260}$	$\frac{2.3125}{2.3135}$	$\frac{2.5000}{2.5010}$
Timing Dimension Before Top Dead Center for Horizontal Engines	H $\frac{.060}{.070}$	H $\frac{.060}{.070}$	H $\frac{.060}{.070}$	H $\frac{.030}{.040}$	H $\frac{.090}{.100}$	H $\frac{.090}{.100}$	H $\frac{.090}{.100}$	H $\frac{.060}{.070}$	H $\frac{.030}{.040}$

Fig. 13-30. A table of specifications in the service manual will provide data concerning piston height (timing dimension before top dead center) for use in timing the opening of breaker points. (Tecumseh Products Co.)

Fig. 13-31. A special timing tool can be used to adjust piston to specified location BTDC.

so it can be turned with a wrench. Then, turn the crankshaft in the direction the engine runs until the piston is at TDC. Since proper piston position is between .060 in. and .070 in. BTDC, two graduations or marks (.062 in.) will be required. However, to compensate for any backlash, back the piston down six graduations and bring it back up

four marks. This places the piston exactly two graduations or .062 in. BTDC. Finally, lock the thumbscrew and remove the wrench.

Another method of adjusting piston height involves using a ***depth micrometer.*** With this technique, the micrometer is set for correct piston height BTDC and the piston is brought up to it, Fig. 13-32.

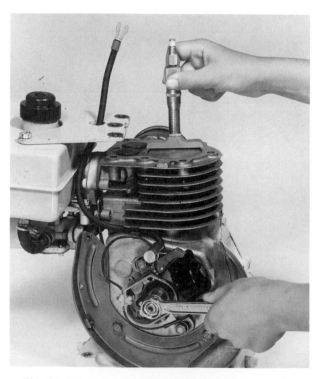

Fig. 13-32. A depth micrometer also can be used to measure piston height BTDC. The piston can be raised or lowered by turning the wrench on the crankshaft nut.

A third method of measuring and setting the piston height is shown in Fig. 13-33. In this method, a special *dial indicator,* which is equipped with an extender leg, is used to reach over the piston top. A tip, which contacts the piston top, is installed on the end of the extender leg. To use this instrument, the correct tip must be installed. Use the small tip for engines with timing dimensions between Top Dead Center (TDC) and .050 in. BTDC. Use the large tip for engines with timing dimensions of between .051 in. BTDC and .150 in. BTDC. If the engine is the four cycle type, make sure to secure the extender leg in position to locate the tip directly over the piston head. Loosen the sleeve screw on the side of the adaptor

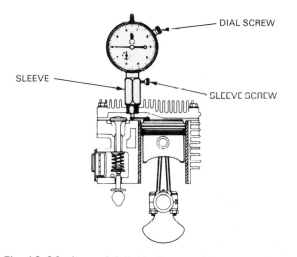

Fig. 13-33. A special dial indicator with an extender leg can be used to determine piston height. Use the correct type of extender leg for the engine being serviced. (Tecumseh Products Co.)

to allow the sleeve to be turned into the threads of the spark plug hole. This will ensure the proper location of the tip. Once the adaptor sleeve is secured in the hole, tighten sleeve screw to prevent the dial from moving up or down, which could give a false reading.

Find Top Dead Center (TDC) by rotating the crankshaft clockwise (when looking at the magneto end of the crank) until the needle on the dial stops and reverses direction. TDC is at the point where the needle stops. Loosen the dial screw and rotate the dial until the zero is lined up with the needle at TDC. Retighten the dial screw to secure it in place, Fig.13-34A.

While watching the needle on the dial indicator, rotate the crankshaft counterclockwise (when looking at the magneto end of the crank) past the specified Before Top Dead Center (BTDC) dimension, Fig. 13-34B (Example: .090 in.). Then rotate the crankshaft back clockwise to the proper dimension, Fig. 13-34C (Example: .080 in.). This will compensate for any slack between the connecting rod and the crankshaft assembly.

TIMING IGNITION SPARK

With the correct breaker point gap set and the piston positioned at the point where spark should occur, the next step is to time the "breaking" of the points. You can do this with a simple, battery-operated continuity tester or an ohmmeter.

First, loosen the two stator adjustment bolts so the stator can be rotated. Disconnect the coil lead

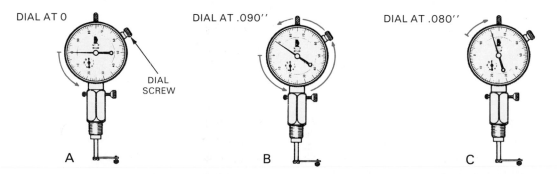

Fig. 13-34. To use the dial indicator and the extender leg to locate piston position: A—Rotate the crankshaft clockwise until the needle stops. Loosen the dial screw and set the zero under the needle. Tighten the dial screw. B—Rotate the crankshaft counterclockwise past the specified BTDC dimension. C—Rotate the crankshaft clockwise to the proper dimension to remove the slack between the connecting rod and the crankshaft.

wire to the points and reinstall the securing nut. Connect one lead of a continuity tester or ohmmeter to the breaker point terminal and the other lead to a good ground location. Rotate the stator until the continuity tester light or the ohmmeter indicates a break in the circuit, showing that the breaker points have opened, Fig. 13-35. Carefully tighten the two stator adjustment bolts and reconnect the coil lead to the breaker point terminal.

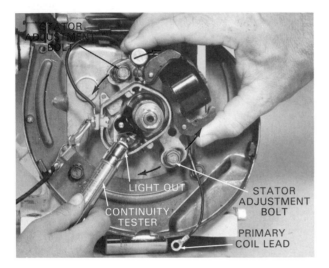

Fig. 13-35. A continuity tester is attached. The stator plate is loosened and turned until the continuity light goes out. This indicates that the points have opened. Then, the stator plate is locked in place. The coil primary lead is disconnected during this test.

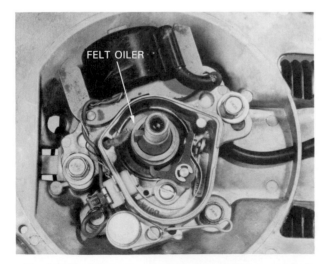

Fig. 13-36. A felt oiler should have several drops of oil to lubricate the cam lobe. Some felt oilers are reversible. (Jacobsen Mfg. Co.)

Before replacing the dustcover over the stator plate, place one or two drops of oil on the felt cam oiler, Fig. 13-36. Lubrication reduces wear on the point wear block. *Do not over oil.* Excess oil will foul the breaker points. Clean the points by sliding a lint-free paper back and forth between the contacts. Manually open the points when removing the paper to eliminate paper fibers from remaining between the contact points. If you have carefully followed the recommended procedure, the engine ignition system will be "in time."

REINSTALLING THE FLYWHEEL

After performing the necessary ignition system testing and service, the flywheel can be replaced. To replace the flywheel, turn crankshaft to place keyway in 12 o'clock position and insert the key. If the shaft uses a Woodruff key, position it as shown in Fig. 13-37. Make sure that the key seats properly in the keyway before starting the flywheel on the shaft.

Next, align the flywheel keyway with the crankshaft key and install the flywheel. Tighten the crankshaft nut to the correct torque. NOTE: Use a strap wrench, Fig. 13-38, or a spanner wrench, Fig. 13-39, to hold the flywheel when torquing the crankshaft nut.

Fig. 13-37. If a Woodruff-type key is used on a crankshaft, it should be placed so that the top of the key is parallel to the centerline of the shaft. (Lawn Boy Power Equipment, Gale Products)

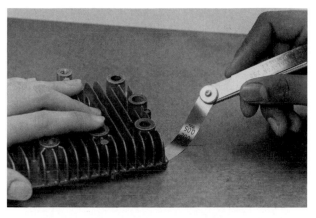

Fig. 13-40. Cylinder head should be placed on a surface plate and tested for flatness with a thickness gage.

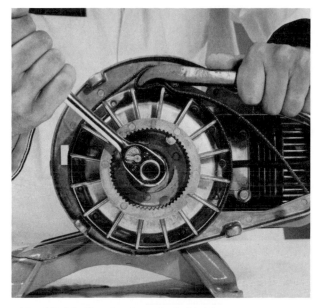

Fig. 13-38. Flywheel can be held with a strap wrench while tightening flywheel nut. (Jacobsen Mfg. Co.)

Fig. 13-39. A spanner wrench also can be used to hold flywheel when torquing flywheel nut.

REINSTALLING THE CYLINDER HEAD

If the cylinder head was removed to set piston position, it should be cleaned of all carbon deposits and checked for flatness by laying it on a surface plate. Use a .002 in. feeler gage between the mating surfaces to detect any warpage, Fig. 13-40.

If cooling fins are broken, the head should be replaced. If there are cracks, nicks, or burrs on the machined surface of the head, it should be resurfaced. Always install a new head gasket and follow the bolt tightening sequence recommended by the manufacturer.

SERVICING BATTERY IGNITION SYSTEMS

Battery operated ignition systems are much like magneto systems. However, they have additional components that require maintenance and service. When servicing a battery ignition system, check for the following problems before beginning an extensive system analysis:

1. Defective or undercharged battery.
2. Corroded or loose terminals and connections.
3. Wrong connections.
4. Cracked insulation or broken wires.
5. A wire "grounding out" in the system.
6. A defective switch.
7. Improperly functioning operator presence system.

The presence of a battery and a starter does not mean that the ignition system is battery operated. Some magneto ignition systems also use these parts. You can identify a battery ignition system by a can-shaped coil, in addition to a battery and a generator.

NOTE: All lawn and garden tractors built after July of 1987 are required to have an operator presence system. Many implements were equipped with these systems prior to this date. If an engine will not start or is "cutting out," check for a malfunction in the operator presence system.

BATTERIES

Storage batteries need regular maintenance to keep them in good operating condition. Remove the battery caps about twice a month and, if necessary, add distilled water to bring the electrolyte above the plates. This prevents *sulfating* (forming of salt-like deposits when air combines with the electrolyte), which will ruin the battery. These deposits are the same as the green deposits that collect around the posts and cable clamps of an automobile battery.

Keep the battery case and terminals clean. Remove the battery cables occasionally and clean all parts thoroughly. Reinstall the cables and coat them with petroleum jelly to retard further corrosion. Keep these connections clean and tight to avoid short circuits and/or voltage loss.

HYDROMETER TEST

If battery condition is doubtful, test it. Use a *hydrometer* to determine if the battery is fully charged. The hydrometer determines charge by measuring the specific gravity (density) of the electrolyte. Remove one of the battery vent caps and draw enough electrolyte into the hydrometer to make the float move freely, Fig. 13-41. The position of the float will correspond to the electrolyte's specific gravity. Additional testing should not be attempted if the specific gravity is less than 1.250. The battery must be recharged.

Maintenance free or *sealed batteries* cannot be checked with a hydrometer because there is no access to the electrolyte. These batteries are liquid filled or have a gel electrolyte and must be tested under load with special equipment.

CHECKING VOLTAGE

Another way of testing a battery is to measure cell voltage. This can be done with an accurate voltmeter, Fig. 13-42, or a high-rate discharge tester. The tester puts a heavy load on the battery for three seconds.

Check each cell in turn. If voltage checks out from 1.95 to 2.08V, the battery is good. If there is a difference of .05V between any cells, replace the battery.

Some batteries with one-piece, hard covers cannot be checked by testing individual cells. In these cases, the hydrometer test is made or an instrument equipped with a carbon pile rheostat is used to test the battery. There is also a tester available that is equipped with probes, which are dipped into the electrolyte. These probes carry an electrical impulse to a voltmeter.

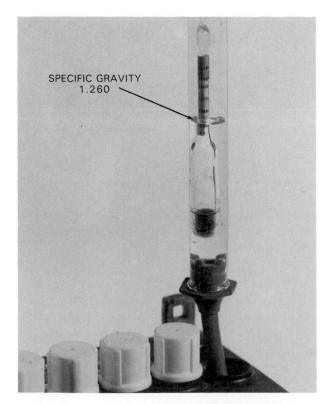

SPECIFIC GRAVITY 1.260

Fig. 13-41. This hydrometer reading shows a fully charged condition of the battery. Specific gravity reading is in the 1.260 to 1.280 range.

BOLENS

Fig. 13-42. An accurate voltmeter can be used to measure battery voltage and cell voltage.

BATTERY RECHARGING

⚠️ Caution must be exercised when charging a battery. Batteries generate *hydrogen* and *oxygen* during charging. These two gases combine to form a highly explosive mixture. Always the following rules:

1. Never check a battery fluid level with a flame.
2. Do not use jumper cables on a battery unless the ignition wiring is disconnected from the battery.
3. Do not use electric welding equipment on an implement with a battery without first disconnecting the wires to the battery terminals.
4. Connect the negative cable last and disconnect it first (in negative ground systems).

If batteries are used in temperatures below 32°F (0°C), it is important to keep them fully charged. A full charge can prevent the electrolyte from freezing and cracking the battery case. In cold temperatures, battery power decreases while the need for engine cranking power increases. Sub-zero temperatures can reduce a fully charged battery's capacity to 30% of its normal power, while increasing the cranking load beyond the warm weather cranking requirements, Fig. 13-43.

Only direct current can be used to recharge a battery. The battery charger automatically rectifies (converts) the alternating current (AC) available from electrical receptacles to direct current (DC). Before charging a battery, add water to bring the electrolyte in the battery cells up to the right level. Make sure the outside of the battery is clean. Connect the positive charger lead to the positive battery terminal and negative lead to negative terminal, Fig. 13-44. Charging times will vary, depending upon battery condition and charging rate. Follow manufacturer's instructions.

Fig. 13-44. The positive charger cable must be attached to the positive battery post and the negative cable must be connected to the negative post. Otherwise, the battery could be ruined. If battery markings are unreadable, dip leads from the battery terminals in a weak solution of sulphuric acid. More gas bubbles will collect around negative lead than the positive lead.

Badly sulfated batteries can sometimes be reclaimed by recharging them very slowly. This reconverts the sulfate to electrolyte.

Most new batteries are dry-charged. To place this type of battery in service, add the electrolyte solution according to manufacturer's instructions. Some newly activated batteries require a short period of charging; others can be placed in service immediately.

WIRING

Electrical system *wiring* must have good insulation between all points of connection. Wires

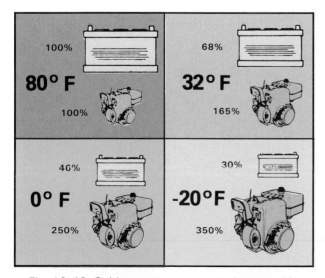

Fig. 13-43. Cold temperatures can reduce cranking power of battery, while increasing the cranking load.

should be securely fastened and connecting points should be free of corrosion, rust, and oil. Loose and corroded connections can severely diminish battery potential. A pinhole in a wire's insulation can cause electricity to leak and "ground out" on the engine frame. This condition can be amplified if water or oil is present on the insulation. A wire that is grounding out can make starting impossible. It can also cause an engine to run erratically. The thickness of copper wire is expressed in gage numbers. The larger the gage number, the smaller the diameter of the wire. The smaller the number, the larger the diameter of the wire, Fig. 13-45. Six gage wire (minimum) must be used for the starter circuit. Sixteen gage wire (minimum) is generally used for the charging circuit. (20 amp system requires 14 gage wire.) Eighteen gage wire (minimum) is required for the magneto circuit (ground circuit).

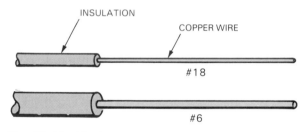

Fig. 13-45. *The larger a wire's gage number, the smaller the wire.*

AMMETERS

Some battery ignition systems are equipped with an **ammeter,** which is used to measure the rate of current flow from the alternator to the battery. If no current flow is indicated by the ammeter, remove the meter from the circuit and check all components in the system. If the system is operating properly, use an ohmmeter to check continuity across the ammeter terminals. If continuity does not exist, the ammeter is faulty and should be replaced.

SWITCHES

Switches are used to control many functions in a battery ignition system. The switch is a common point where most of the wiring comes together on an implement such as a small riding tractor. Many varieties of switches are available. When a switch fails, replace it according to the manufacturer's specifications. *Never substitute an automotive switch for small engine applications.*

In some switching applications, current is very high and normal switches are too small to continuously "make" and "break" the circuits without burning their electrical contacts. In these situations, a solenoid should be installed in the circuit. A *solenoid* is a heavy-duty, electromagnetic switching mechanism that is used to handle large amounts of current. It consists of a heavy metal strip or disk, which is activated by an electromagnet. The metal strip or disk connects two contact points and "makes" or "breaks" the electrical circuit. Because the metal strip (disk) is heavier than most electrical switch contacts, it will not pit or burn away due to arcing, Fig. 13-46.

Switches and solenoids that are used for small engine applications should be waterproof to prevent shorting, which could occur if the unit is left out in the rain or washed with a garden hose. Several common switches are shown in Fig. 13-47.

Most switches and solenoids can be tested with a standard ohmmeter or other type of continuity tester, Fig. 13-48.

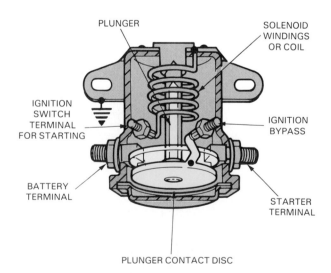

Fig. 13-46. *Study the construction of this starter solenoid. As the plunger moves, it pulls the disk in to contact the battery terminals. When contact is made, the starter is energized.*

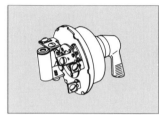

Multiple contact switches can be turned to different positions to open or close complex circuits (combination lighting and variable speed control circuits).

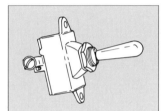

Toggle switches are simple on-off switches used to control auxiliary circuits.

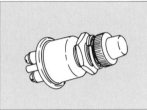

Push button switches are momentary switches that are pushed in one direction to open or close a circuit.

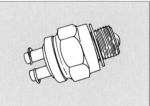

Pressure switches sense high or low pressure conditions and close a circuit to provide audible and/or visual warning signals.

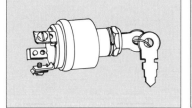

Ignition key switches must have a key inserted to turn them on and off.

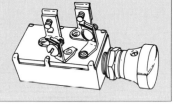

Push-pull switches are used for light switches and emergency switches.

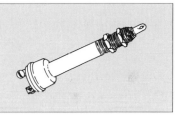

Cutout switches are used to break electrical circuits during emergencies (if operator involuntarily leaves implement).

Fig. 13-47. There are many switches used in a battery ignition system. (Deere & Co.)

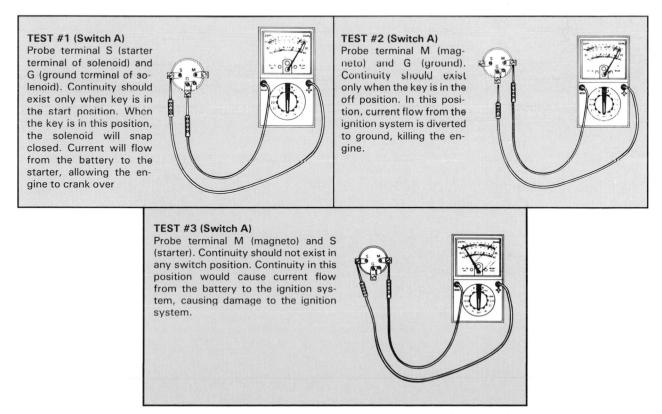

TEST #1 (Switch A)
Probe terminal S (starter terminal of solenoid) and G (ground terminal of solenoid). Continuity should exist only when key is in the start position. When the key is in this position, the solenoid will snap closed. Current will flow from the battery to the starter, allowing the engine to crank over

TEST #2 (Switch A)
Probe terminal M (magneto) and G (ground). Continuity should exist only when the key is in the off position. In this position, current flow from the ignition system is diverted to ground, killing the engine.

TEST #3 (Switch A)
Probe terminal M (magneto) and S (starter). Continuity should not exist in any switch position. Continuity in this position would cause current flow from the battery to the ignition system, causing damage to the ignition system.

Fig. 13-48. This solenoid is being checked for continuity with an ohmmeter. Follow manufacturer's instructions when testing switches and solenoids. (Tecumseh Products Co.)

CHECKING THE DC STARTER-GENERATOR CIRCUIT

There are certain steps to follow when checking output or troubleshooting the direct current (DC) starter-generator circuit found on some small engines.

1. Disconnect all equipment, place the transmission in neutral (if applicable), and turn on the ignition switch.
2. If the generator warning light comes on or the ammeter shows a discharge, these units are working and the battery is supplying current.
3. Start the engine. If the generator warning light stays on or the ammeter continues to show a discharge, look for trouble in the generator circuit.
4. If the system is charging, check out the starting circuit. There are four separate starting circuit checks:
   a. Test the ground connection at the starter-generator or at the battery with a voltmeter. If the negative battery cable goes to ground, attach the negative voltmeter lead to the starter-generator mounting frame, and clip the positive voltmeter

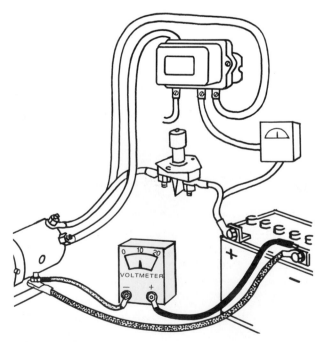

Fig. 13-49. A voltmeter can be used to check battery and starter-generator system ground connections.

lead to the negative post of battery, Fig. 13-49. Reverse this procedure if the system is a positive ground.

If the voltmeter shows about 10V, move on to the next step. A lower reading indicates a poor ground. Clean and tighten all ground connections. Replace the ground cable if it is worn or corroded. Operate the starter and check the voltage.

   b. If the voltage is still low, check the circuit between the battery and the starter switch. To do this, leave the negative cable grounded to the mounting frame. Move the positive lead to the switch terminal nearest battery. Then, turn on the starter switch.

A low reading on the voltmeter indicates a poor connection between the battery and the starter switch. If the engine will not crank, clean and tighten the cable connections. Then turn on starter switch and recheck voltage.

   c. Check the starter switch. Leave the negative lead of voltmeter attached to the starter-generator frame. Move the positive lead to the switch terminal nearest the starter-generator. Turn on the switch. If there is little or no voltage, the starter switch is not closing the circuit. Repair or replace. If the voltage is still low, go to the next step.

   d. Move the positive lead to the armature post of the starter-generator. Leave the negative lead grounded. The starter should operate and the voltmeter should read about 11V. If the engine does not start and the voltage is normal, the starter is faulty. A no-start condition and little or no voltage indicates a loose or broken connection between the starter switch and the starter-generator connection. Clean, tighten, and inspect or replace wiring. Recheck voltage.

COMMUTATORS AND BRUSHES

All DC starters and generators have **commutators** and **brushes,** which occasionally need service.

Start by cleaning the unit's metal housing. Avoid getting cleaning solvent on insulated wiring.

Check for worn bearings at both ends of the armature shaft. (You can feel play with your hands, and worn bearings are usually noisy when operating.)

If there is a cover band, remove it. A ring of solder along the inside of the band indicates that the unit has overheated. Further repair must be done by an experienced technician.

If the generator has no cover band, remove the long bolts running through the housing and pull off the end plate nearest the commutator. The commutator consists of a group of bars arranged in a cylinder-like fashion inside the generator or starter. Spring-loaded brushes rub on the commutator, Fig. 13-50. Check the brushes for wear. They should move freely and press firmly against the commutator.

A generator will have either two or three brushes. A starter-generator will have two. If the brushes are worn to half of their original length or the clips are resting on the brush holders, replace the brushes. If brushes are binding in the holders, wipe the holders with a clean, dry rag.

Check and tighten all electrical connections. Inspect the commutator for damage and/or wear. If the bars are rough and out of round, the armature will have to be chucked in a lathe and turned to a smooth finish. After turning, the mica (insulating material) between the commutator bars should be undercut. Follow all of the manufacturer's specifications.

If the commutator is only dirty and glazed, clean it with fine sandpaper while the engine is running slowly. Do not leave any wires disconnected during this operation or you will burn out the generator!

You can clean the starter commutator while the starter is turning over the engine. Disconnect the spark plug wire so the engine does not accidentally start. Wrap No. 00 sandpaper over the squared end of a stick and move the sandpaper back and forth on the commutator until all dirt and glaze is removed. *Never use emery cloth or solvent. Emery will cause arcing. Solvents will soften the insulation between the bars.*

If necessary, install new brushes. If they do not seat squarely, pull sandpaper back and forth between brush and the commutator. Let the sandpaper work the brush down to the shape of the commutator. Blow out the dust and replace the band.

Polarize the generator at the regulator by placing one end of a jumper wire on the battery terminal. Momentarily, touch the other end of the wire to the generator terminal. This is only necessary if you have disconnected any of the wire leads to the generator.

If you do not do this and the polarity has reversed, you may burn out the generator, damage the cutout relay points in the regulator, or run down the battery.

MAINTAINING AN ALTERNATOR

Small engine *alternators* come in many different sizes. Maintenance generally involves lubrication (on some units, but not on others) and inspection. Periodically, check that brushes, slip rings, and bearings are in good condition. If the battery is run down, test alternator output and, if necessary, test separate parts of the alternator. Check battery polarity first. Is the proper terminal going to ground? If not, reverse them.

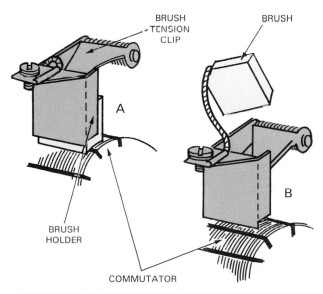

Fig. 13-50. Brushes are held against the commutator by a brush clip. A—Brush is worn down so clip rides on holder. B—Brush is removed for replacement. Never pull on brush leads.

CAUTION: Never reverse polarity when servicing the battery. It could burn out alternator diodes and damage wiring by overheating. Do not try to polarize an alternator. It is NOT necessary. Polarity of an alternator cannot be lost or reversed.

Most small engine alternator systems use permanent magnets that are attached to the inner rim of the flywheel. These magnets are similar to the magneto magnets. The *stator assembly* consists of a series of coils that are mounted on a circular plate and attached to the engine inside the flywheel. The alternator produces alternating current (AC), which is converted to direct current (DC) to charge the battery. This is accomplished through the use of a rectifier or diode arrangement, which is located in the circuit between the alternator and the battery. The size of the magnets and stator coils determines the output current of the alternator. Fig. 13-51 illustrates an alternator ring and a regulator-rectifier.

Dual, unregulated alternator systems charge the battery with DC current and provide AC current for operating the lights and other accessories. Fig. 13-52 shows a wiring diagram for a dual, unregulated alternator system using a diode in the direct current circuit. The current output is limited by the construction of the coils in the stator. Therefore, no regulator is used. Current to the electrical accessories is available only when the engine is running, and the brightness of the lights varies with the speed of the engine. The AC and DC units are separate, so the load on one does not affect the other.

Another common type of alternator system is the unregulated AC-only type. In these systems, the alternator is only used to operate the lighting

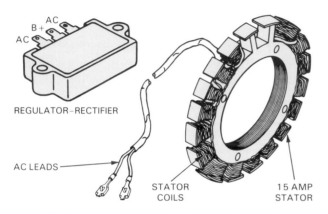

Fig. 13-51. One type of alternator ring and a regulator-rectifier.

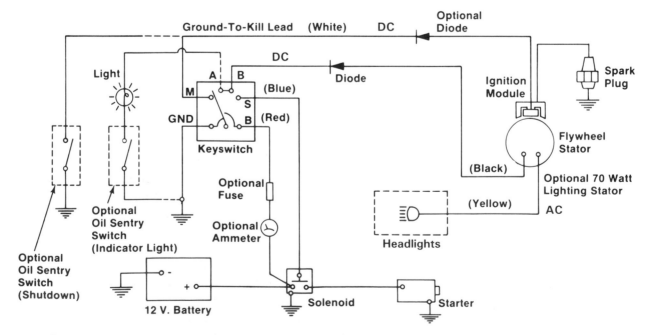

Fig. 13-52. This wiring diagram is for a dual-purpose alternator. It provides AC for headlights and DC for battery charging.

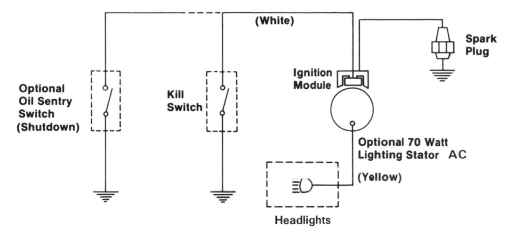

Fig. 13-53. Wiring diagram for a manual-start engine with an AC-only lighting stator for the headlights. A battery is not used in this system. (Kohler Co.)

system, Fig. 13-53. Current is available only when the engine is running and the brightness of the lights varies with engine speed. A battery is not used in these systems. The troubleshooting chart in Fig. 13-54 points out common alternator system problems.

ALTERNATOR OUTPUT TESTS

A few common tests can be performed to determine the output of most small engine alternators. A volt/ohm/ampere (VOM) meter is recommended for carrying out these tests, Fig. 13-55. When checking alternators, make the tests in the following sequence:

1. Test alternator output.
2. Test diode(s) or regulator-rectifier (if equipped).

Most meter receptacles and test lead connector ends are color coded to ensure correct test lead attachment, Fig. 13-56. Before testing the alternator's output (volts, amps), use an accurate tachometer to temporarily adjust the engine speed to the rpm specified in the test instructions. CAUTION: Upon completion of the alternator output test, always readjust the engine rpm to its correct no-load governed speed as specified in the engine service manual.

Fig. 13-56 illustrates a typical output test for a 9 amp regulated alternator. This system provides alternating current to a regulator-rectifier.

The regulator-rectifier converts the AC to DC and regulates the current to the battery. The charging rate will vary with engine RPM and temperature. When testing the regulator-rectifier for

ALTERNATOR SYSTEMS TROUBLESHOOTING

Battery Not Charging:
Engine RPM too low. Inline fuse ''blown'' (if equipped). Defective battery. Loose, pinched, or corroded battery leads. Open, shorted, or grounded wires between output connector and battery. Defective diode (open or shorted). Defective or improperly grounded regulator-rectifier. Diode installed incorrectly (reversed). Damaged battery (shorted battery cells). Excessive current draw from accessories. Low magnetic flux or damaged alternator magnets.
Battery In State Of Over Charge:
Severe battery vibration (missing or broken tie-down straps). Battery rate of charge not matched to alternator output. Damaged battery (shorted battery cells). Defective regulator.
Headlamps Not Working:
Inline fuse ''blown'' (if equipped). Defective headlamps. Loose or corroded wires. Open, shorted, or grounded wires between output connector and headlamps. Low magnetic flux or damaged alternator magnets.

Fig. 13-54. Alternator system troubleshooting chart.

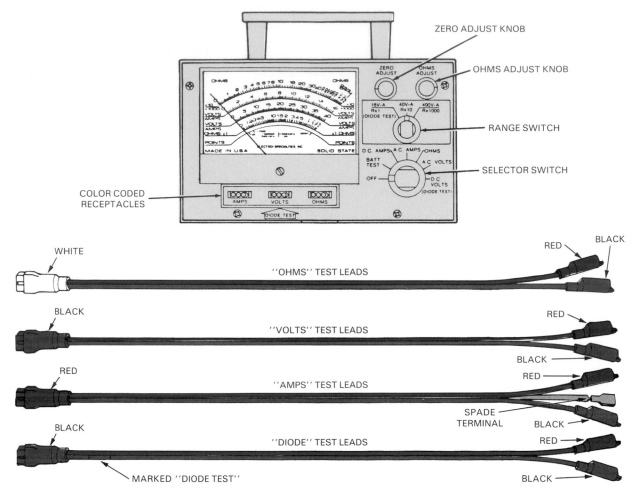

Fig. 13-55. A volt/ohm/ampere meter (VOM) can be used when performing alternator tests. Note color-coded test leads. (Kohler Co.)

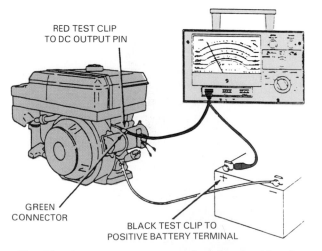

Fig. 13-56. Typical test connections for checking the output of a 9 amp, regulated alternator. Output for this particular model should not be less than 40 volts. (Briggs and Stratton Corp.)

output, a 12 volt battery with a minimum charge of 5 volts is required. There will be no charging output if battery voltage is below 5 volts. See Fig. 13-57.

NOTE: When testing a regulator-rectifier, connect test leads before starting engine. Be sure connections are secure. If a test lead vibrates loose while the engine is running, the regulator-rectifier may be damaged.

DC-only and dual-circuit alternator systems use a *diode* to convert AC to DC. The diode is located in the output wire on most alternators. A typical test hookup for one type of diode is shown in Fig. 13-58. After the test is made in one direction, the leads are reversed and the diode is tested again. The meter should only show a reading in one direction. If meter shows a reading in both

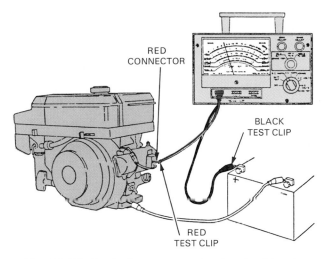

Fig. 13-57. Typical test connection for checking the output of one type of regulator-rectifier. Typical output should range from 3 to 9 amps, depending on battery voltage. (Briggs and Stratton Corp.)

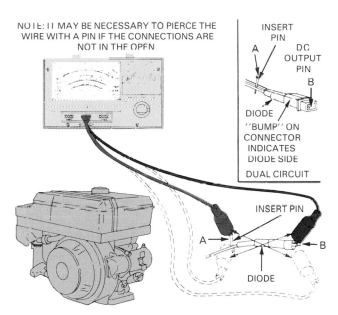

Fig. 13-58. Testing a diode for proper operation. Meter should show a reading in one direction only. (Briggs and Stratton Corp.)

directions, the diode is defective (closed). If the meter does not show a reading (needle movement) in either direction, the diode is defective (open).

NOTE: Replacement diode harnesses are available. When installing a new harness, use rosin-core solder. Use shrink tubing or electrician's tape on connections. *Do not use crimp connectors.*

VOLTAGE REGULATOR SERVICE

Do not attempt to service *voltage regulators* until the engine has run for about 20 minutes. This will allow the temperature of the regulator to stabilize. Then check:

1. Condition of the ground strap running to the engine. Connections must be clean and tight.
2. Appearance of the breaker points. They tend to wear in a concave pattern and should be filed as necessary. Wipe filed points with a clean, lint-free cloth soaked in carbon tetrachloride. Never use emery cloth or sandpaper on regulator points.

ADJUSTING CURRENT VOLTAGE

There are two points of adjustment in the current voltage regulator, Fig. 13-59. Disconnect battery, then:

1. Check the armature air gap. Push down on the armature until the contact points just touch. Measure the air gap and check it against the manufacturer's specifications (usually .075 in.). Adjust by loosening the contact

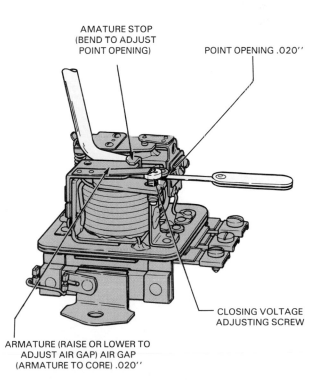

Fig. 13-59. Current-voltage adjustments on typical voltage regulator. (Kohler Co.)

mounting screws and raising or lowering the contact bracket. Align and adjust the points and retighten the screws. This is always done before adjusting the voltage setting.

2. Adjust voltage setting by turning the adjusting screw. Clockwise adjustment will increase the voltage setting. Counterclockwise adjustment will decrease the setting. Replace the cover. Run the engine to stabilize the voltage and recheck.

CUTOUT RELAY SERVICE

The generator regulator usually has a cutout relay unit with three adjustments, Fig. 13-60. Disconnect battery, then:

1. Measure and adjust the air gap. Place finger pressure on the armature, directly above core until points close. Measure gap and adjust to manufacturer's specifications. Generally, this is about .020 in. Retighten screws.

2. To adjust the point opening, bend the armature stop. See the service manual for the correct gap.

3. Adjust the closing voltage by turning the screw clockwise to increase tension and voltage; counterclockwise to decrease tension and voltage. Closing voltage must be at least .5V less than current voltage regulator unit setting.

DISTRIBUTORS

Small engines with more than one cylinder and a battery ignition system use a *distributor* to bring spark to the right cylinder at the right time. During tune-ups, the distributor cap should be removed and inspected for cracks, carbon tracking, and pitted contacts.

The rotor also requires inspection. This is the small plastic "arm" that is mounted on top of the distributor shaft and revolves during engine operation. The metal tab on top of the rotor must make good contact with the metal inset in the center tower of the distributor cap. Therefore, the firing end of the rotor should not be worn or irregular.

The breaker points should be filed flat or replaced as described for magneto points. Spark timing can be set by rotating the distributor. Usually, the distributor is clamped in place to lock in the ignition timing adjustment.

IGNITION TIMING

To set the ignition *timing* on the engine shown in Fig. 13-61, the number one spark plug must be removed. Next, while turning the engine over slowly by hand, hold your finger over the spark plug hole to determine when the piston is on the compression stroke. Then, with the piston at TDC, the rotor should be approximately in line with a timing notch in the distributor housing. Loosen the clamp and rotate the distributor until the points just begin to open. Hold the distributor in place and retighten the clamp. This procedure is accurate enough to start and operate the engine.

More accurate timing can be done with a neon *timing light*. In this method, the timing light is aimed at a fixed mark on the engine block or the engine shroud. When the correct timing is reached, a mark on the flywheel will appear to be aligned with the fixed mark when the engine is at

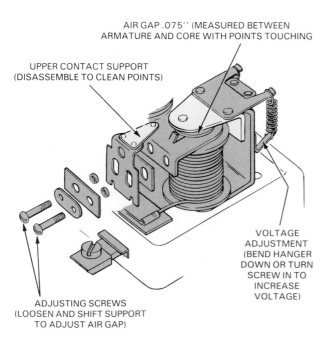

AIR GAP .075″ (MEASURED BETWEEN ARMATURE AND CORE WITH POINTS TOUCHING

UPPER CONTACT SUPPORT (DISASSEMBLE TO CLEAN POINTS)

VOLTAGE ADJUSTMENT (BEND HANGER DOWN OR TURN SCREW IN TO INCREASE VOLTAGE)

ADJUSTING SCREWS (LOOSEN AND SHIFT SUPPORT TO ADJUST AIR GAP)

Fig. 13-60. Points of adjustment on a cutout relay are shown. To clean the points, loosen the upper contact support.

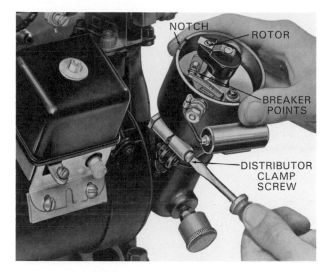

Fig. 13-61. With the number one piston at TDC, the rotor should point to the notch in the distributor housing. To time ignition, loosen clamp screw and rotate distributor housing until breaker points close. Then, turn distributor in the opposite direction until the points begin to open and retighten clamp screw. (Wisconsin Motors Corp.)

idle. When using a timing light, follow manufacturer's recommendations.

LUBRICATION

Distributors should be lubricated at several points. Put a small amount of oil or grease in the reservoir or cup provided for the shaft, a film of grease on the breaker cam, and a drop or two of oil on the pivot for the breaker points. *Use care; too much lubricant can cause the points to burn.*

SUMMARY

The ignition systems in small engines require periodic inspection and maintenance. Each ignition system component should be checked for proper operation and replaced if defective.

When an engine is hard to start, a new spark plug may seem to solve the problem. A new plug may require 5000 volts to fire. After many hours of use, the same plug may require 10,000 volts to fire. Changing the plug means that less voltage is needed to fire the plug and, therefore, the engine is easier to start. Often, however, a less obvious problem has caused the plug to fail.

You can analyze the quality of combustion by examining the carbon deposits on a spark plug. Normal combustion produces a beige or gray-tan deposit. Plugs can be cleaned in a spark plug cleaning machine. When gapping spark plugs, only use leaf-type feeler gages on new plugs. Wire-type thickness gages are recommended for use on worn plugs. When installing plugs, tighten only to the torque specified by the manufacturer.

A basic ignition test can help determine the condition of the ignition system. Remove the plug wire from the plug. Hold the wire 3/16 in. from top of the plug and pull the starter cord. If a spark is produced between the plug and the wire, the ignition is good. If a spark does not appear between top of the plug and the wire, hold the wire near the base of the spark plug and pull the cord. If a spark occurs, the plug is failing under compression. If a spark does not occur, the problem is in the magneto system.

In some magneto systems, the gap between the coil or trigger module and the flywheel is adjustable. In some systems, the ignition components are completely contained under the flywheel. A knock-off tool or a wheel puller can be used to remove the flywheel from the crankshaft.

After the flywheel is removed, all magneto parts should be checked for defects. Look for dents, cracks, and gouges. Make sure all wire insulation is in good condition. An ohmmeter can be used to check the condition of coils and condensers. In many cases, a special ignition tester must be used to check the condition of ignition system components.

In breaker point ignition systems, the points must be adjusted, the piston height must be set, and the ignition spark must be timed. Follow manufacturer's recommendations. An engine can be timed using a continuity tester or an ohmmeter. After performing necessary ignition system service, the flywheel can be replaced.

If the cylinder head was removed to adjust piston height, it should be cleaned of carbon deposits and checked for flatness. A new head gasket should be used and the head bolts should be tightened as specified by the manufacturer.

Battery operated ignition systems operate much like magneto systems. However, they contain many additional components that require service. Before beginning extensive tests, check for a defective battery, corroded or loose terminals, incorrect connections, faulty wires, defective switches, or a malfunctioning operator presence system.

DC generators and starters occasionally require service. Check for worn bearings and overheating. Inspect the commutator and brushes for wear. Check and tighten all electrical connections. Polarize DC generator if wire leads are disconnected.

Alternators require occasional lubrication and inspection. Make sure brushes, slip rings, and bearings are in good condition. A volt/ohm/ampere meter can be used to determine the output of most small engine alternators. Check service manual for proper meter connections.

If the charging system is equipped with a voltage regulator, adjust the unit per manufacturer's instructions. An engine should run for approximately 20 minutes before conducting regulator tests. The will allow the temperature of the regulator to stabilize.

Multi-cylinder small engines use a distributor to send spark to the appropriate cylinder at the correct time. During service, the distributor cap should be removed and checked for cracks, carbon tracking, and pitted contacts. Distributors may require lubrication in several locations. In addition to the distributor cap, the rotor should also be checked for wear.

Spark timing is set in distributor ignition systems by rotating the distributor. When ignition timing meets specifications, the distributor clamp must be tightened

KNOW THESE TERMS

Ignition system tune-up, Spark plug, Magneto system, Spark plug gap, Spark plug cleaning machine, Compression test, Leaf-type feeler gage, Wire-type thickness gage, Solid state system, Flywheel, Knock-off tool, Wheel puller, Continuity, Timing tool, Depth micrometer, Dial indicator, Battery operated ignition system, Sulfating, Hydrometer, Maintenance free battery, Hydrogen, Oxygen, Wiring, Ammeter, Switches, Solenoid, Commutator, Brush, Alternator, Stator Assembly, Diode, Voltage regulator, Distributor, Timing, Timing light.

REVIEW QUESTIONS–CHAPTER 13

1. Which of the following determines the amount of voltage induced by the magnets when the engine is running?
   a. Breaker points.
   b. Spark plug.
   c. Condenser.
   d. Engine speed.
2. What color should the internal porcelain insulator be on a normal, used spark plug?
3. What five conditions cause spark plug deposits?
4. If a torque specification is not given, what is the procedure for tightening a new spark plug?
5. When performing an ignition system test, spark does not occur between the high tension lead and the plug tip when the engine is cranked. However, spark does occur when the lead is held near base of the plug. The problem lies with the:
   a. Magneto.
   b. High tension lead.
   c. Coil.
   d. Plug.
6. Name two tools used to remove flywheels from small engine crankshafts.
7. If the key that positions the flywheel is deformed or partly sheared, the engine will most likely:
   a. Lose the flywheel.
   b. Be out of time.
   c. Run exceptionally fast.
   d. Burn fuel excessively.
8. The three major steps in timing a small engine are as follows: adjust the piston height BTDC, rotate the stator plate until the points open, set the breaker point gap. List these steps in the correct order.

9. What are the six basic things to check for before starting an extensive battery ignition system analysis?

10. What two gases are generated during battery charging.

11. Why is it important to keep batteries fully charged in temperatures that are below freezing?

12. Why is engine cranking load increased significantly in cold weather?

13. Name the single component that indicates for certain that an engine has a battery ignition system:
    a. Battery.
    b. Distributor.
    c. Canister-type coil.
    d. Starter.
    e. Generator.

14. What kind of tester can be used to test switches?

15. What type of current is generated by an alternator?

16. Explain why emery cloth should not be used to clean a commutator of a generator or a starter.

17. Under what conditions would you polarize an alternator.

18. Two electronic components that convert alternating current to direct current are _____ and _____ .

19. Why should you put only one or two drops of oil on a felt cam oiler?

20. Why should a wooden table or an insulated bench top be used when performing electrical tests?

SUGGESTED ACTIVITIES

1. Perform the complete timing procedure on a small engine.

2. Demonstrate the proper technique for removing, inspecting, and replacing a flywheel.

3. Examine some old spark plugs. Attempt to analyze engine condition by their appearance. Clean, gap, and test them.

4. Make a cutaway from a defective coil. Open a condenser and examine its construction.

5. Place a bad plug in an engine. Test for spark at the plug tip and, then, at the base. Explain the difference.

6. Demonstrate the proper procedure for charging a battery. Be sure to follow all safety rules when working with batteries and chargers.

7. Perform continuity tests on several types of switches.

8. Test several electrical system components with the test equipment described in this chapter and in the Useful Information section of this text. If other equipment is available, follow manufacturer's instructions for the equipment at hand.

Cutaway view of a two-cylinder, horizontal shaft, four cycle engine. This engine features a foam air filter, a float-type carburetor, and a pressure lubrication system. (Onan)

CHAPTER 14

CYLINDER RECONDITIONING

After studying this chapter, you will be able to:
☐ Inspect engines for problems.
☐ Describe the procedure for removing an engine from an implement.
☐ List the steps involved in disassembling an engine.
☐ Measure a cylinder for wear and out-of-roundness.
☐ Explain the procedures involved in reboring a cylinder.
☐ Summarize the reasons for honing a cylinder.

When you repair an engine, it is best to work in a clean, well-lighted area. Tools should be clean and close at hand.

⚠ When using tools, safety rules should be observed. For example, safety glasses should be worn to protect the eyes. Oily and gasoline-soaked rags should be placed in flameproof cans. Heat or flames should never be allowed near solvents or other materials that will burn. A work place is not safe unless you use good judgment while working.

Some engines are easier to work on if they are mounted on an engine stand. Other engines in need of major repairs can be removed from the implement and torn down right on the workbench, Fig. 14-1.

You will need a *service manual* for the engine you are working on. Every engine make and model is built to certain dimensions and specifications that are different from other engines. The kind of job you do depends a great deal upon the care and accuracy used in following the engine service manual.

Fig. 14-1. Small gasoline engines are easier to work on when removed from implement.

ENGINE INSPECTION

It is good practice to look for causes of engine problems even before removing the engine from the implement. Loose or broken engine mounts, misaligned pulleys, or unevenly worn drive belts can cause excessive vibration. Wet oil on the outside of the engine may mean that there are loose parts, leaking gaskets, leaking oil seals, or a cracked casting.

Wires may need to be disconnected before the engine can be removed. If so, make flags of masking tape for the wire ends. Identify them with matching numbers, Fig. 14-2. This will prevent damage from wrong connections and will save time during reassembly.

Cylinder Reconditioning   243

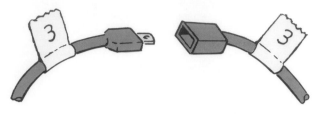

Fig. 14-2. If wires are coded with masking tape, they can be properly identified and reconnected.

DISASSEMBLING THE ENGINE

Before removing the engine, take out the spark plug. It could fire accidentally or break during disassembly of other parts. When working on lawn mowers, snow throwers, and other implements with exposed moving parts, never touch the blades or driven parts until the spark plug has been removed from the engine.

After the engine has been removed, the starter unit can be unbolted. It may be a simple pulley-rope starter, a retractable rope starter, or an electronic starter. Usually, only a few bolts fasten the starter unit to the engine. See Fig. 14-3.

The exhaust manifold pipe and muffler can be taken off next. Set them aside and out of the way.

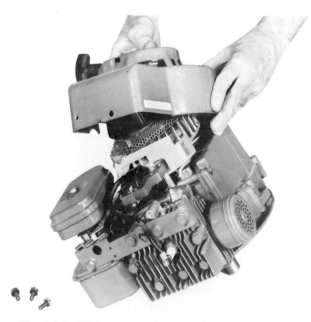

Fig. 14-3. With the engine out of the implement, the starter unit can be removed.

The carburetor and intake manifold pipe also can be removed, Fig. 14-4. It often helps to sketch the carburetor and linkage locations. This can save time in reassembly. Check all gasket surfaces for defects. On some engines, the fuel tank is fastened to the carburetor. In any case, the tank can be removed, and the fuel lines can be disconnected.

The air shroud, blower housing, and baffles can be removed to uncover the flywheel. Flywheel removal is explained in Chapter 13. Always use the proper puller to prevent damage to the crankshaft or the flywheel. After the flywheel is removed, the magneto components can be disassembled and set aside.

Fig. 14-4. When removing carburetor and fuel tank, sketch carburetor linkage hookup for later reference.

ORGANIZE YOUR WORK

Organizing the job saves time and effort. At this point, the outer parts have been taken off the engine. They should be set aside in their own groups. Magneto, flywheel shroud, and starter parts should be in one group; carburetor, fuel tank, and exhaust manifold in another. See Fig. 14-5.

Keeping the work clean is part of good organization. Outside surfaces can be cleaned at this

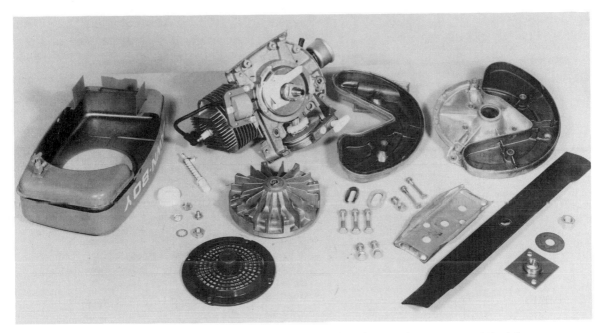

Fig. 14-5. A clean workbench and an organized work procedure can speed the job.

point of tear-down, Fig. 14-6. Inside surfaces will be cleaned later. Grass clippings and other debris should be removed by scraping and brushing the fins and housings before using cleaning fluid.

A safe engine cleaning solvent should be used to remove grease, oil, and grit. Some parts, like the coil and condenser, may be cleaned by wiping with a clean cloth moistened with solvent. This is better than total immersion. Never use solvents that burn easily or those that may be harmful to humans.

As each part is washed, wipe it with a clean cloth and set it aside to dry. If the workbench is oily and greasy, clean it before doing disassembly work on it. Hands should be washed often to keep dirt off of the cleaned parts.

CYLINDER INSPECTION

Several types of cylinders may be found on small engines. One type is the separate cylinder block shown in Fig. 14-7. Another variation is the integral (one-piece) cylinder block and crankcase, Fig. 14-8.

If the engine is of the separate cylinder type, remove the bolts and lightly tap the cylinder block with a soft leather hammer, Fig. 14-7. A slow, smooth pull will remove the cylinder from the piston assembly. An alternate procedure is to remove the connecting rod cap and pull the cylinder and piston from the crankcase as a unit. Then, pull the piston from the cylinder.

The piston must be removed from the top of the cylinder in the integral cylinder block and crankcase type engine. First, remove the cylinder head. Next, take the crankcase cover from the

Fig. 14-6. Grass and other debris should be removed from cooling fins before using solvent.

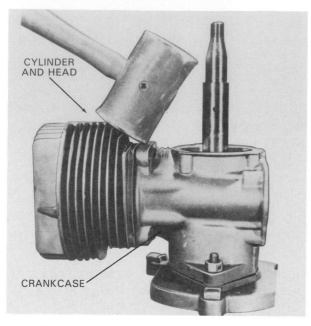

CYLINDER AND HEAD

CRANKCASE

Fig. 14-7. After bolts are removed, a light tap with a soft hammer will loosen cylinder block from crankcase. (Lawn-Boy Power Equipment, Gale Products)

PISTON

CRANKSHAFT

CAMSHAFT

CONNECTING ROD CAP

Fig. 14-8. Piston may be removed from cylinder if there is no noticeable ridge at top of cylinder above the ring wear area. (Wisconsin Motors Corp.)

side of the block and remove the connecting rod cap. Then, push the piston out of the top of the cylinder as shown in Fig. 14-8. If there is a heavy ridge around the inside top of the cylinder, the piston rings will not pass the ridge without damage. Special *ridge reaming tools,* Fig. 14-9, are available for cutting the ridge to make this part of the cylinder the same diameter as the worn portion below it.

Fig. 14-9. Maximum cylinder wear takes place near the top of the cylinder. The unworn portion at the top of the cylinder (ring ridge) must be removed. Cut it flush to the cylinder wall with a ridge reamer. (Tecumseh Products Co.)

With the piston out, inspect the cylinder block. Look for areas of scuffing or scoring on the walls. Check for nicks or grooves in gasket surfaces. Inspect it for chipped or broken fins. Examine head bolt holes and spark plug holes for damaged or stripped threads.

A worn cylinder has a narrow, unworn ridge at the top. The bottom of this ridge indicates the extent of the top piston ring travel. Right below the ridge is the area of most cylinder wear. The wear will be the greatest on two opposite sides of the cylinder, 90° from the crankshaft centerline. The cylinder, therefore, wears into an oval shape. This increased wear is due to several things:

- Less lubrication at this portion of the cylinder wall.
- The diluting effect of raw gas on the engine oil.

- The pressure that builds up behind the rings at their highest position.

Below the point where cylinder diameter is the greatest, cylinder wear lessens rapidly. Because of this wear pattern, there is a gradual taper toward the bottom of the ring travel. Below ring travel there is almost no wear. This area is well lubricated and only receives light wall pressure from the piston skirt.

CYLINDER MEASUREMENT

If the cylinder wall looks smooth and free of scuff and score marks, you are ready to measure the cylinder for wear and out-of-roundness.

Finding the amount of *cylinder taper* is the first important measurement in determining cylinder condition. First, measure the cylinder diameter below the ring travel. Then, measure it just below the ring ridge. The difference between these measurements is an accurate indication of the amount of cylinder taper. The taper measurements should be taken both parallel and at right angles to the crankshaft to determine the greatest amount of wear and out-of-roundness.

The amount of taper allowed before reboring is required depends on engine design, its general condition, and the type of service in which it is used. There are no rules that apply to all engines regarding cylinder taper. The manufacturer, however, may set a limit. Beyond a certain point, the engine manufacturer will advise reboring or cylinder replacement.

Out-of-roundness for small engines is generally limited to .005 in. or .006 in. Beyond this limit, engine performance is greatly reduced.

Several methods can be used to measure cylinders. Fig. 14-10 shows an *inside micrometer* equipped with an extension handle. This precision instrument must be carefully adjusted to cylinder size. It must give the exact diameter of the cylinder.

Fig. 14-11 shows a *telescoping gage* being used to measure cylinder size. The gage head is spring loaded to expand when the thumbscrew is released. Once located in the cylinder, the gage is locked in place by tightening the thumbscrew. Then, it is removed from the cylinder, and an outside micrometer is used to measure the length of the telescope head. This measurement represents the cylinder size (diameter).

There is another convenient tool designed for measuring small engine cylinders. Fig. 14-12 shows the various sections of the *set gage* part of this tool.

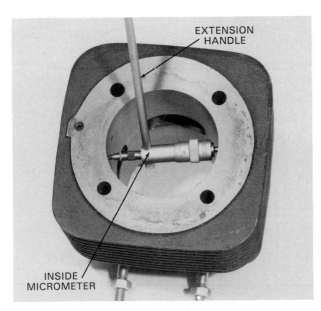

Fig. 14-10. An inside micrometer can be used to measure cylinder diameter. (Kickhaefer Mercury)

Fig. 14-11. A telescoping gage can also be used to measure cylinder diameter. Gage is adjusted to cylinder size and is measured with an outside micrometer. (Deere & Co.)

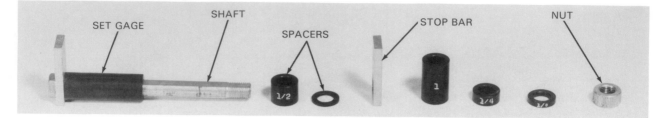

Fig. 14-12. Parts of a disassembled set gage are shown. Gage is preset to a specified size.

To use the tool, look up the standard cylinder size in the engine manual. Then place the right spacers on the shaft. Place the stop bar on the shaft with the remaining spacers and the nut.

The *cylinder gage* attaches to the set gage as shown in Fig. 14-13. The needle can be set to zero by turning the knurled ring to the left. Lock the ring with the thumbscrew and remove the cylinder gage from the set gage. The extension sleeves are used only when necessary.

When the cylinder gage is adjusted properly (set zero at specified diameter), put it into the cylinder, Fig. 14-14. The centering bar automatically centers the gage in the cylinder. Since the gage is spring loaded, it is free of the measuring spindle. The amount indicated by the needle is cylinder diameter oversize in thousandths of an

Fig. 14-14. Cylinder gage is self centering and will provide a direct reading. Needle deflection from zero indicates the amount of wear in thousandths of an inch.

inch. Take readings in three or four directions across the diameter. Check these measurements against manufacturer's specifications.

Cylinder service needs are determined by cylinder condition. If the bore is not damaged, and if taper and out-of-round readings are within specified limits, only a light deglazing with a fine emery cloth may be needed. The engine block must be thoroughly washed afterward.

REBORING THE CYLINDER

There are several ways to repair a cylinder that shows too much wear. Repair procedures depend on engine type. Some engines have chrome-plated, aluminum cylinders, Fig. 14-15. Worn or

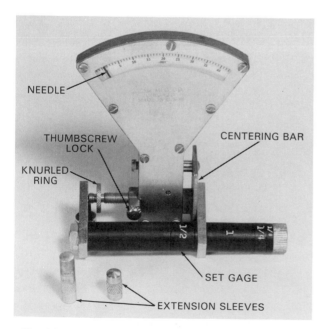

Fig. 14-13. When located in set gage, cylinder gage is adjusted to read zero by turning the knurled ring.

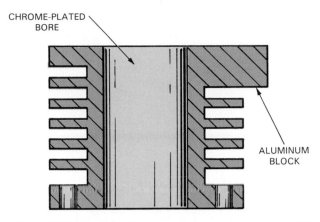

Fig. 14-15. Special cylinder construction is shown with a chrome-plated surface.

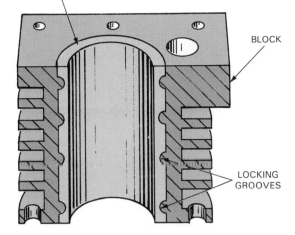

Fig. 14-17. Some cast iron cylinders are cast in place in the block.

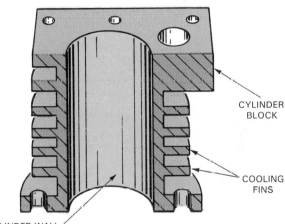

Fig. 14-18. Most cast iron cylinders are part of the engine block.

damaged cylinders of this type should be thrown away and replaced with a whole new cylinder. Other engines have a flanged *cylinder sleeve* that can be removed and replaced with a new sleeve, Fig. 14-16. The cast-in sleeve, Fig. 14-17, or the solid cast iron cylinder, Fig. 14-18, can be rebored to a larger size.

Two problems must be solved when *reboring* a cylinder. The first is to resize and maintain original alignment while producing a round, straight bore. The second is to produce the correct cylinder wall finish.

Cylinders are usually rebored in .010 in. steps. If the cylinder is being rebored for the first time,

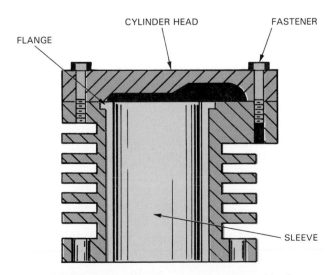

Fig. 14-16. Pressed-in sleeve can be secured by forming a flange on the upper end. Cylinder head will hold sleeve in place.

its diameter will be increased .010 in. over the standard size. If this does not "clean up" the cylinder (remove imperfections), the next step is .020 in. over the standard size. When replacing pistons and rings, order .010 or .020 in. over standard size to match the new cylinder bore.

The *boring machine* shown in Fig. 14-19 is designed for use on small engines. In setting up the machine, the engine block or cylinder block can be clamped in place below the boring head. Fig. 14-20 shows the cutter being adjusted to the correct diameter for the new bore. A built-in micrometer provides accurate setup to .0001 in. (one ten-thousandth of an inch).

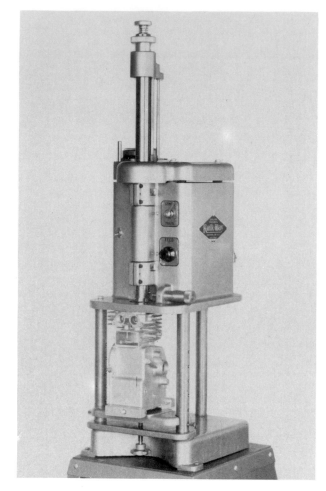

Fig. 14-19. A small engine cylinder boring machine is prepared for operation. Engine block is clamped in place. (Cedar Rapids Engineering Co.)

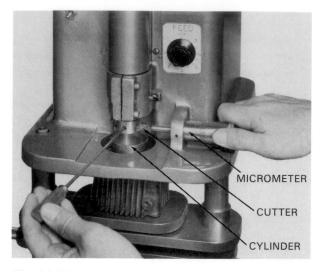

Fig. 14-20. Using a built-in micrometer gage, the cutter on this boring machine is set to give a precise bore diameter. (Cedar Rapids Engineering Co.)

An electric motor drives the spindle to rotate the cutter. The feed rate controls the distance the cutter advances into the bore during each cutter revolution and can be changed by moving the feed dial. See Fig. 14-20.

Boring the cylinder will produce a straight, round bore. However, the boring operation does not produce a satisfactory surface finish. The boring tool leaves microscopic furrows and surface fractures, so it is recommended that an overall minimum of .0025 in. of stock should be left for finish honing.

HONING THE CYLINDER

Honing is an abrasive (sandpaper-like) finishing process that removes the boring tool marks and surface fractures. Honing also produces the desired cylinder wall finish. The smoothness of the finish depends upon the grit size of the stones used.

Ideally, a newly honed cylinder will wear smooth at just about the same rate the new piston rings. When this "break-in" process is complete, both the cylinder walls and the piston rings will be smooth and should last for hundreds of hours of engine operation.

Fig. 14-21 illustrates a typical cylinder honing tool for small engine work. It has two abrasive stones and two guides to keep the tool aligned with the cylinder. The stones can be removed and replaced with new ones as necessary.

An electric drill is used to rotate the cylinder hone. In operation, the assembly is slowly and steadily moved in and out of the cylinder, Fig. 14-22. Stones should not be permitted to extend out of the cylinder end as uneven wearing of the stones may occur. For best results, the honing process should produce a fine surface pattern, like that shown in Fig. 14-23.

After reboring and honing the cylinder, use a piece of fine emery cloth to remove any burrs that may have developed around the ports. Clean the cylinder walls and the block thoroughly with kerosene and clean rags. Apply a light coat of SAE 10 oil to the cylinder to prevent rust.

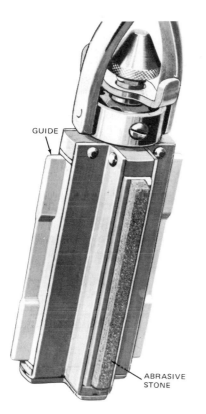

Fig. 14-21. A typical cylinder honing tool has two abrasive stones and two guides. (Sunnen Products Co.)

SUMMARY

Engine work should be performed in a clean, well-lighted area. An engine stand can be used if necessary. Always refer to a manufacturer's service

Fig. 14-23. A fine cross-hatched surface pattern is created by the in-and-out motion of the revolving hone.

manual for exact dimensions and specifications.

Before removing the engine from an implement, look for causes of engine problems. If wires need to be disconnected before removing engine, flag ends with masking tape. Remove the spark plug to avoid accidental firing.

After the engine has been removed, unbolt the starter unit, exhaust manifold pipe, muffler, carburetor, intake manifold pipe, fuel tank, and fuel lines. Check all gasket surfaces for defects. Remove the air shroud, blower housing, and baffles to expose the flywheel. Use a proper puller to remove the flywheel. After the flywheel is detached, remove the magneto components and set them aside.

Clean the outer engine parts by scraping off loose debris and soaking the components in solvent. Some parts, such as the coil and the condenser, should not be immersed in solvent.

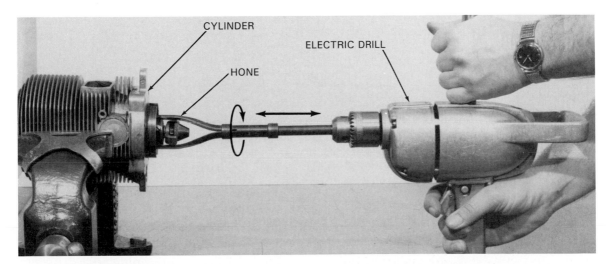

Fig. 14-22. A cylinder boring tool leaves microscopic furrows and surface fractures. These are removed and the cylinder is brought down to the correct size by honing. (Kickhaefer Mercury)

If an engine is the separate cylinder type, remove bolts and tap the block with a soft leather hammer. Slowly pull the cylinder from the piston assembly. In an integral cylinder block and crankcase, the piston must be removed from the top of the cylinder.

When the piston is removed, inspect the cylinder walls for scuffing and scoring. Check gasket surfaces for nicks and grooves. Examine the head bolt holes and spark plug holes for damaged threads.

If the cylinder wall looks smooth, measure it for wear and out-of-roundness. Check for excessive cylinder taper.

Cylinder service is determined by the cylinder condition. If the bore is within specifications, only a light deglazing is necessary.

If the cylinder is worn excessively, repair will depend on engine type. Chrome-plated aluminum cylinders must be replaced. Cast-in sleeve and solid cast iron cylinders can be rebored. Cylinders are usually rebored in steps of .010 in. The cylinder must be honed after boring. Honing is an abrasive finishing process that removes boring tool marks and produces a desirable cylinder wall finish.

KNOW THESE TERMS

Service manual, Ridge reaming tool, Cylinder taper, Out-of-roundness, Inside micrometer, Telescoping gage, Set gage, Cylinder gage, Cylinder sleeve, Reboring, Boring machine, Honing.

REVIEW QUESTIONS–CHAPTER 14

1. When should the small engine mechanic begin looking for engine defects or problems?
2. Excessive vibration could be caused by:
   a. Loose engine mounts.
   b. Pulleys out of line.
   c. Worn drive belts.
   d. All of the above.
   e. Only a and c.
3. Before removing an engine from an implement, it is safe practice to:
   a. Drain the oil.
   b. Turn off the ignition switch.
   c. Remove the spark plug.
   d. Crank the engine slowly to remove all the fuel from the cylinder(s).
4. When you remove a carburetor and tear it down for repair, what will help you remember how it goes back together?
5. In what ways can you save time and effort when reconditioning an engine?
   a. Consult the operator's manual.
   b. Keep workbench and engine parts clean.
   c. Organize your work by grouping certain engine parts.
   d. All of the above.
6. Explain how the piston is removed if there is a heavy ridge at the top of the cylinder in an integral cylinder block and crankcase type engine.
7. The location of cylinder diameter measurements should be:
   a. At the very top and bottom of the cylinder.
   b. Taken 90° to each other at the center of the cylinder.
   c. Below the ridge and at the bottom of the cylinder to determine taper.
   d. Only below the ridge, but at 90° to each other.
8. What do you call the special tool designed to remove cylinder ridges?
9. After a cylinder is rebored, it must be:
   a. Lapped.
   b. Reamed.
   c. Deglazed.
   d. Honed.

SUGGESTED ACTIVITIES

1. Design a well-organized workbench and tool panel for repairing small engines.
2. Completely disassemble an engine that has been in service for many hours and is in need of reconditioning.
3. Inspect disassembled parts for wear or damage. Discuss possible causes.
4. Measure a worn cylinder with an inside micrometer or other measuring tool. Record readings for taper and out-of-round.
5. Rebore and hone or deglaze a cylinder.
6. Ream a cylinder ridge.

CHAPTER 15

PISTONS AND PISTON RINGS

After studying this chapter, you will be able to:
- ☐ Describe piston and piston ring construction.
- ☐ Differentiate between compression rings and oil control rings.
- ☐ Explain the purpose of ring end gap.
- ☐ Identify common types of piston damage and list possible causes.
- ☐ Summarize what happens during piston ring wear-in.
- ☐ Explain the purpose of a piston pin.

In reconditioning small gasoline engines, pistons and piston rings are critical service items. Generally, reboring and/or honing of the cylinders is necessary, followed by the thorough inspection and repair of the parts closely fitted to them.

To do the job well, the small engine technician must understand the stresses to which a piston and its rings are subjected. He or she also must know the kinds of materials they are made from. Lastly, he or she must know what to do to put these parts back into top shape and to reassemble them for efficient engine operation.

The condition of the rings and pistons can be learned by observing and inspecting the parts during disassembly of the engine. The need for service is evidenced by low compression, blow-by, oil pumping, and fouled plugs.

The piston slides up and down in the cylinder. It sucks in the air-fuel mixture, compresses it, and then carries the force of the burning fuel to the crankshaft through the connecting rod. On the final stroke of the engine cycle, the piston pushes burned gases from the cylinder.

In normal operation, the piston travels up and down in the cylinder more than a thousand times per minute. Subjected to heat, pressure, and friction, the piston must be lightweight, strong, and properly fitted. It has a lot of work to do.

PISTON CONSTRUCTION

Pistons can be made of aluminum or steel. Aluminum is by far the most popular metal for this application. The surface may be coated with a special break-in finish (tin or other coating). Sometimes pistons are chrome plated for installations where they operate directly on aluminum-alloy cylinder walls.

The type of piston often used in a four cycle engine is shown in Fig. 15-1. The head is quite thick, giving this hardworking part strength and resistance to overheating. The area below the head has grooves for the piston rings. The full-diameter

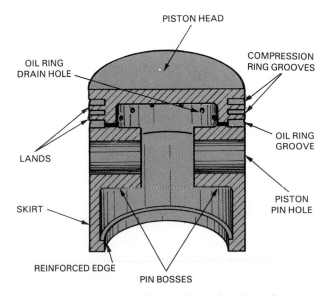

Fig. 15-1. A typical four cycle engine piston is cut away to show construction details.

ridges between the grooves are called **lands**. The wall or bottom of the oil ring groove is either slotted or pierced with holes. Oil wiped from the cylinder wall by the oil ring flows through these holes and back into the oil sump. The **sump** is the low area of the engine block where the oil collects.

Pistons may have grooves for one to four rings. Generally, the two cycle engine piston has one or two grooves. Both are compression ring grooves. Four cycle engine pistons will generally have three grooves, two for compression rings and one for an oil control ring.

The section of the piston surrounding the piston pin hole is called the **pin boss**. It is thick and often reinforced with cast-in webs. The **piston skirt** is the part of the piston below the bottom of the lower ring groove. The skirt is designed to be as light as possible to hold down the weight of the assembly.

The skirt actually guides the piston and keeps it from tipping from side to side. Portions of it may be cut away for lightness. Also, in some two cycle engines, portions may be cut away to allow the air-fuel mixture to pass through the piston skirt into other parts of the cylinder.

PISTON FIT

The piston is subjected to high temperatures, causing it to expand during operation. To allow for this increase in size, there must be a specific amount of clearance between the piston skirt and cylinder wall.

The cylinder also expands, but not as much as the piston. Normal clearance must be great enough to allow for lubrication and piston expansion. Different engines have different clearances. The amount depends upon engine design and use. Most small engines call for .003 in. to .005 in. piston-to-cylinder wall clearance, Fig. 15-2.

CAM-GROUND PISTONS

When the designer wants the smallest possible clearance between the piston skirt and the cylinder wall, skirts are often cam ground to an elliptical (oval) shape, Fig. 15-3A. The oval shape of a **cam-ground piston** allows the **thrust surfaces**

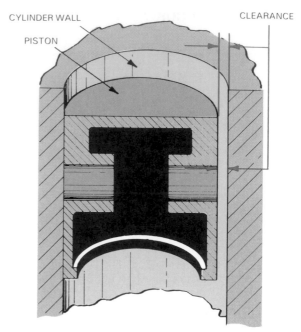

Fig. 15-2. Sufficient clearance must be allowed between piston skirt and cylinder wall to permit adequate lubrication and to allow for expansion of parts due to high temperatures.

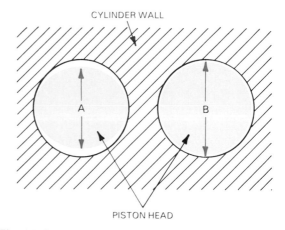

Fig. 15-3. Exaggerated top views of a cam ground piston as it would fit in a cylinder. A–Cold. B–Hot. Arrows indicate piston pin position.

(sides of skirt forced against cylinder during compression and firing) to fit more closely, even when cold. As the piston heats up, the diameter across the thrust surfaces remains constant and the piston enlarges parallel to (in the same direction as) the piston pin. See Fig. 15-3B. These exaggerated views illustrate how a cam-ground piston expands to a round shape as it becomes hot.

PISTON THRUST SURFACES

During the compression stroke, the pressure of the confined air-fuel mixture forces the piston toward one side of the cylinder. See Fig. 15-4A. When the crankshaft throw passes TDC, burning and rapidly expanding gases push hard on the piston, forcing it against the opposite side of the cylinder. See Fig. 15-4B.

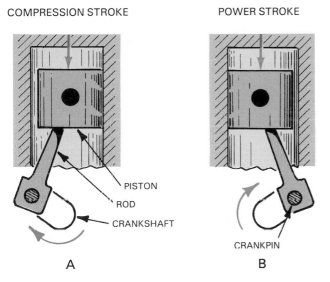

Fig. 15-4. Thrust surfaces of piston must resist heavy side pressure against cylinder walls. A–Upstroke. B–Downstroke.

In each instance, the sides of the piston forced against the cylinder wall are called *thrust surfaces*. These surfaces are at right angles (90°) to the centerline of the crankshaft and piston pin.

If the piston has too much clearance in the cylinder, side thrust during the compression and firing strokes will make it move, or *slap*, from one side of the cylinder to the other. As it moves sideways, the piston will tend to tip or cock in the cylinder. This "loose fit" can be very harmful to the piston and rings. The piston must fit the cylinder properly to avoid slapping.

PISTON HEAD SIZE

The piston head receives the brunt of combustion heat, so it runs hotter than the skirt and expands more. Because of this, the head of the

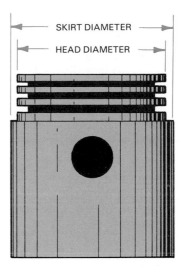

Fig. 15-5. Piston head receives greatest heat and is sometimes made smaller to compensate for expansion.

piston often is made with a smaller diameter than the skirt. Fig. 15-5 shows an exaggerated view of a piston with the head smaller than the skirt. The actual difference is only a few thousandths of an inch.

PISTON HEAD SHAPE

Piston heads are manufactured in many different shapes, depending on the type of small engine and its use. On four cycle engines, the piston head can be flat, domed, or wedge-shaped. Pistons used in two cycle engines generally are flat when used with a loop-scavenging design. Cross-scavenging designs use a raised baffle or deflector head piston, Fig. 15-6.

PISTON RING DESIGN

All small engine pistons must have clearance for lubrication and expansion. At the same time,

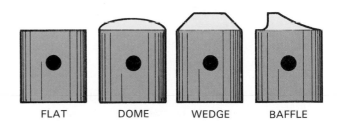

Fig. 15-6. Small gasoline engine piston heads are manufactured in a wide variety of shapes.

they must have **rings** to help do the job of sealing the cylinder(s). Without piston rings, the piston could not compress the fuel charge properly. Also, burning gases would leak out between the sides of the piston and the cylinder wall.

In performing their job, the piston rings ride against the cylinder wall, separated from it only by a thin film of oil. The rings rub freely against the sides of the ring grooves, which hold the rings squarely to the bore and force them to slide up and down the cylinder with the piston, Fig. 15-7. Since the ring face is in steady contact with the cylinder walls, Fig. 15-7, an effective seal is formed.

Fig. 15-8 shows a compression ring in its groove. The sides of the ring groove are flat, parallel, and smooth. The ring has proper side clearance. In operation, expanding gases force the ring

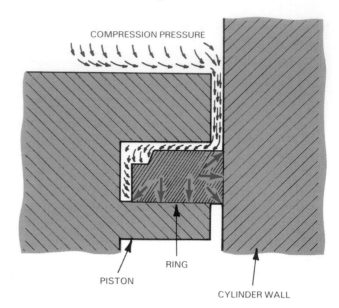

Fig. 15-8. Combustion chamber pressure forces ring against cylinder wall and bottom side of groove.

down against the lower side of the groove. At the same time, gases behind the ring force it against the cylinder wall. These forces help to form a good seal.

PISTON RING CONSTRUCTION

Piston rings are made of cast iron or steel. Both types may be plated with chrome or other long-wearing materials, Fig. 15-9. Most pistons use cast iron compression rings. Steel, when used, generally goes into the construction of the oil control ring. In some installations, a cast iron center spacer-scraper may be combined with steel side rails.

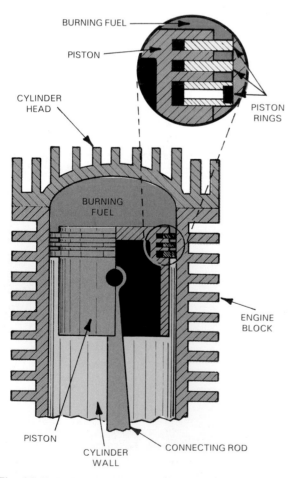

Fig. 15-7. Piston rings form a seal between the piston and cylinder wall.

Fig. 15-9. Piston rings are often plated with chrome or other materials to reduce wear.

RING TENSION

To permit the piston rings to expand and contract under varied temperatures and operating conditions, the rings are cut through at one place at the time of manufacture. See Fig. 15-10. The size of this opening between the ends of the ring (with piston and rings in the cylinder) is called the ring end gap. Although a great number of gap joint designs have been used in an effort to seal against gas leakage, the plain butt joint is the most common.

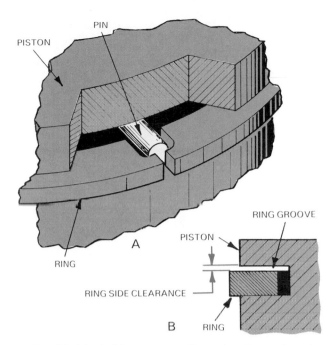

Fig. 15-10. End gap is cut through the piston ring to permit the ring to enter the cylinder and still exert tension on the cylinder wall.

Fig. 15-11. A—Many two cycle engines have pinned rings to prevent ring rotation. B—Ring side clearance allows movement, admits lubricating oil, and permits expansion of parts due to heat.

In another design feature, the outside diameter of a piston ring is made slightly larger than cylinder bore diameter. This causes the ring to exert force on the cylinder wall when installed. This force is called **ring tension.** Because the outside diameter of the ring is larger than the cylinder bore diameter, the ends of each ring in a single-piston set must be squeezed together to get the piston assembly into the cylinder.

RING MOVEMENT

Piston rings are free to move inward and outward in their respective grooves in the piston. See Fig. 15-11B. In addition, the rings will gradually work their way around (float) in the grooves, unless each ring is pinned in place, as shown in Fig. 15-11A. In most four cycle engines, the rings float. Some two cycle engines have the rings pinned in position. This is to prevent the ring ends from catching on the edge of the intake or exhaust ports and cutting into the cylinder wall.

Floating rings must be installed with the ring end gaps staggered to avoid gap alignment and possible oil flow through the series of gaps to the combustion chamber. **Pinned rings** are held in position by a short pin manufactured into the ring groove in the piston. The ring ends are cut out to straddle the pin, Fig. 15-11. Obviously, the pin prevents rotation of the ring around the groove.

RING SIDE CLEARANCE

Piston rings must have the right amount of **side clearance**, which permits them to move in and out in the groove while exerting tension on the cylinder wall. Side clearance also provides for adequate lubrication and heat expansion. Fig. 15-12 shows how the groove is cleaned with a special tool having different size scrapers that are pulled around the grooves. Fig. 15-13 illustrates how side clearance is checked with a feeler gage.

COMPRESSION RINGS

Generally, the first and second rings from the top of the piston are compression rings. **Compression rings** are designed to provide a strong seal, keeping the compressed fuel mixture and the burning gases above the piston by preventing

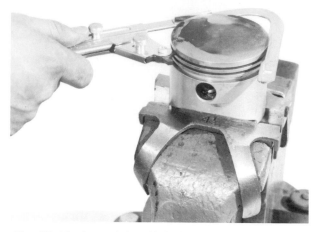

Fig. 15-12. A special tool is being used to clean carbon from ring grooves of piston. Do not widen or deepen any part of grooves.

Fig. 15-13. Side clearance of groove is checked with a piston ring and a thickness gage.

passage between the piston and the cylinder wall. Compression ring shapes vary somewhat with scraper grooves, beveled faces, and grooves or bevels on the inner side of the ring. See Fig. 15-14.

The various bevels and grooves are designed to create an internal stress in each compression ring. The stress causes the ring to twist slightly in its groove during the intake stroke of the piston. The twisting action places the lower edge of the ring, rather than the face, in contact with the cylinder wall. This allows compression ring to act as a mild scraper to aid in oil control. See Fig. 15-15A.

On the compression and exhaust strokes (four cycle engine), the rings are in a tipped position and tend to slip lightly over the oil film on the cylinder, Fig. 15-15B. On the power stroke, the

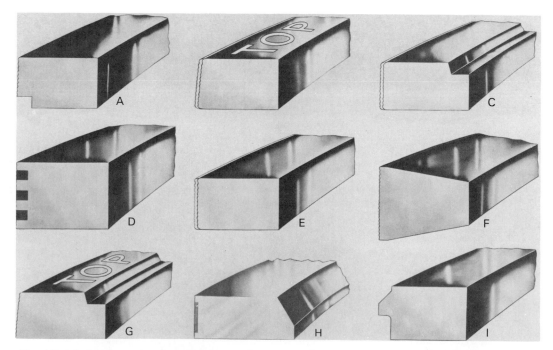

Fig. 15-14. Compression ring shapes. A—Outer groove. B—Chrome-plated, tapered face. C—Inner groove, chrome face. D—Ferrox-filled, grooved face. E—Plain chrome face. F—Keystone. G—Inner groove, tapered face. H—Inner chamfer, molybdenum-filled, grooved face. I—Scraper face. (Perfect Circle Products)

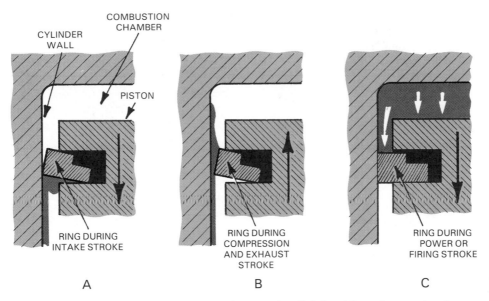

Fig. 15-15. Inner groove causes ring to twist slightly, aiding oil control and compression. It also reduces wear.

pressure of the gases forces the ring flat so that the entire edge bears firmly against the cylinder wall. Maximum sealing is provided during this critical time, as in Fig. 15-15C.

OIL CONTROL RINGS

The *oil control rings* are designed to remove surplus oil from the cylinder walls. They do this through a light scraping action against the walls. The ring and groove are both slotted and perforated (having holes). Oil trapped by the ring passes through the slots or holes of both the ring and the groove, Fig. 15-16. It then flows down inside the piston, where it drops into the oil pan or the crankcase. A three-piece oil control ring with a hump-type, spring-steel expander is shown in Fig. 15-17.

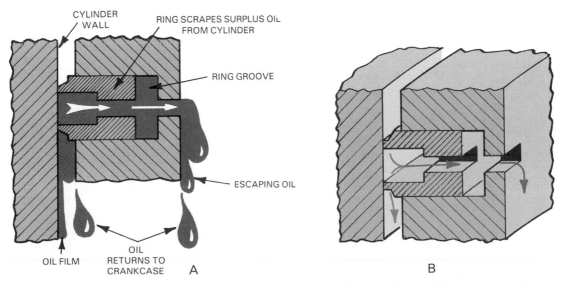

Fig. 15-16. An oil control ring removes surplus oil from cylinder walls. A–Oil being removed. B–Path of oil during removal.

Fig. 15-17. Three-piece oil control ring with flat steel (hump-type) expander in place. (Perfect Circle Products)

RING END GAP

The inside diameter of a piston ring is always made smaller than the piston's diameter. This being the case, each ring must be expanded to get it over the piston head and into the ring groove. The amount of *end gap* is critical (vital to success of rebuilding operation), so manufacturer's specifications should be followed. As a rule of thumb, however, allow .004 in. of end gap for every inch of cylinder diameter. For example, the minimum end gap for a 2 1/2 in. cylinder is .010 in.

Too much end gap will allow the gases to leak between the ring ends. Too little gap is even more serious. When the rings heat up in service, they will expand and close up. If the rings continue to heat and expand, they will break and score the cylinder wall.

To measure ring end gap, place the ring in the cylinder. Then, turn a piston upside down and push the ring to the lower end of the cylinder. When the ring reaches the proper depth, remove the piston and try various size feeler gages in the gap until one fits, Fig. 15-18. If the ring gap is too small, dress the butt ends of the ring with a file, Fig. 15-19.

When piston ring-to-groove side clearances and end gaps are satisfactory, install the rings on the piston, Fig. 15-20. Next, use a piston ring compressor to compress the rings flush with the

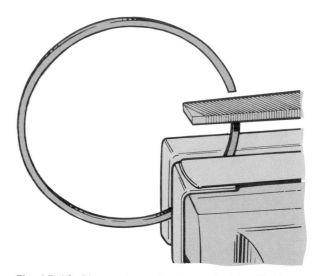

Fig. 15-19. Ring ends can be dressed with a file if end gap is too small. Use copper vise jaws to protect ring.

Fig. 15-18. Ring end gap is measured by pushing ring into cylinder with an inverted piston. Then, piston is removed and a thickness gage is used to measure gap.

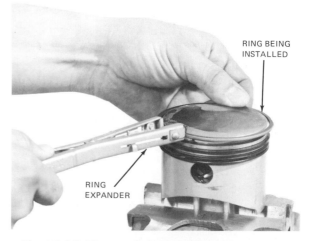

Fig. 15-20. Rings must be expanded with a special tool. Otherwise, danger of ring breakage is increased. (Tecumseh Products Co.)

Fig. 15-21. A ring compressor is used to squeeze ring ends together while the piston is pushed into cylinder. (Kubota Tractor Corp.)

grooves, Fig. 15-21. Hold the compressor firmly against the top of the block and use a wooden dowel or hammer handle to tap the piston out of the compressor and into the cylinder. Once free of the compressor, the rings will maintain firm contact with the cylinder wall.

RINGS AND CYLINDER WALLS

When cylinders wear, they become larger at the top. This is due to the wearing action of burning gases and to the dust and grit brought in with the air-fuel charge. Lack of lubrication in the upper cylinder also increases wear in the upper area of ring travel.

Even though the cylinder walls have a tapered wear pattern, the rings must stay tight against them. When taper becomes greater than .008 or .010 in., ring tension on the cylinder wall is lost and the end gaps open wide to allow oil to pass through to the combustion chamber, Fig. 15-22.

When the piston is at the bottom of its stroke, the rings are forced inward by the walls, Fig. 15-23. When the piston travels to the top of the cylinder, the rings must constantly expand to maintain contact with the ever-widening cylinder, Fig. 15-22.

When the engine is running at high speed, the piston can move to the top of a badly tapered cylinder and back down again so quickly that the rings do not have a chance to expand. This leaves the rings completely free of the walls at the top of the stroke. An engine with cylinder walls in this condition would burn large amounts of oil. It is possible, too, that the piston would tip and slap, causing ring wear and damage.

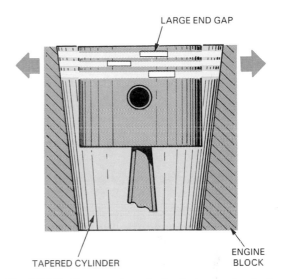

Fig. 15-22. Piston at top of a badly tapered cylinder. Rings must expand a great distance to stay in touch with cylinder walls.

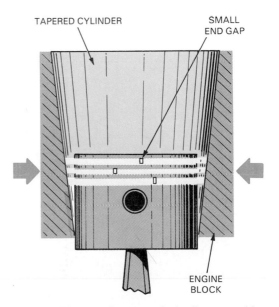

Fig. 15-23. Piston at bottom of a badly tapered (exaggerated) cylinder. Rings are forced into grooves.

Once the tapered condition exceeds .010 in., it can only be corrected by boring the cylinder oversize or by replacing the unit.

PISTON DAMAGE

Pistons can be damaged in many ways. They should be cleaned, examined, and measured after removal. The accompanying illustrations are good examples of conditions to look for.

Most piston and cylinder damage can be traced to one or more of the following causes: lack of oil, use of the wrong oil or oil-fuel mixture, use of the incorrect type of gasoline, foreign particles in the cylinder, overheating caused by clogged cooling fins, excess carbon buildup in the cylinder exhaust ports, and improperly fitted rings and pistons.

Fig. 15-24 shows scoring and light scuffing of both a piston and rings due to overheating. This occurs when high friction and combustion cause temperatures to approach the melting temperature of the piston materials.

Check and correct:
- Dirty cooling shroud and cylinder head fins.
- Lack of cylinder lubrication.
- Poor combustion.
- Improper bearing or piston clearance.
- Overfilled crankcase causing fluid friction.

Fig. 15-25 illustrates a piston with rings stuck and broken from lacquer, varnish, and carbon buildup, caused by abnormally high operating temperatures.

Fig. 15-25. Piston rings are stuck due to lacquer, varnish, and carbon buildup, resulting from high temperatures.

Check and correct:
- Overloading the engine.
- Ignition timing.
- Lean fuel mixture.
- Clogged cooling fins.
- Wrong oil.
- Low oil.
- Stale fuel.

Fig. 15-26 reveals vertical scratches on the ring faces and the piston, caused by the presence of abrasive particles.

Check and correct:
- Damaged or improperly installed air cleaner.
- Air leaks between air filter and carburetor.
- Leak in gasket between carburetor and block.

Fig. 15-24. Piston scored and scuffed due to overheating. (Tecumseh Products Co.)

Fig. 15-26. Vertical scratches on ring faces and piston are caused by abrasive materials entering the engine.

- Air leak around throttle shaft.
- Improperly cleaned cylinder bore after reconditioning engine.

Compare the oil ring (bottom ring) in Fig. 15-27 with the one in Fig. 15-26. The rails of the oil ring in Fig. 15-27 are worn down to the drain holes, and the ring surface is flat. This type of wear results from extended use and, possibly, from abrasives in the engine.

Check and correct:
- Rails worn down.
- Low ring tension.

Fig. 15-27. Piston has extremely worn rings because of long use and possible abrasives. (Tecumseh Products Co.)

The piston in Fig. 15-28 has a burned top land, resulting from detonation. Detonation is abnormal combustion which causes too much pressure and excessively high temperatures in the combustion chamber. Detonation is sometimes referred to as carbon knock, spark knock, or timing knock. It occurs when the air-fuel mixture ignites spontaneously and interferes with the normal combustion flame front.

Check and correct:
- Lean fuel mixture.
- Low octane fuel.
- Over-advanced ignition.
- Engine lugging.
- Excessive carbon deposits on piston and cylinder head (increasing compression).

Fig. 15-28. Burned top land results from detonation.

- Milled cylinder head (increasing compression ratio).

Preignition is the burning of the air-fuel mixture before normal ignition occurs. Preignition creates a pinging sound, resulting from a severe internal shock. It is accompanied by vibration, detonation, and power loss. If allowed to continue, it could cause severe damage to the piston, rings, and valves. See Fig. 15-29.

Check and correct:
- Internal carbon deposits remaining incandescent (red hot).
- Spark plug heat range too high.
- Spark plug ceramic shell broken.

Fig. 15-29. A hole burned through piston head was caused by preignition.

- Thin edges on valves or elsewhere in combustion chamber too hot.

If the connecting rod and piston are not aligned, a diagonal wear pattern will show on the piston skirts, Fig. 15-30. This condition can occur, along with poor ring contact, if the cylinder is bored at an angle to the crankshaft.

Check and correct:
- Rapid piston wear.
- Uneven piston wear.
- Excessive oil consumption.

Fig. 15-30. A diagonal wear pattern indicates improper alignment of connecting rod and piston. (Tecumseh Products Co.)

Piston damage can be caused by foreign objects carelessly left inside an engine during reconditioning. Fig. 15-31 shows the results of a needle bearing that became embedded in a piston. Do not be careless. Do the job with painstaking care from the start to final assembly.

PISTON RING WEAR-IN

After the small engine is reconditioned, a short wear-in period occurs. *Wear-in* is the process in which the face of each ring wears off until it fits perfectly against the cylinder wall. To help the rings seat quickly, the face is covered with microscopic grooves. During the first few hours of op-

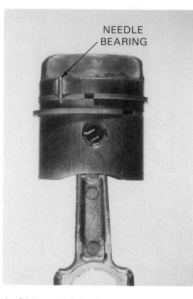

Fig. 15-31. Objects left inside engine can cause serious damage. A needle bearing is shown embedded across ring. (Jacobsen Mfg. Co.)

eration, these grooves rub against the cylinder wall and all high spots are worn off.

As the grooves wear away, the faces of rings and the cylinder wall become very smooth. Under normal operating conditions, very little wear occurs beyond this point.

PISTON PINS

A *piston pin* is used to secure the connecting rod to the piston. These pins are made of case-hardened steel and are ground to exact size. They may be hollow or solid.

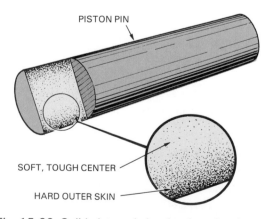

Fig. 15-32. Solid piston pin has hardened and ground surfaces.

A typical solid piston pin is shown in Fig. 15-32. Note the finer grain size of the metal at the surface. This is due to a heat treatment that provides a hard, durable case with a soft, tough center.

Many different piston pin assemblies have been used. The full-floating pin arrangement shown in Fig. 15-33 is free to turn in the rod as well as in the piston bosses. When both the connecting rod and piston are of aluminum alloy, the pin can operate directly against this material. If the rod is steel, either a bronze bushing or a needle roller bearing is used in the rod. The piston bosses also may have bronze bushing inserts for the pin. The snap rings in each boss prevent the pin from rubbing on the cylinder surface. Fig. 15-33 shows a full-floating pin with a steel connecting rod. This setup requires the use of a bushing or bearing.

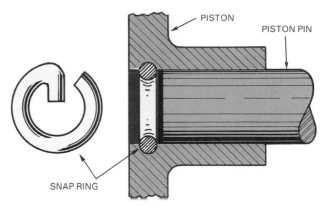

Fig. 15-34. Snap rings keep full-floating pin in place in piston.

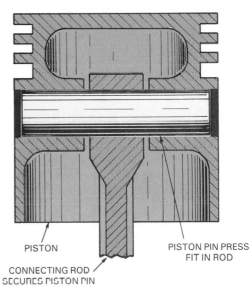

Fig. 15-35. This piston pin is pressed into connecting rod, but is allowed to turn in pin bosses of piston.

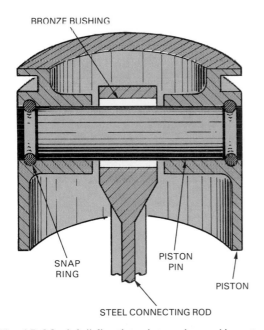

Fig. 15-33. A full-floating piston pin used in a steel connecting rod requires use of a bushing.

Retaining *snap rings* are compressed and placed in grooves in the piston pin bosses, Fig. 15-34. Some piston pins are a tight, press fit in the connecting rod, Fig. 15-35. The pin may turn in the piston bosses, bushings, or needle bearings, depending on the type of construction used.

REMOVING PISTON PINS

If snap rings are used, they must be removed first. Some are formed, spring-steel wire; others are stamped flat spring steel. Fig. 15-36 shows one method of removing the wire-type snap rings with a screwdriver. Needle nose pliers can also be used. Removal of stamped snap rings usually requires specially designed snap ring pliers.

Be careful when removing or replacing snap rings. They can slip out of the pin boss or the jaws of the pliers and cause severe eye injuries. Wear safety glasses when working with snap rings.

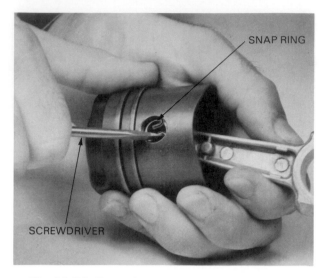

Fig. 15-36. Some rings can be removed from their grooves as shown. Always wear safety glasses for this operation. (Jacobsen Mfg. Co.)

If the piston pin is full floating and somewhat worn, it may slide out easily. On the other hand, it may be quite snug. A soft-faced mallet and a dowel rod can be used to tap the pin out, Fig. 15-37. Be careful not to hit the piston. Let the pin fall into a soft cloth.

Fig. 15-37. Tapping out piston pin requires the use of a soft hammer and dowel rod to avoid damaging the piston or the pin boss.
(Lawn-Boy Power Equipment, Gale Products)

Press-fit pins must be removed with a mechanical or hydraulic press. Refer to the engine manual for the proper support of the piston. If needle bearings are used, be careful not to lose them or misplace them.

MEASURING PISTON PINS AND BOSSES

When the piston pin has been removed, measure its outside diameter with a micrometer. Make a note of the measurement. Next, measure the inside diameter of the bosses using a small hole gage. Expand the gage in the boss until it gently contacts the inner surfaces. Withdraw the gage and measure it with the micrometer. Then, subtract the pin diameter from the boss diameter. The difference must be within the limits specified for the engine being serviced.

If the pin-to-boss clearance is excessive, a new pin and bushing or an oversize pin and reaming of the bosses are needed. Most piston pins fit snugly in the piston. If the pin turns by hand, inspect it for wear.

SUMMARY

The condition of the piston and the rings can be observed by inspecting the parts during disassembly of the engine. Low compression, blow-by, oil pumping, and fouled plugs all signal the need for piston and ring service.

Pistons can be made of aluminum or steel. Pistons have grooves for one to four rings. Two cycle engine pistons have one or two compression ring grooves. Four cycle engines have two compression ring grooves and one oil control ring groove.

Pistons expand during operation. Therefore, a specific amount of clearance between the piston skirt and the cylinder wall must be present.

A piston's thrust surfaces are forced against the sides of the cylinder wall. The thrust surfaces are at right angles to the centerline of the crankshaft and the piston pin.

The piston head is often smaller than the skirt because it receives the brunt of combustion heat and expands more than the rest of the piston. Pistons are manufactured in many different shapes, depending on their application.

Piston rings seal the cylinder. Without rings, the piston could not compress the fuel charge properly. Rings also keep burning gases from leaking between the sides of the piston and the cylinder wall. Piston rings ride against the cylinder wall, separated only by a thin film of oil.

Generally, the first and second rings from the top of the piston are compression rings. These rings are designed to keep the compressed fuel mixture and burning gases from escaping between the piston and the cylinder wall. Oil control the rings remove surplus oil from the cylinder walls.

Pistons can be damaged in many ways. They should be cleaned, examined, and measured after removal. A piston pin is used to secure the piston to the connecting rod. The pin may be held in place by a snap ring or can be press fit in the connecting rod.

KNOW THESE TERMS

Piston, Lands, Sump, Pin boss, Piston skirt, Cam-ground piston, Thrust surfaces, Piston rings, Ring tension, Floating rings, Pinned rings, Side clearance, Compression rings, Oil control rings, Ring end gap, Tapered wear pattern, Wear-in, Piston pin, Snap rings.

REVIEW QUESTIONS–CHAPTER 15

1. Explain why the piston head diameter may be less than the skirt diameter.
2. If a piston has too much clearance in the cylinder, it will produce a motion and sound called:
   a. Knocking.
   b. Slapping.
   c. Pinging.
   d. Hammering.
3. Piston rings are made of:
   a. Cast iron.
   b. Steel.
   c. Aluminum alloy.
   d. Cast iron, aluminum alloy.
   e. Cast iron, steel.
4. Name the tool used to squeeze the piston rings together so they can be installed in the cylinder.
5. Briefly name the results of the following conditions:
   a. Excessive ring end gap.
   b. Lack of ring end gap.
6. Pinned piston rings would most likely be found in:
   a. Four cycle engines.
   b. Two cycle engines.
7. Piston rings can be one of two basic types. Can you name them?
8. Piston pins are prevented from contacting the cylinder wall by use of:
   a. Cotter pins.
   b. Setscrews.
   c. Tapered pins.
   d. Snap rings.

SUGGESTED ACTIVITIES

1. Remove the rings from a piston, clean the piston and measure it.
2. Measure ring end gap and side clearance.
3. Using an old ring, demonstrate the method of dressing ring ends with a file to increase ring end gap.
4. Recondition piston bosses by reaming, and replace the worn piston pin with an oversize pin.
5. Replace piston rings with a ring expander.
6. Using ring compressor, replace reconditioned piston assembly in the cylinder.

Servicing a valve assembly on an overhead valve engine. (Tecumseh Products Co.)

CHAPTER 16

RODS, BEARINGS, AND VALVES

After studying this chapter, you will be able to:
☐ Describe the function of the connecting rod and the bearings.
☐ Define bearing spread and bearing crush.
☐ Differentiate between friction bearings and antifriction bearings.
☐ Summarize the function of the crankshaft.
☐ Service conventional and overhead valve assemblies.
☐ Explain the operation of ports, reeds, and rotary valves.
☐ Describe the purpose of the camshaft.
☐ Explain the purpose of an automatic compression release.

Like pistons and piston rings, connecting rods, bearings, and valves are used in areas of the engine that demand close "fits." Due to high temperatures, however, some clearance must be allowed for part expansion. While there are differences between makes of engines, maintenance of the rods, bearings, and valves is much the same for all.

Special attention must be given to four cycle engines because they contain more parts that require service. Rod and bearing service is the same for both two cycle and four cycle types. The valve system of four cycle engines (major area of difference) will be covered in detail near the end of the chapter.

CONNECTING RODS AND BEARINGS

The *connecting rod* attaches the piston to the crankshaft. The upper end of the connecting rod has a hole through which the piston pin is passed. The lower end contains a large bearing that fits around the crankshaft journal, Fig. 16-1.

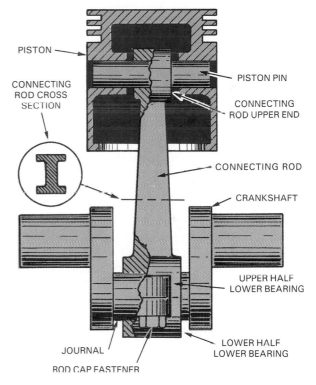

Fig. 16-1. Connecting rod attaches piston to crankshaft. Bearings are used at both ends of rod to reduce friction.

The lower end of the connecting rod is usually split when friction bearings are used. *Friction bearings* use smooth, sliding surfaces to reduce friction between moving parts. The place at which the halves separate is called the parting line. When needle or roller bearings are used, the rod end can be split or solid. See Fig. 16-2.

A variety of connecting rods, both split and solid, are pictured in Fig. 16-3. Roller bearings, needle bearings, and precision inserts are also shown.

FRICTION-TYPE ROD BEARINGS

There are three types of friction bearings commonly used in the big end of connecting rods. See Fig. 16-4.

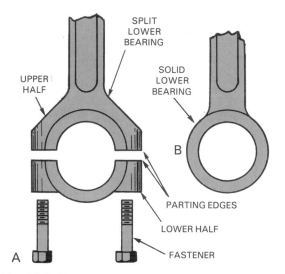

Fig. 16-2. Two types of connecting rod designs (crankshaft end). A—Split construction. B—Solid construction.

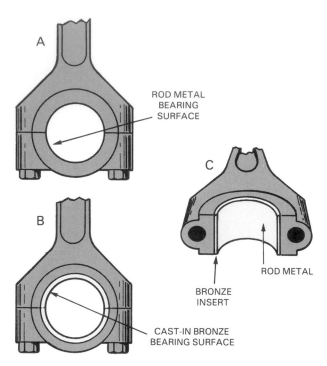

Fig. 16-4. Friction-type connecting rod bearings. A—Rod metal forms bearing surface. B—Bronze bearing is cast into rod metal. C—Replaceable precision insert bearing.

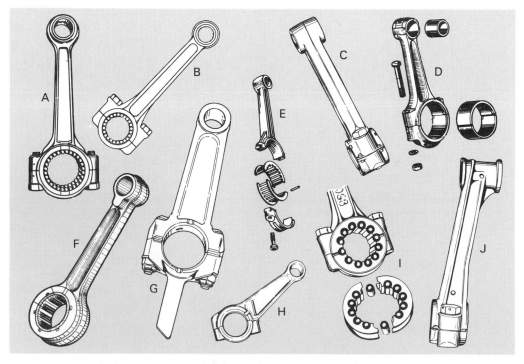

Fig. 16-3. Small engine rods and bearings. A and B—Split rod, roller bearings. C—Split rod, cast-in bearing. D—Split rod, precision inserts. E—Split rod, needle bearings. F—Solid rod, paired roller bearings. G—Split rod with dipper, precision inserts. H—Rod split at an angle, precision inserts. I—Split rod, roller bearings. J—Split rod with offset cap, cast-in bearing.

- Rod metal (used when rod is made of aluminum alloy).
- Bearing bronze (cast into rod end, bored, and finished).
- Removable precision insert bearings (steel shells lined with various materials).

The thin lining material on removable bearing inserts can be lead-tin babbitt, aluminum, or copper-lead-tin. Fig. 16-5 shows a steel-backed insert (1) that is coated with cast babbitt (2). This type of bearing is called a "precision" insert because it is made to an exact size for proper fit.

Bearing inserts are kept from turning in the rod end by a locating tab on the parting line edge of each insert. The tab fits into a slot in the rod itself. Fig. 16-6 illustrates this tab and slot arrangement.

Fig. 16-5. Construction of a typical precision insert bearing. 1–Steel back. 2–Cast babbitt (about .004 in. thick). (Clevite Corp.)

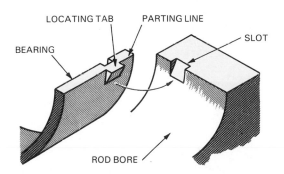

Fig. 16-6. Locating tabs prevent precision inserts from turning.

BEARING SPREAD

The diameter across the parting surfaces of insert bearing halves is slightly larger than the diameter across the curve machined into the rod and rod cap. This condition is called *bearing spread*. To seat the insert, the ends must be forced down and snapped into place. Never press down in the center of the insert to seat it in the rod bore. The correct amount of bearing spread gives tight insert-to-bore contact around the entire bearing and provides support and alignment. It also helps to carry heat away through the rod and bearing cap and holds the bearing in place during assembly.

BEARING CRUSH

When precision inserts are snapped into the rod bore, the ends will protrude slightly above the parting surface. See Fig. 16-7A. This built-in design feature is called *bearing crush*. Generally, bearing crush varies from .001 in. to .002 in.

When the rod cap is installed and drawn into place, the insert ends meet first and force the insert halves tightly against the rod bore. This provides firm support for the insert. The forced fit makes the insert round and, through close metal-to-metal contact, allows heat to be carried away through the rod. Fig. 16-7B shows how radial pressure is exerted against the rod bore.

INSERTS ARE MATCHED

Precision inserts must be kept in matched pairs. Never mismatch bearing inserts. Always

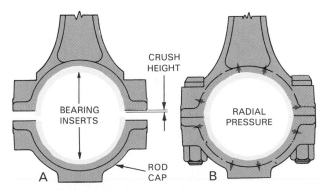

Fig. 16-7. Effect of bearing "crush." A–Rod and cap separated. B–Rod and cap drawn together, creating radial pressure on inserts. (Sunnen Products Co.)

use the exact size needed. Bearings cannot be made larger or smaller in the shop. Standard and various undersizes are available.

ANTIFRICTION BEARINGS

Many small gasoline engines use an antifriction bearing in the big end of the connecting rod. *Antifriction bearings* use rollers or balls to reduce friction between moving parts. Often, needle rollers are used. These roller elements can be held together by a roller cage or separator. See Fig. 16-8A. The rollers can also be left free as in Fig. 16-8B. Antifriction bearing assemblies are hardened and ground to an exact size. They must fit accurately, but still have some clearance for expansion.

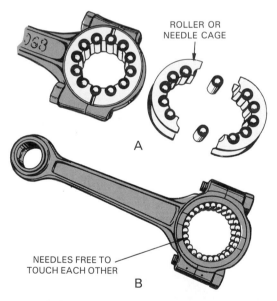

Fig. 16-8. Two types of connecting rod needle bearings. A–Caged needles. B–Free needles. (Evinrude Motors)

During manufacture, the *rod cap* is bolted into position on the rod. Then, the assembly is bored to an exact size. It is important, therefore, that the rod cap is always put back in its original position. If the cap is turned 180°, the upper and lower halves can be offset. This error in assembly will cause bearing and shaft failure. See Fig. 16-9A.

Connecting rods are usually marked with either a line, punch mark, number, special boss, or

chamfered edge to show correct cap alignment during assembly. If a "mark" is not apparent, punch mark the rod end and cap for later reference, Fig. 16-9B. NOTE: Caps must never be switched from one rod to another.

ROD BOLT LOCKING DEVICES

To stop connecting rod bolts or cap screws from loosening in service, locking devices are used. One common device is a thin sheet metal strip with locking tabs, Fig. 16-10. The cap screw is inserted through holes in the locking strip, holding it in place against the rod. After the cap

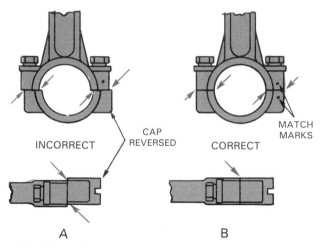

Fig. 16-9. Connecting rod cap installation. A–If cap is turned 180°, rod bore will be offset. B–Match marks on rod and cap signal correct assembly.

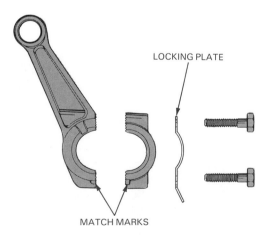

Fig. 16-10. A locking plate is often used between connecting rod cap and cap screws. (Tecumseh Products Co.)

screw is tight, the metal tabs are bent up against the flat sides of the screw head, Fig. 16-11.

Self-locking nuts, lock washers, and specially shaped cap screws are also used to prevent loosening. The final tightening of the cap screws is especially important. Always use a torque wrench to tighten rod fasteners to the exact torque specified by the manufacturer.

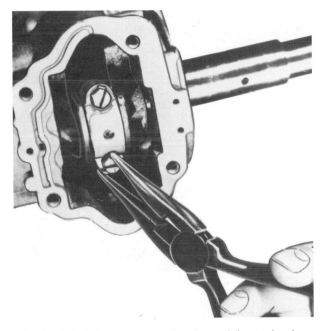

Fig. 16-11. After cap screw has been tightened, tab on locking plate is bent up against head of screw to lock it in place. (Lawn-Boy Power Equipment, Gale Products)

CRANKSHAFT

The *crankshaft* converts the reciprocating (back and forth) motion of the piston into rotary (circular) motion. It transmits engine torque to a pulley or gear, so that some object may be driven by the engine. The crankshaft also drives the camshaft (on four cycle engines), supports the flywheel, and, in many engines, operates the ignition system.

Crankshafts can be made of cast or drop forged steel. One-piece and multi-piece crankshafts are used. Fig. 16-12 shows a typical one-piece small engine crankshaft. A multi-piece crankshaft is shown in Fig. 16-13.

The *crankshaft throw* is the offset portion of the shaft measured from the centerline of the main bearing bore to the centerline of the connecting rod journal. The connecting rod journal is commonly referred to as the "crank throw" or "crankpin."

CRANKSHAFT BALANCE

To help offset the unbalance created by the force of the reciprocating mass (connecting rod, piston, and crankpin), *counterweights* are added to the crankshaft. By placing these weights opposite the crankpin, engine vibration is greatly reduced. The counterweights are usually forged as

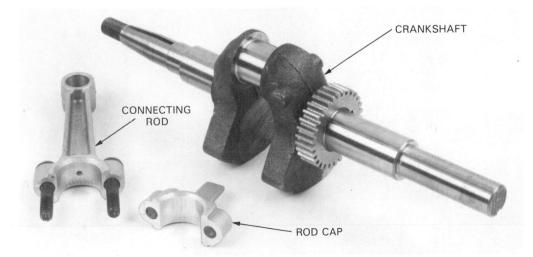

Fig. 16-12. Single-piece crankshafts are most popular in small gasoline engine applications.

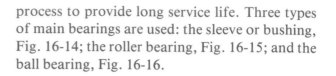

CONNECTING ROD JOURNAL
OR CRANKPIN

MAIN
BEARING
JOURNAL

MAIN
BEARING
JOURNAL

COUNTERWEIGHT

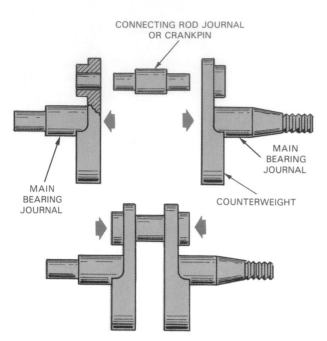

Fig. 16-13. Multi-piece crankshafts have various parts pressed together under heavy pressure.

an integral part of the crankshaft, Figs. 16-12 and 16-13.

CRANKSHAFT MAIN BEARINGS

The crankshaft is supported by one or more **main bearings.** Often, the main bearing journal surfaces are hardened by an induction hardening

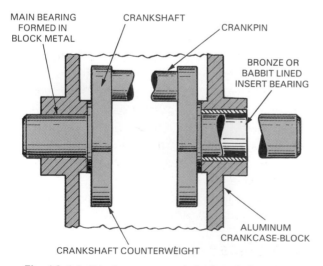

Fig. 16-14. Friction-type crankshaft main bearings. Shaft at left uses bore in aluminum crankcase as a bearing surface. Shaft at right uses a precision insert bearing.

process to provide long service life. Three types of main bearings are used: the sleeve or bushing, Fig. 16-14; the roller bearing, Fig. 16-15; and the ball bearing, Fig. 16-16.

CRANKSHAFT CLEARANCES

To allow space for lubricant between the moving parts, as well as to provide room for expansion when heated, crankshaft bearings must have

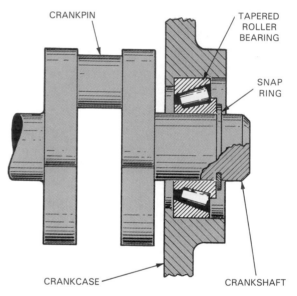

Fig. 16-15. Typical use of tapered roller bearing as a crankshaft main bearing.

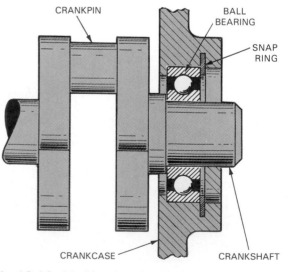

Fig. 16-16. A ball bearing also can be used as a crankshaft main bearing.

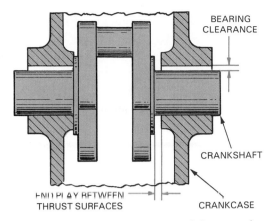

Fig. 16-17. Crankshaft bearings and thrust surfaces must have some clearance (end play) to provide space for lubricant and for heat expansion.

a slight end clearance, Fig. 16-17. Shaft movement from end to end is controlled by the bearing adjustment when tapered roller bearings or ball bearings are used.

With friction bearings, a thrust surface on the shaft rubs against a similar surface on the crankcase. A precision insert main bearing may have a thrust flange for the crank to rub against. In some applications, a bronze thrust washer is used.

Clearances will vary with engine type, design, and use. Bearing and thrust surface clearances are critical. They must be held to exact tolerances as recommended by the manufacturer.

Fig. 16-18 illustrates the method of measuring the bearing surfaces on a crankshaft with a mi-

Fig. 16-18. A micrometer is required to accurately measure bearing surface diameter on a crankshaft.

crometer. A measurement must be taken in at least two positions 90° to each other. If any of the dimensions are smaller than specified, or if there are any score marks, the bearing surfaces should be reground. Basically, wear and taper should not exceed .001 in.

MEASURING BEARING CLEARANCE

Bearing clearance is the space between the inner bearing surface and the crankshaft main or rod journal. When checking bearing clearance, use a special compressible plastic material called *Plastigage.* This material is color coded and selected according to the recommended clearance range. It comes in a thin, round strand, which is stored in a paper package.

To use Plastigage, select the correct color for the specified clearance. Cut a piece of plastic equal to the width of the bearing and lay it across the bearing surface. Torque the bearing cap in place. Then, remove the cap and compare the compressed width of the plastic with the comparison chart printed on the Plastigage package. Clearance is given alongside the matching marks on the chart. In effect, the wider the plastic, the less clearance there is.

If bearing clearance is too great, undersize inserts will have to be used. If the crank journal is worn, it will require grinding to clean it up. After grinding the journal, recheck the clearance with a Plastigage.

On many small engines, the main bearings are simply machined bores in the crankcase halves or pressed inserts. Plastigage will not measure wear in these bearings. To check clearance, first measure the crankshaft diameter with a micrometer. Next, measure the inside diameter of the main bearing with a telescoping gage, Fig. 16-19. Lock the gage, remove it from the bearing, and measure the setting with a micrometer. The difference between the shaft reading and the bearing reading is the amount of bearing clearance.

CRANKCASE SEALS

Crankcase seals prevent leakage of oil from the areas where the crankshaft and crankcase come

Fig. 16-19. Measuring diameter of a pressed main bearing with a telescoping gage. (Deere & Co.)

together. The shell of the seal makes fixed contact with the crankcase, while the knife edge of the sealing lip rubs lightly against the crankshaft. See Fig. 16-20.

Seals are made of neoprene, leather, graphite, or other materials, depending on how they are used. A typical crankcase seal has a steel outer shell with a neoprene center. A small coil spring keeps the sealing lip in constant contact with the shaft it seals, Fig. 16-21.

Note in Fig. 16-20 that the sealing lip must face the fluid being sealed in. In this application, it faces the crankcase. In this way, the pressure of the oil will tend to force the lip against the shaft. If the seal is installed backwards, oil pressure will force the sealing lip away from the shaft and oil leakage will occur.

When removing the crankcase cover from the crankcase and crankshaft, as in Fig. 16-22, place tape over the keyway. This will keep the sharp keyway edges from cutting the neoprene oil seal.

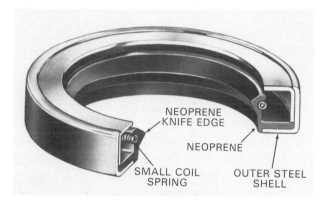

Fig. 16-21. This crankshaft oil seal has an outer steel shell with a neoprene center. A small coil spring produces contact pressure. (Chicago Rawhide Mfg. Co.)

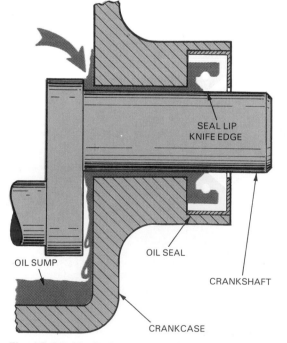

Fig. 16-20. Typical neoprene oil seal has sealing lip with sharp edge, providing increased pressure and reduced friction.

Fig. 16-22. Taped keyway edges will protect oil seal when cover is removed. (Deere & Co.)

Fig. 16-23 shows how a press is used to push the old seal out of the backplate. In Fig. 16-24, the seal is being readied for installation. A liquid sealant is applied to the outside of the shell of the seal before pressing it in place in the backplate. Often, seals can be replaced by tapping them into the bore of the backplate with a special driving tool, Fig. 16-25.

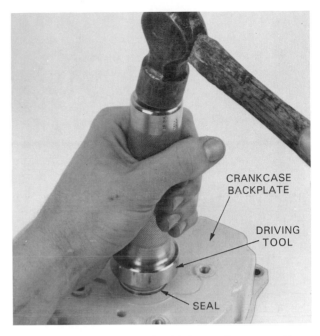

Fig. 16-25. Use a special driving tool to tap seal in place in housing. (Deere & Co.)

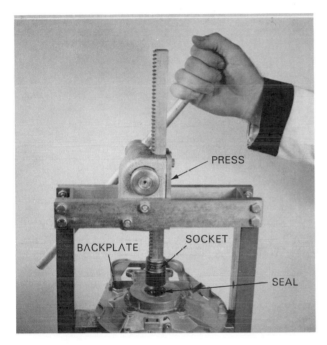

Fig. 16-23. On some engines, an arbor press can be used to push out old oil seals. Press with care to avoid damaging housing. (Jacobsen Mfg. Co.)

VALVE SERVICE

Four cycle engines contain **poppet valves**, which are subjected to tremendous heat. The normal operating temperature of the exhaust valve exceeds 1000°F. To withstand this heat, high-quality, heat-resistant steel must be used and the correct operating clearances must be maintained.

REMOVING VALVE ASSEMBLY

The engine valve assembly includes the valve, valve spring, and one or more retainers, Fig. 16-26.

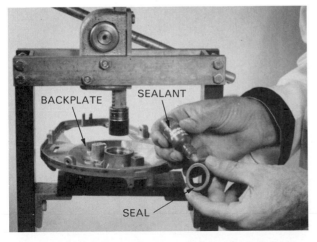

Fig. 16-24. Apply sealing compound to shell of seal before pressing seal into bore of housing.

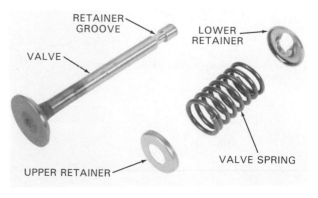

Fig. 16-26. A poppet valve assembly is used in four cycle engines.

Rods, Bearings, and Valves 277

The locking-type retainers are called *valve keepers.* Once the cylinder head has been removed, remove the valve by compressing the valve spring with a compressor, Fig. 16-27. Remove the valve spring retainer from groove in the valve stem using a pair of pliers, Fig. 16-28. Then, slowly release pressure on the spring and remove the compressor. The valve can be pulled out the top and the spring taken from the side.

INSPECTING VALVES AND SEATS

When the valves have been removed, clean them with a power-operated wire brush and inspect them for the following defects:
1. Eroded, cracked, or pitted valve faces, heads, or stems.

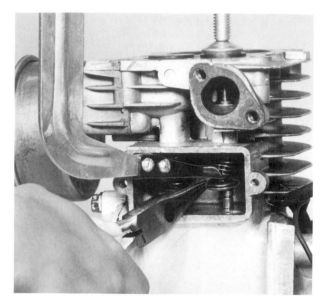

Fig. 16-28. When spring is fully compressed, use a pliers to remove valve keepers. Then, release valve compressor and remove valve, spring, and retainers.

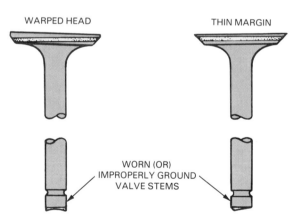

Fig. 16-29. Warped valve head, thin margin, or worn stem may mean refacing or replacement is needed.

2. Warped head, Fig. 16-29.
3. Worn or improperly ground valve stems, Fig. 16-29.
4. Bent valve stems.
5. Margin less than 1/64 in.
6. Partial seating.

Heavy carbon deposits on intake valves sometimes cause faulty valve operation by restricting the flow of fuel into the cylinder. If any serious defects are observed, the valve should be replaced. In any case, valve faces should be machined to a smooth, true finish.

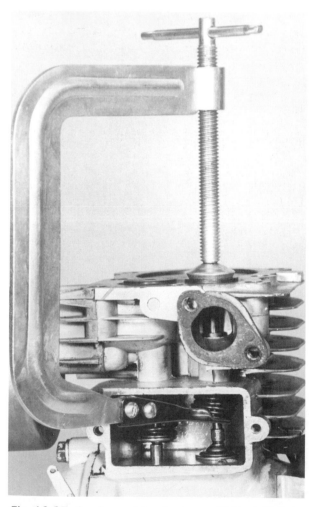

Fig. 16-27. A valve spring compressor "squeezes" the spring to uncover keepers in the valve stem.

INSPECTING VALVE SPRINGS

Through overheating and extensive use, *valve springs* can lose their elasticity and become distorted (warped or bent). Check each spring for squareness and proper length with a square and a surface plate, Fig. 16-30. Replace all springs that are badly distorted or reduced in length. Test spring tension and compare with the specifica-

tions in the engine manual, Fig. 16-31. Lack of spring tension can "valve flutter" or incomplete closing and sealing of the valves.

VALVE GUIDES

Valve guides align and "steer" the valves so that they can open fully and close completely. Guide-to-valve stem clearance must not exceed tolerances, since this would permit the valve to tip. Tipping causes the valve face to strike the seat at an angle, allowing hot combustion gases to escape. Some clearance is required, however, to allow for heat expansion and lubrication. Generally, guide-to-valve stem clearance should run about .002 to .003 in.

Valve guides can be a replaceable insert or an integral part of the block. See Fig. 16-32. Replaceable guides are cast iron. The old guide must be driven out and the new guide pressed in place. Integral guides can be reamed to fit a valve with an oversize stem.

Fig. 16-30. Use a square and a surface plate to check a valve spring for proper length and squareness. (Deere & Co.)

Fig. 16-31. A special tester is available to test a valve spring for adequate tension.

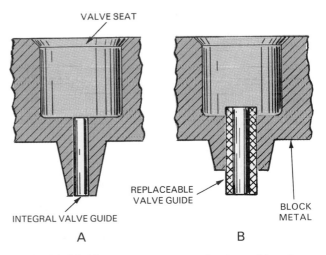

Fig. 16-32. There are two types of valve guides. A–Bored in block. B–Pressed in block. Guides that are pressed in the block can be replaced.

INSPECTING VALVE GUIDES

Valve guides must be cleaned before inspection. A special cylindrical wire brush, driven by a power drill, is made for this job. After cleaning the guide, measure the bore with a small hole gage, Fig. 16-33. Expand the gage until it lightly

Rods, Bearings, and Valves 279

Fig. 16-33. Valve guide diameter can be measured with a small hole gage. (Deere & Co.)

touches the sides of the bore. Remove the gage and measure it with a micrometer.

Next, measure the valve stem diameter with a micrometer, Fig. 16-34. Subtract the stem diameter from the guide diameter to find the precise amount of clearance. Compare this with the clearance specified.

VALVE GUIDE REAMING

If the clearance between the stem and the guide exceeds the allowable limit, enlarge the guide to the next oversize dimension with an adjustable

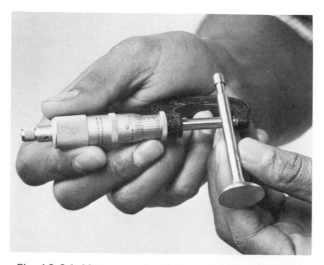

Fig. 16-34. Measure valve stem diameter to determine valve stem-to-guide clearance.

Fig. 16-35. Use an adjustable reamer to recondition and resize integral valve guides. Replacement valves must have oversize stems. (Deere & Co.)

reamer, Fig. 16-35. Select and install a valve with the correct oversize stem. Do not enlarge the tappet guides, because oversize tappet stems are seldom available.

VALVE SEAT ANGLE AND WIDTH

The correct *valve seat angle* is necessary for proper valve seating. Valve seats are generally cut to a 45° angle, although 30° seat angles are used in a few engines. Follow all of the manufacturer's recommendations.

The *valve seat width* is important for effective valve system operation. The seat must be wide enough to prevent cutting into the valve face. It also must provide enough contact area to provide for adequate heat dissipation. On the other hand, the seat must not be too wide. If it is, carbon will pack between the seat and the valve face, holding the valve off the seat. A valve that fails to seat produces a rough-running engine and will quickly warp and burn. Specified seat widths range from .030 to .060 in. (1/32 to 1/16 in.), Fig. 16-36.

Some valve seats are ground to an angle of 46° and the valve face is ground to an angle of 45°, or vice versa. The 1° variation produces a hairline contact that gives fast initial seating. Some manufacturers believe that, upon heating, the valve will form a perfect seal. The difference in the angle between the valve face and the valve seat is called an **interference fit,** Fig. 16-37. Valve seat

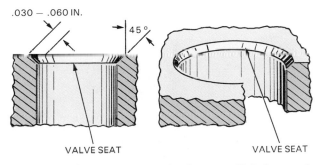

Fig. 16-36. A typical integral valve seat. Hole for seat is bored in the block metal. Note valve seat angle and valve seat width.

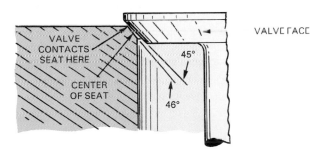

Fig. 16-37. A 1° difference between valve face and valve seat provides better seating.

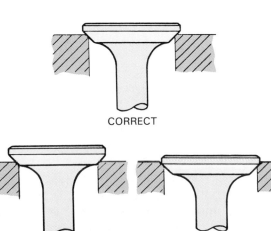

Fig. 16-38. Comparison of correct and incorrect location of seating area on valve face. (Deere & Co.)

contact must be near the center of the valve face, Fig. 16-38.

Valve seats should be cut with a special cutting tool, such as the one shown in Fig. 16-39. If the carbide cutter blades on this tool are damaged or become dull, they can be easily replaced. Cutting action is controlled by turning the tool as cutting takes place. Fig. 16-40 illustrates the proper procedure to follow when using a valve cutter and pilot. Fig. 16-41 shows how to correct poor valve seat geometry.

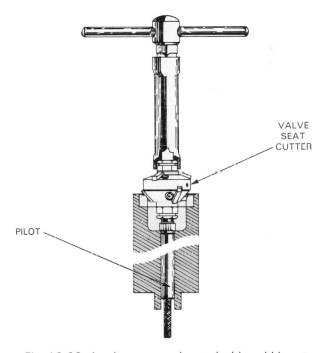

Fig. 16-39. A valve seat cutting tool with carbide cutting blades. This tool is used to recondition valve seats by hand. (Neway Mfg. Co.)

LAPPING VALVES

Good used valves may be reseated by **lapping.** Some manufacturers do not recommend lapping. When a valve expands, it does not seat in the same place that it does when it is cold. Therefore, any benefits from the lapping process will cancel out. Nevertheless, the procedure can be followed if you wish to seat valves this way.

Some engine manufacturers recommend hand lapping of the valve seats. Lapping compound is available from engine parts distributors. Some

Valve Seat Cutter Kit
GENERAL INSTRUCTIONS

SELECTION AND USE OF PROPER PILOT

A. SOLID PILOTS.

1. Select a pilot same diameter (fractional or metric) as valve guide.

2. Insert pilot in valve guide, twisting slightly, until very snug. Pilot shoulder should not touch valve guide. (Fig. 1)

   • If small, try next size larger.

   • If too large, try next smaller size.

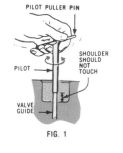

FIG. 1

DETAILED CUTTING INSTRUCTIONS

A. Slowly lower cutter to valve seat. DO NOT DROP CUTTER.

B. Turn clockwise and apply very light pressure. Release the down pressure at end of each cut. Make one or two turns with no pressure.

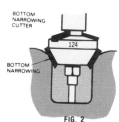

FIG. 2

C. BOTTOM NARROWING CUT.

1. Cut lightly with narrowing cutter (usually 60°).

2. Cut until a fine continuous line is formed with valve seat. (Fig. 2)

D. TOP NARROWING CUT.

1. Cut lightly with narrowing cutter (usually 15°). For engines with hemispheric combustion chambers, use 30°/31°.)

2. Cut until seat width is slightly less than required. (This operation LOWERS THE SEAT.)

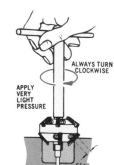

FIG. 3

E. FINAL SEAT CUT.

1. Cut lightly, with seat cutter (usually 31° or 46°).

2. Cut seat to proper width. This should take only a few turns (Fig. 3)

F. INSPECT SEAT.

1. Remove pilot, using pilot puller pin. (Fig. 4)

2. Insert valve in valve guide.

3. Tap valve slightly up and down in the guide (holding it with fingers top and bottom — above and below the cylinder head). Do this until seat contact ring shows on the **valve face.**

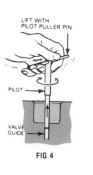

FIG.4

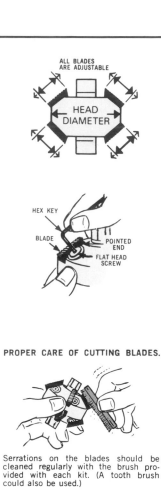

PROPER CARE OF CUTTING BLADES.

Serrations on the blades should be cleaned regularly with the brush provided with each kit. (A tooth brush could also be used.)

PROPER CARE OF CUTTERS AND PILOTS.

The tools should be kept in tool case. When removed from case, they should be placed on a cloth or pad. These precision instruments will last a long time if reasonable care is used.

Fig. 16-40. Proper procedures for using a valve seat cutting tool. (Neway Mfg. Co.)

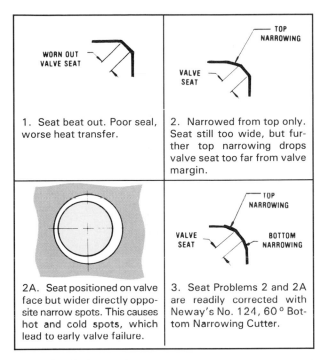

WORN OUT VALVE SEAT	TOP NARROWING VALVE SEAT
1. Seat beat out. Poor seal, worse heat transfer.	2. Narrowed from top only. Seat still too wide, but further top narrowing drops valve seat too far from valve margin.
	TOP NARROWING VALVE SEAT BOTTOM NARROWING
2A. Seat positioned on valve face but wider directly opposite narrow spots. This causes hot and cold spots, which lead to early valve failure.	3. Seat Problems 2 and 2A are readily corrected with Neway's No. 124, 60° Bottom Narrowing Cutter.

Fig. 16-41. Poor valve seat geometry can be corrected with proper reconditioning procedures. (Neway Mfg. Co.)

suppliers package it in two-compartment canisters. One compartment contains a coarse silicon carbide abrasive combined with a special grease. The second compartment contains a finer compound. The condition of the valve will dictate which grade to use.

If the coarse lapping compound is used, follow up with the finer compound. Fig. 16-42 shows how to apply the compound to the valve face only. The compound should not be allowed to contact the valve stem or guide. Next, a lapping tool is attached to the valve head by means of a suction cup, Fig. 16-43. The tool shown has a spring-loaded piston in the handle to help create suction. With the tool attached, the valve is placed in the guide and twirled back and forth, Fig. 16-44.

The lapping process is complete when a dark gray, narrow band, which is equal to the seat width, can be seen all the way around the valve face. Do not lap more than is necessary to show a complete seat.

After lapping, thoroughly clean the valve and valve seat chamber so that none of the abrasive finds its way into the engine. The best way to get the cleaning job done is to turn the engine upside down and wash the chamber with solvent from the bottom.

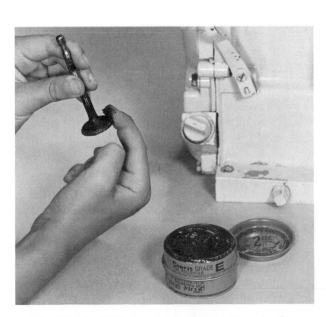

Fig. 16-42. Apply lapping compound to valve face before lapping face to seat.

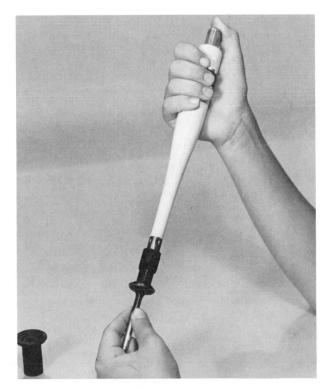

Fig. 16-43. To use a lapping stick, attach stick to valve head with suction cup. (Power-Grip Co.)

Rods, Bearings, and Valves 283

Fig. 16-44. Lap valve to seat by twirling lapping stick between palms of hands. Lift lapping stick and valve occasionally to increase cutting action of compound.

VALVE LIFTER-TO-STEM CLEARANCE

Valve clearance refers to the space between the end of the valve stem and the top of the valve lifter when the valve is closed. The amount of clearance needed depends upon engine design and use. Due to hotter operation, the exhaust valve often requires more clearance than the intake valve. Clearances of around .008 in. for the intake valve and .012 in. for the exhaust valve are fairly common. Follow manufacturer's specifications.

When there is too little valve clearance, the valve may be held open when the valve stem heats up and lengthens (expands). As a result, engine performance is poor and both the valve face and valve seat will burn, Fig. 16-45. Insufficient clearance can also alter valve timing, making it too far advanced.

Too much valve clearance, on the other hand, will make valve timing late and reduce valve lift.

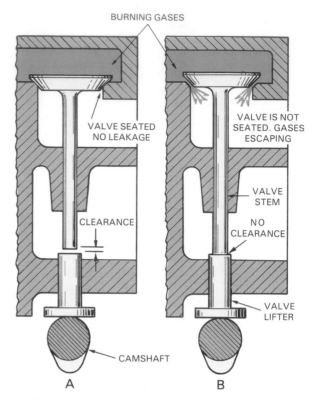

Fig. 16-45. Valve clearance setting is essential to good engine performance. A—Correct clearance permits valve to seat. B—Lack of clearance keeps valve open.

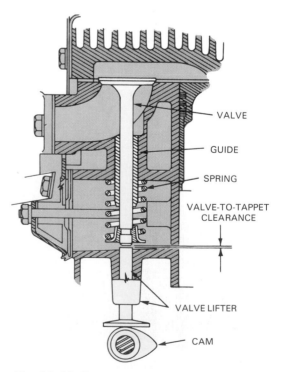

Fig. 16-46. Components of a complete valve train. (Kohler Co.)

This results in sluggish engine performance. It also can cause rapid lifter wear because of the pounding action involved. Under these conditions, the engine will be noisy and the valve could break. Fig. 16-46 shows a complete small engine valve train.

Fig. 16-47. During refacing operation, valve rotates slowly while being fed back and forth across the abrasive wheel. (Sioux Tools, Inc.)

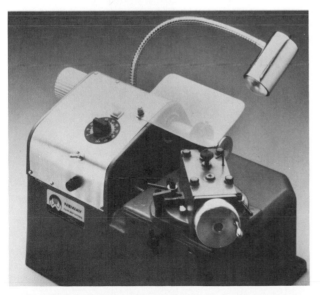

Fig. 16-48. A valve lathe is specially designed to cut and true the valve face with a carbide cutting tool. (Neway Mfg. Co.)

REFACING VALVES

Valve refacing is done on a specially designed grinder, Fig. 16-47. The valve is revolved while being fed over an abrasive wheel. The collet that holds the valve is adjusted to achieve the desired face angle. Coolant flows over the valve head during grinding to reduce heat and produce a good surface finish. The infeed wheel is used to precisely control the amount of material being ground from the valve face. In some cases, a lathe can be used to reface valves. See Figs. 16-48 and 16-49.

Fig. 16-49. Closeup of a valve face being turned and trued on a valve face lathe.

When valve lifters become concave or deformed, they can be ground flat as in Fig. 16-50. Likewise, the end of the valve stem can be dressed and shortened to produce correct valve clearance, Fig. 16-51.

ADJUSTING VALVE CLEARANCE

When a valve has been refaced, it rides lower in the guide, and, therefore, valve-to-tappet clearance is reduced. If the engine does not have adjustable tappets, the end of the valve stem must be ground to obtain correct clearance. To check clearance, turn the camshaft until the lobe is away

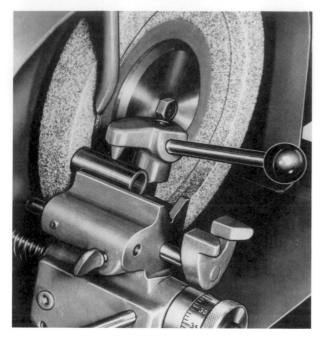

Fig. 16-50. Machine the head of a valve lifter as shown.

Fig. 16-51. Grinding the end of the valve stem trues the end surface and removes stock for clearance adjustment.

Fig. 16-52. Valve stem-to-lifter clearance is checked with a thickness gage. (Deere & Co.)

OIL DRAINBACK HOLES

Fig. 16-53. Make sure that all drainback holes are open before installing breather cover.

from the tappet. Hold the valve against its seat while testing clearance with a thickness gage, Fig. 16-52. If there is too little clearance, remove the valve and grind .001 or .002 in. off of the stem. Repeat the clearance check and grinding operation until the clearance is correct.

After the valves and seats have been properly reconditioned, place each valve in its respective guide. Use a valve compressor to compress the spring and, then, install the keepers. Make sure

that the oil drainback holes are open before reinstalling the valve breather cover, Fig. 16-53.

OVERHEAD VALVE SYSTEMS

Many four cycle engines use an *overhead valve system*, which transmits motion through *pushrods* and *rocker arms* to open and close valves, Fig. 16-54. The overhead valve design improves volumetric efficiency and eliminates combustion chamber hot-spots, which can cause cylinder distortion. Overhead valve systems increase fuel efficiency by as much as 25% and improve engine service life.

OVERHEAD VALVE SYSTEM DISASSEMBLY

The following procedure should be observed when disassembling an overhead valve system:

1. Remove engine accessories that interfere with the removal of the valve cover.

2. Remove the valve cover bolts and the cover, Fig. 16-55. The rocker arm assembly should now be exposed.

3. Remove the rocker arm locking screws and nuts as shown in Fig. 16-56. Remove the rocker arms. Some engines have rocker arms that pivot on a rocker arm shaft, Fig. 16-57. To remove, loosen the adjusting screws, remove the retaining rings, and slide the rocker arms off of the shaft.

4. When the rocker arms are removed, lift out the pushrods, Fig. 16-58. (Note the location of each pushrod and do not interchange them when reassembling the engine.) After cleaning the pushrods, check them for

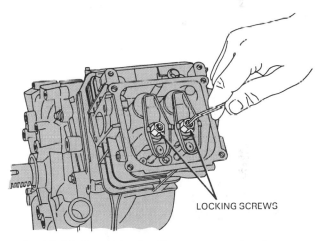

Fig. 16-56. Removing locking screws and rocker arms. (Briggs and Stratton Co.)

Fig. 16-54. Simplified overhead valve system. (Deere & Co.)

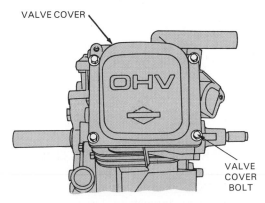

Fig. 16-55. Typical valve cover. To remove cover, loosen bolts on each corner. (Briggs and Stratton Co.)

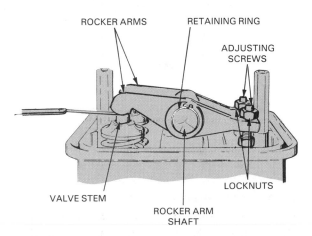

Fig. 16-57. Rocker arms on a rocker arm shaft. Loosen adjusting screws, remove retaining rings, and slide rocker arms off shaft. (Tecumseh Products Co.)

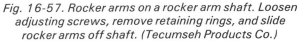

Rods, Bearings, and Valves   287

straightness by rolling them on a flat, machined surface. If they are bent, replace them. Do not try to straighten pushrods.

5. Remove the cylinder head bolts and the cylinder head, Fig. 16-59.

REMOVING VALVES

To remove the valves in an overhead valve system, locate the cylinder head on a workbench and place small wooden blocks under the valve faces to hold them in place. If there are wear buttons or caps on the valve stems, remove them.

NOTE: When removing valves and valve springs, identify the parts to prevent interchanging them during reassembly.

On some engines, you can compress the valve springs by pressing on the spring retainers with your thumbs. Push the spring retainer toward the large end of its slot and release pressure, Fig. 16-60. Remove the retainer, spring, and valve stem seals. Discard the seals.

If the valve springs cannot be compressed with your thumbs, a special valve spring compressing tool may be necessary, Fig. 16-61. In this illustra-

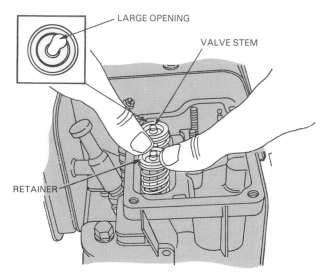

Fig. 16-60. Compress valve spring and move retainer to large opening to release. A wooden block placed under the valve head prevents valve movement. (Briggs and Stratton Co.)

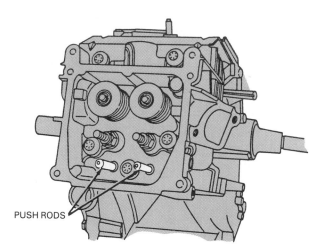

Fig. 16-58. Remove pushrods by lifting them out of holes. Check for straightness by rolling the rods on a flat surface. (Briggs and Stratton Co.)

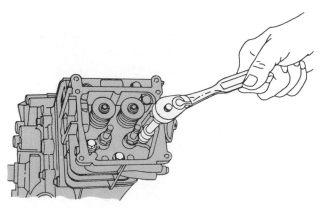

Fig. 16-59. Remove cylinder head by removing head bolts evenly and in slight increments. (Briggs and Stratton Co.)

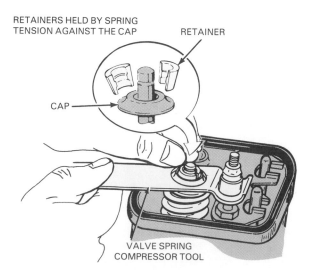

Fig. 16-61. Compressing valve spring with a valve spring compressor to remove split-type retainers. (Tecumseh Products Co.)

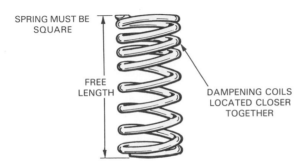

Fig. 16-62. Valve spring with dampening coils. Measure for correct length and tension. (Tecumseh Products Co.)

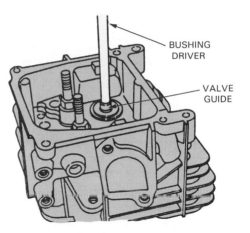

Fig. 16-63. Pressing valve guide out of cylinder head with a bushing driver or a flat punch. (Briggs and Stratton Co.)

tion, split-type retainers are used to secure the valve springs to the valve stems.

When removing the valve springs, note that the coils are closer together on one end of the spring than on the other. These are called *dampening coils* and they should be located opposite the valve cap and retainers, Fig. 16-62.

SERVICING OVERHEAD VALVES, SEATS, AND GUIDES

The valves, seats, and guides used in overhead valve systems are serviced in the same way as those in conventional systems. Valves should be cleaned and resurfaced to a 45° angle (or a 30° angle) on a valve grinding machine.

Valve seats can be reconditioned with a valve seat cutting tool. Valves should be lapped if recommended by the manufacturer. Thoroughly clean lapping compound from valve seats and faces. Inspect, measure, and test valve springs. Replace any parts that do not meet specifications.

Measure intake and exhaust valve guides. See Fig. 16-33. If dimensions are not within specifications, the guides must be replaced. To remove worn guides, use a bushing driver or flat-ended pin punch. Support the cylinder head and press the guides out, Fig. 16-63. When pressing new guides into a cylinder head, press only to the specified depth, Fig. 16-64. This dimension will vary from one engine model to another.

INSTALLING OVERHEAD VALVES

Before starting assembly, inspect valve stems for foreign material and burrs, which can cause

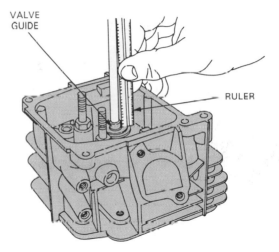

Fig. 16-64. Pressing new valve guides into cylinder head. Press to specified height above hole as shown. (Briggs and Stratton Co.)

sticking and damage the new stem seals. Coat the valve stems with valve guide lubricant. Do not allow the lubricant to contact the valve face, valve seat, or end of the valve stem.

Place the cylinder head on a workbench and support the valve faces with wooden blocks. Place the valve springs over the valve stems and set the retainers on the springs. Compress the springs and install the retainers, Figs. 16-65 and 16-66. If stem seals are used, place them over the stems as required. Do not attempt to install the rocker arms until after the cylinder head is installed on the engine.

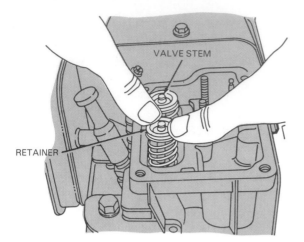

Fig. 16-65. Installing valve spring retainers.
(Briggs and Stratton Co.)

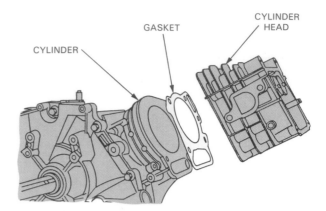

Fig. 16-67. Use a new gasket when installing the cylinder head on the block. Head and block surfaces must be clean. (Briggs and Stratton Co.)

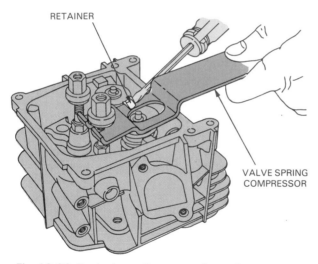

Fig. 16-66. Replacing split-type retainers. A magnetized screwdriver or a bit of grease helps to place retainer onto the valve stem recess. (Briggs and Stratton Co.)

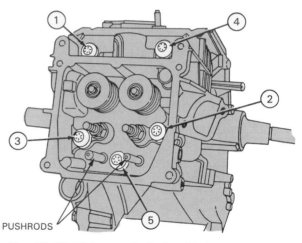

Fig. 16-68. Tighten cylinder head bolts with a torque wrench in proper sequence and in gradual increments to avoid head warpage. Place pushrods in proper holes.

INSTALLING CYLINDER HEAD

The mating surfaces of the cylinder and the cylinder head should be completely clean. Use a new head gasket and place the cylinder head on the cylinder, Fig. 16-67. *Never use gasket cement or sealer on a head gasket.* Lubricate the cylinder head bolt threads with oil. Install the bolts through the head and into the cylinder block holes. Tighten the bolts evenly by hand and, then, use a torque wrench to tighten them to the proper torque specifications, Fig. 16-68. Torque the head bolts in sequential increments to avoid caus-

ing the cylinder head to warp. Finally, place the pushrods into their respective holes.

INSTALLING ROCKER ARMS

Place the rocker arms on the studs and install the rocker arm nuts. Turn the nuts until they just touch the rocker arms. Carefully rotate the crankshaft to verify proper pushrod operation.

ADJUSTING VALVE CLEARANCE

Proper clearance between the rocker arm and the valve stem is essential. Too much clearance will reduce volumetric efficiency. Too little clearance can cause valve burning or warpage.

Before checking valve clearance, position the piston as recommended by the manufacturer. To accomplish this, simply rotate the crankshaft until the piston reaches the position specified by the manufacturer. Top Dead Center may be the correct piston position for some engines; others may require the piston to be a certain distance beyond Top Dead Center. Always check specifications. If necessary, the distance past Top Dead Center can be measured through the spark plug hole with a ruler, dial indicator, or similar tool. Once the piston position is attained, place the proper feeler gage leaf between the rocker arm and the valve stem, Fig. 16-69. Check engine specifications for the required clearance. Some engines require equal clearance for both intake and exhaust valves. However, some engine manufacturers use a different metal for exhaust valves than intake valves, and, therefore, the coefficient of thermal expansion is not the same for each valve. Therefore, clearances must be different for each valve.

Turn the locking/adjusting nut clockwise to reduce clearance or counterclockwise to increase clearance. The feeler gage should drag slightly when pulled out. Hold the adjusting nut with a wrench and tighten the locking screw slightly. Recheck clearance with the feeler gage. If necessary, readjust until correct clearance is obtained. Tighten the locking screw.

Replace the valve cover and gaskets, Fig. 16-70. Tighten the valve cover bolts to the recommended torque setting. Overtightening the valve cover bolts can warp the cover flanges, causing oil to leak. Be careful! See Fig. 16-71.

PORTS, REEDS, AND ROTARY VALVES

Two cycle engines generally use *porting* of the cylinder wall (instead of poppet valves) in the fuel-feed and exhaust systems. Porting basically consists of two holes (ports) in the cylinder wall. One port admits the air-fuel mixture and the other port allows exhaust gases to escape.

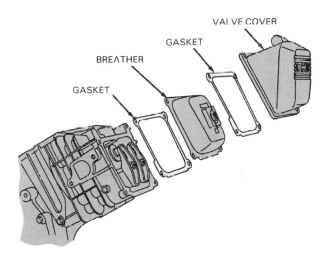

Fig. 16-70. Replacing valve cover. All mating surfaces should be clean and new gaskets should be used. (Briggs and Stratton Co.)

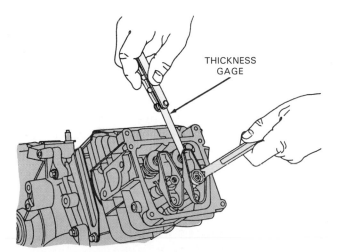

Fig. 16-69. Adjusting rocker arm-to-valve stem clearance. Very slight drag should be felt on the thickness gage. (Briggs and Stratton Co.)

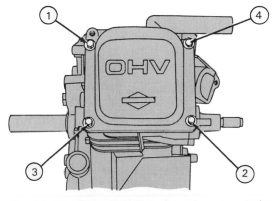

Fig. 16-71. When installing the valve cover bolts, tighten them in the proper sequence. Tighten to specified torque to avoid warping the cover flanges. (Briggs and Stratton Co.)

Ports have no service requirements other than keeping them free of carbon. Unlike poppet valves, they cannot be replaced since they are part of the cylinder. *Reed valves* are used in two cycle engines to control fuel flow from the carburetor to the crankcase, which serves as a second combustion chamber. A reed valve, operating on vacuum, opens during the intake stroke of the piston and closes before the start of the power stroke.

A reed valve is usually mounted on a plate to separate it from the carburetor. Fig. 16-72 shows a reed valve assembly used in an outboard engine.

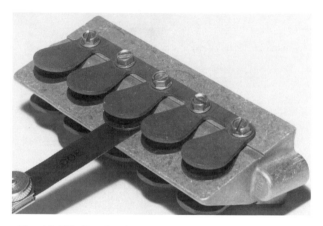

Fig. 16-72. Reed valves are not apt to stick shut if they are adjusted to stay open from .005 to .010 in. when at rest.

The openings through the reed plate and the reeds must be kept clean. The surface of the plate must be smooth so that there is a good seal when the reed closes against it. If the reed valve has lost its springiness or tension, install a new one. High-speed, two cycle engines use reed stops to prevent distortion and damage to the reeds, Fig. 16-73. If the reeds are bent, they must be replaced.

Some two cycle engines use rotary valves (instead of reeds) to control air-fuel intake. These valves are generally attached to the end of the crankshaft, although some run on a separate shaft geared off of the crankshaft. In operation, a valve rotates against a wear plate. Both the valve and plate have holes in them. When the holes align, an air-fuel charge enters the crankcase.

Inspect rotary valve ports for wear or damage. Replace defective parts. Keep openings clean and see that the surface of the wearplate is smooth and flat. If the wearplate or the mating surface of the valve is pitted or bent, replace the assembly. If a spring holds the parts in contact, check it against specifications for proper length and tension.

CAMSHAFTS AND GEARS

The *camshaft* of an engine is designed to operate the valves. A single camshaft is used in most

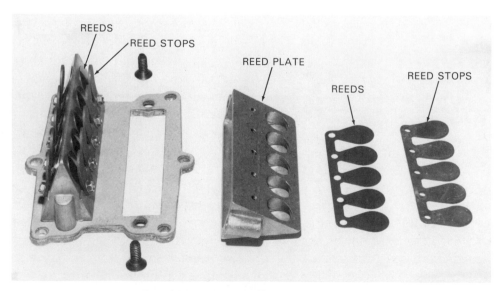

Fig. 16-73. A partially disassembled reed valve assembly shows ease with which parts can be replaced.

small engines, with a cam (lobe) for each valve. When the camshaft rotates, the lobe of the cam lifts the valve from its seat.

Camshafts are made of steel or cast iron. The surface of the shaft is hardened to improve wearability. The ends of the camshaft may turn in bearings or in the block metal. See Fig. 16-74A. Some small engine camshafts are hollow and have a second shaft running through them, Fig. 16-74B. With this setup, the inner shaft is fixed and the hollow camshaft revolves on it.

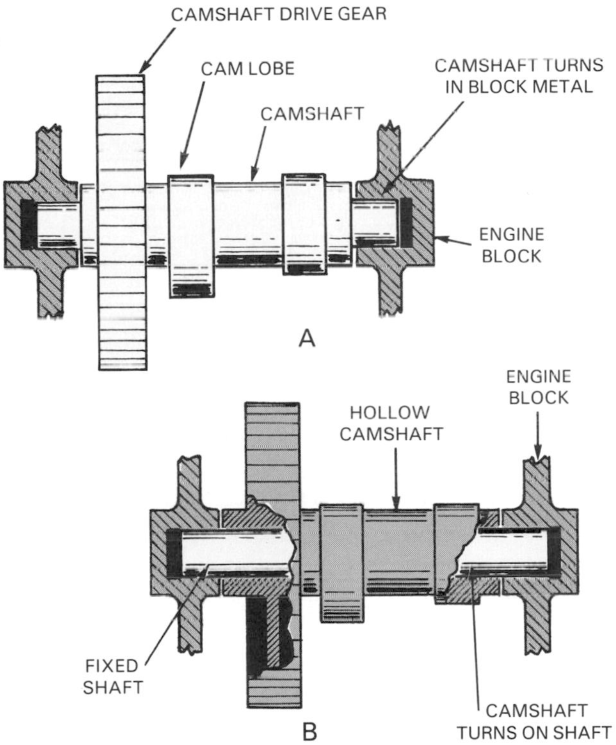

Fig. 16-75. Camshaft gear is meshed with crankshaft gear so that timing marks are aligned. Camshaft turns at half crankshaft speed. (Deere & Co.)

Fig. 16-74. Typical small engine camshafts. A–Solid camshaft. B–Hollow camshaft turning on a fixed shaft.

Most small gasoline engines use gears to turn the camshaft. A gear on the crankshaft meshes with and drives a gear on the camshaft. Since the camshaft gear is exactly twice the size of the crankshaft gear, it runs at half crankshaft speed, Fig. 16-75. Also note that the timing mark on the cam gear is aligned with the keyway on the crankshaft. This is the correct procedure for timing valve operation to the crankshaft on this particular engine.

AUTOMATIC COMPRESSION RELEASE

To make hand cranking easier, some small engines have an *automatic compression release* mechanism on the camshaft. This device lifts the exhaust valve slightly during cranking and releases part of the compression pressure.

One manufacturer's compression release mechanism is pictured in Fig. 16-76. In view A, the camshaft is at rest and springs are holding the flyweights in. In this position, the tab on the larger flyweight sticks up above the base circle of the exhaust cam, holding the exhaust valve partially open. In view B, the tab prevents the exhaust lifter from resting on the cam.

After the engine starts and its speed reaches about 600 rpm, centrifugal force overcomes spring pressure and the flyweights move outward. Movement of the flyweights causes the tab to be retracted, and the exhaust valve seats fully. See views C and D in Fig. 16-76. The flyweights remain in this position until the engine is stopped.

Fig. 16-77 shows the compression release mechanism in the starting position (A) and the running position (B). Automatic compression release is one of the many advances in small engines that ease the chore of engine start-up.

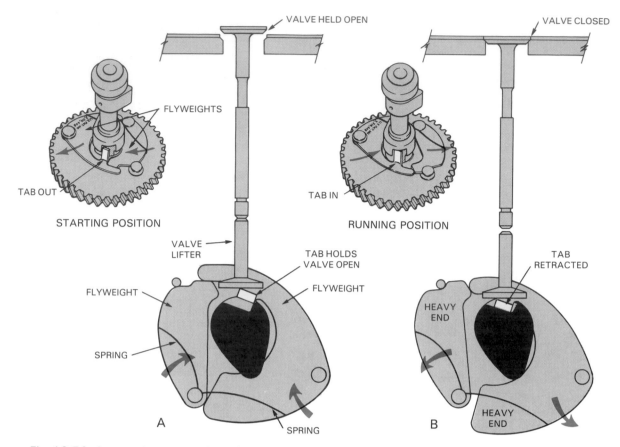

Fig. 16-76. Automatic compression release makes cranking easier. A–Tab is out, preventing valve from closing completely. B–When engine starts and reaches 600 rpm, the flyweights move out, the tab retracts, and the valve functions normally. (Kohler Co.)

Fig. 16-77. Automatic compression release. A–Starting position. B–Running position.

SUMMARY

The connecting rod attaches the piston to the crankshaft. There are three types of friction bearings commonly used in the large end of the con-necting rod: rod metal bearings, cast bronze bearings, and precision insert bearings. Many small engines use antifriction bearings in the large end of the connecting rod.

The crankshaft converts the reciprocating motion of the piston into rotary motion. To help offset the unbalanced condition created by the force of reciprocating mass, counterweights are added to the crankshaft. The crankshaft is supported by one or more main bearings. Crankcase seals prevent leakage of oil from areas where the crankshaft and crankcase come together.

A four cycle engine's valve assembly includes the valve, the valve spring, and one or more retainers. After valves are removed, clean and inspect them for defects. Valves with serious defects must be replaced. Valve springs should be checked for squareness, length, and tension. Replace all springs that are not within specifications.

Check valve guides with a small hole gage. If clearance between guide and stem exceeds the allowable limit, enlarge the guide with an adjustable reamer. A new valve with an oversize stem must be installed.

Valve seats are generally cut to a 45° angle. Seat contact must be near the center of the valve face. A valve seat cutter is used to recondition seats. Good used valves can be reseated by a hand-lapping process.

Valve clearance refers to the space between the end of the valve stem and the top of the lifter. Valves must be closed when measuring clearance. If there is too little clearance, a valve may be held open when the stem expands.

Valve refacing is done on a specially designed grinder. Valve clearance is reduced when a valve is refaced. Therefore, the tappets must be adjusted or the valve stem end must be ground to obtain the correct clearance.

Overhead valve systems transmit motion through pushrods and rocker arms to open and close the valves. These systems improve volumetric efficiency and eliminate hot spots.

Two cycle engines generally use intake and exhaust ports instead of poppet valves. Ports have no service requirements other than keeping them free from carbon.

The camshaft is designed to operate the valves. When the camshaft rotates, the lobe of the cam lifts the valve from its seat. To make hand cranking easier, some engines have an automatic compression release mechanism on the camshaft. This device lifts the exhaust valve during cranking to release compression pressure.

KNOW THESE TERMS

Connecting rod, Friction bearings, Bearing spread, Bearing crush, Antifriction bearings, Rod cap, Crankshaft, Crankshaft throw, Counterweights, Main bearings, Bearing clearance, Plastigage, Crankcase seals, Poppet valves, Valve keepers, Valve springs, Valve guides, Valve seat angle, Valve seat width, Interference fit, Lapping, Valve clearance, Valve refacing, Overhead valves, Pushrod, Rocker arm, Porting, Reed valve, Camshaft, Automatic compression release.

REVIEW QUESTIONS–CHAPTER 16.

1. Properly fitted friction bearing ends protrude slightly above the parting surface of the connecting rod cap. This characteristic products what is commonly called:
   a. Bearing crush.
   b. Bearing spread.
   c. Bearing seat.
   d. Bearing swell.
2. Bearing caps must never be _____ _____ when being replaced on the rods.
3. What tool must always be used to tighten rod caps?
4. Name three types of crankshaft main bearings.
5. What do you call the special plastic substance used to measure bearing clearance?
6. When removing the crankcase cover from the engine:
   a. Always pry it loose with screwdrivers.
   b. Pry out the oil seal first
   c. Place tape over the keyway to protect the oil seal.
   d. Hold the cover while hammering on the crankshaft end.
7. When replacing oil seals, the knife edge of the seal lip should face:
   a. Fluid being sealed.
   b. Away from fluid being sealed.
8. Valve margins should not be allowed to be less than:
   a. 1/64 in.
   b. 1/32 in.
   c. 1/16 in.
   d. 3/32 in.
9. Valve seats that are too wide will:(select correct answers)
   a. Cause valves to stick closed
   b. Cause valves to stick open.
   c. Transfer too much heat to the block.
   d. Warp and burn.
10. What is the 1° difference between the valve face angle and the valve seat angle called?
11. Name the process of placing abrasive compound on the valve face and twirling it back and forth in the valve seat.

12. Too little valve clearance will cause the valve to:
    a. Break.
    b. Be noisy.
    c. Burn.
    d. Open late.
13. The camshaft revolves at:
    a. Twice crankshaft speed in all engines.
    b. Four times crankshaft speed in four cycle engines.
    c. One-half crankshaft speed in four cycle engines.
    d. One-half crankshaft speed in two cycle engines.
14. An automatic compression release is used on some engines to:
    a. Make cranking easier.
    b. Prevent knocking due to excessively high compression.
    c. Control speed.
    d. Prevent overheating.
15. What are the advantages of overhead valves?
16. Why should parts be identified as they are disassembled?
17. How should pushrods be checked for straightness?
18. When the coils are closer together on one end of a valve spring than on the other, they are called _____ _____.
19. When new or reconditioned valves are being installed in guides, what should be placed on the valve stems first?
20. What tool is used to check valve clearance?
21. Why should valve cover screws never be tightened excessively?

SUGGESTED ACTIVITIES

1. Measure crankshaft bearing clearances with Plastigage and telescoping gages.
2. Install new main and rod bearing inserts. Observe rules of cleanliness and torque tighten rod bolts to specified value.
3. Replace oil seals in the crankcase.
4. Grind valves on a grinding machine or turn and true valve faces on a valve lathe.
5. Recondition old valve seats with a valve seat cutter.
6. Lap valves into seats after the valve faces and seats are reconditioned by grinding.
7. Test valve springs for length, straightness, and tension.
8. Ream valve guides to fit an oversize valve stem with proper clearance.
9. Adjust valve lifter to valve clearance by grinding valve stems or adjusting tappets.
10. Adjust valve clearances in an overhead valve assembly.
11. Time the camshaft to the crankshaft.

CHAPTER 17

SMALL GAS ENGINE APPLICATIONS

After studying this chapter, you will be able to:
☐ List features to consider when purchasing a lawn mower.
☐ Summarize basic lawn mower maintenance procedures and safety precautions.
☐ Describe proper method for storing a lawn mower for long periods of time.
☐ List the features to consider when purchasing a chain saw or an edger/trimmer.
☐ Summarize the maintenance, safety, and storage procedures for chain saws and edger/trimmers.

This chapter will examine a few of the more common small gas engine powered implements. It will discuss purchasing considerations, safety features, service warnings, and maintenance methods. Since there are hundreds of specialized powered implements, you should always study the owner's manual before operating and servicing these devices.

⚠ Lawn mowers have caused a great number of injuries. The more severe injuries were lacerations to hands and feet and injuries from objects thrown from under the mower housing or out of the discharge opening.

Recently, laws have been enacted forcing manufacturers to provide certain safety devices on lawn mowers to help prevent accidents.

Like many other implements, there is not an age, skill, or intelligence requirement for using a lawn mower, However, young children should never be allowed to operate a power mower. The human body is no match for a sharp steel blade rotating at hundreds of miles per hour. Objects have reportedly been thrown a distance of one quarter of a mile by a power mower.

A grass discharge chute guard is shown in Fig. 17-1. It is an important safety device that can prevent lawn mower injuries.

PURCHASING A LAWN MOWER

Because of the large variety available, certain considerations should be made when purchasing a power mower. The operator's strength may determine whether the mower should be the push type or the self propelled type. *Rotary mowers* (those having horizontally rotating blades) for large areas have large diameter blades with large, heavy engines and housings. Also, safety and accessory items may add to mower weight. Remember, a mower may roll easily on a smooth

Fig. 17-1. Grass discharge chute guards, also called stone guards, are designed to deflect objects toward ground. Guards must always be kept in place and in good working order.

floor. However, it may be difficult to push on a deep, rough lawn when the grass catcher is loaded with grass.

If you have a super-quality, "showpiece" lawn that is very level, you may want to consider a *reel-type mower* (blades rotate down vertically). This is the type used on golf course greens. They produce a high-quality job but are more difficult and expensive to maintain. Special equipment is needed to sharpen the reel. Adjustments of the reel blades to the cutter bar are critical, and occasional readjustments are necessary as the blades wear. Cutter height is controlled by raising or lowering the rollers.

For the average size yard, a rotary-type mower with blade diameter of 22 inches is usually satisfactory. For small yards, a 20 inch mower is more maneuverable and takes less storage space. The length that the grass is cut to is controlled by raising or lowering the entire mower through adjustments on the wheels. See Fig. 17-2.

PUSH-TYPE AND SELF-PROPELLED MOWERS

Selecting a push-type or self-propelled mower is a matter of personal preference. Engines for self-propelled mowers have to rotate the blade and drive mechanism to the wheels, Fig. 17-3. This requires additional engine horsepower and a drive mechanism to the wheels. In some mowers,

power is transmitted to the wheels through shafts and gears; others use a belt and pulley system, Fig. 17-4. Greasing bearings and gears and replacing belts are additional maintenance requirements for a self-propelled mower.

Fig. 17-3. Self-propelled mowers are commonly driven by friction rollers that engage lawn mower wheels.

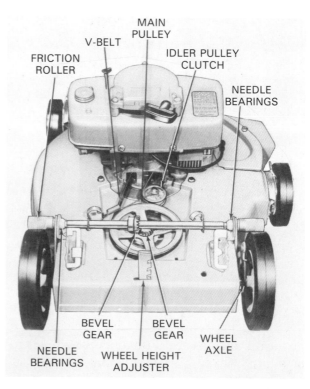
Fig. 17-4. This self-propelled mower uses a V-belt pulley system with an idler pulley clutch to tighten the belt. The main pulley drives the bevel gears to turn the drive shaft. Friction rollers turn the mower wheels.

Fig. 17-2. Each wheel on this mower has an adjuster to raise or lower housing and blade for desired length of grass. Always stop engine before making these adjustments.

TWO CYCLE OR FOUR CYCLE ENGINE

As discussed earlier in this text, two cycle engines have fewer moving parts than four cycle engines and require a proper mixture of oil and gasoline. See Chapter 11 for information on mixing oil and fuel. Four cycle engines retain the oil in the crankcase and only gasoline is placed in the fuel tank. Crankcase oil must be drained and replaced at recommended intervals.

Both types of engines are quite reliable but be aware of the type you are purchasing. If it is a two cycle engine, purchase some two cycle oil, Fig. 17-5A. You should also have an approved type of gasoline can that is labeled so you do not confuse it with another can of plain gasoline. Without the oil in the fuel, the two cycle engine will be ruined in only a few moments of running.

Fig. 17-5. A–Two cycle engine oil. B–Fuel stabilizer for extended storage. Both can be obtained from local implement dealers.

OPTIONAL FEATURES

One optional feature for mowers is side-bagging, rear-bagging, or no bagging at all. Grass bags may be fabric or molded plastic containers. Side-bagging mowers tend to extend the width of the mower but may be a little easier to remove and install. If mowing in narrow quarters, a rear bag may be slightly more maneuverable, particularly if the mower must be backed out of the space.

Using either a side or rear bag reduces thatch (compacted dead grass cuttings) in the lawn.

Fig. 17-6. Never have engine or blade running when grass catcher is removed. On this mower, blade brake and clutch stop blade while allowing engine to continue running.

Thatch, if allowed to accumulate, tends to smother the lawn and create growth problems.

⚠ WARNING: To prevent serious injury, always stop the engine or blade when emptying the grass catcher. Never start the engine or blade without the grass catcher in place, Fig. 17-6.

Grass cuttings are excellent additions for the compost pile or garden. Some prefer to use a bagless mower and rake the grass cuttings after mowing. Again, the cuttings must be removed. Some mowers have blades and housings designed to cut and mulch (cut into fine particles) grass and leaves for composting.

Another optional accessory is a *dethatcher blade,* Fig. 17-7. This can be purchased to fit any mower and is installed in place of the cutting blade. There are spring-like fingers attached to the blade that reach down into the lawn. As the blade turns, the fingers rake the thatch up to the surface. The thatch is then collected in the grass bag or hand raked later.

CONVENIENCES

Most manufacturers try to make bag removal and replacement as easy as possible. The bag usually must be removed a number of times before mowing is completed. This means stopping the

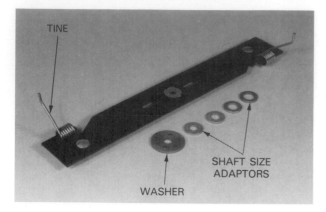

TINE

SHAFT SIZE
ADAPTORS

WASHER

Fig. 17-7. A dethatcher blade uses spring tines (pointed prongs) to rake out dead grass. Adaptors with various hole sizes permit bar to fit on any mower shaft diameter.

mower, removing the bag, emptying the bag, replacing the bag, and restarting the mower. This is enough work without having to "wrestle" with a clumsy bag connection. The bag should also be durable and not wear on chaffing points.

Ease of cleaning the mower, particularly underneath, is another convenience. It is good practice to let the mower cool after mowing; then tip it on its side and wash out the residue of grass with a garden hose. See Fig. 17-8. If grass is allowed to build up on the blade and in the housing, it will

dry and become very difficult to remove. Also, cutting efficiency will be lessened. The upper parts of the mower and engine should also be kept clean to maintain engine cooling and proper functioning of the carburetor and governor parts.

Air filter maintenance (cleaning and oiling) should be easily done. See Figs. 17-9 and 17-10.

COVER

AIR FILTER

Fig. 17-9. Air filter must allow large amounts of air to enter the engine easily, while trapping dirt and debris. It should be easily removed for service.

Fig. 17-8. Bottom of mower should be kept clean for efficient cutting. Steel housings should be cleaned and painted occasionally to prevent rust.

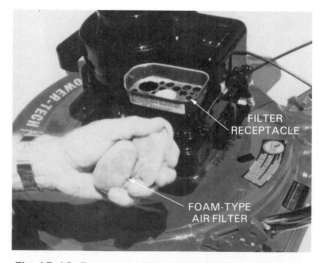

FILTER
RECEPTACLE

FOAM-TYPE
AIR FILTER

Fig. 17-10. Foam-type filter can be easily washed in detergent and water and squeezed dry. Saturate filter in oil and squeeze out excess before replacing.

Also, the spark plug should be easily reached for cleaning or changing as needed, Fig. 17-11. Mufflers, like those on any internal combustion engine, eventually deteriorate to the point where they must be replaced. Ease and cost of replacement may vary, Fig. 17-12.

The "business end" of the mower is the cutting blade. After prolonged use, it will become nicked and dull. There are many styles of blades. To remove the blade for sharpening, the mower is tipped on its side and several bolts are removed, Fig. 17-13. The mower should be set on its side so that it cannot fall upside down.

Fig. 17-13. Before removing a blade, always remove the spark plug wire and tie it away from the plug. Hands and knuckles should be protected with gloves while loosening or tightening bolts.

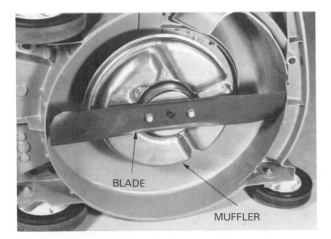

Fig. 17-11. Spark plug should be serviced or replaced at least once each season. A spark plug socket should be used to prevent damage to porcelain insulator.

WARNING: Always remove the spark plug wire and tie it back before tipping the mower. Wear gloves or wrap a cloth around the blade to protect hands from the sharp edges.

Storage and portability may be important. Today, many mowers are designed with handles that fold down so that little space is needed for storage, Fig. 17-14. If transporting the mower in a car, this may also be convenient so that you can shut and latch the trunk lid.

Fig. 17-12. Muffler on this mower requires removal of blade for replacement. It is tuned to the engine for quietness and efficiency.

Fig. 17-14. If storage space is limited, mower handle should fold down or be removed easily.

ENGINE STARTING

Various mechanical means have been devised for starting small engines. The simplest method has been the ***independent rope starter***. A rope with a knot in one end is wrapped around the flywheel pulley and given a quick pull. This design is no longer in use.

The ***recoil starter*** is common today. It utilizes a rope, a ratchet mechanism, and a rewind spring. When the rope is pulled, the ratchet engages the flywheel and rotates the crankshaft. When the engine starts, the ratchet disengages from the flywheel. The rewind spring retracts and recoils the rope for the next starting. The proper technique for starting is shown in Fig. 17-15.

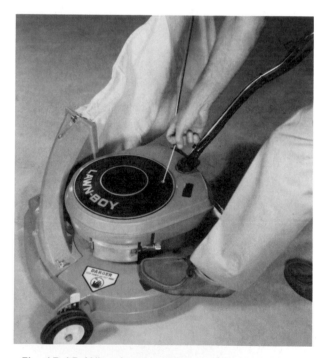

Fig. 17-15. When hand starting a vertical-pull engine, place one foot on the deck and the other foot away from the mower. Pull the rope briskly.

Inertia starters are also available. The inertia starter utilizes a coil spring attached to a ratchet mechanism and a crank or lever. The coil spring is wound tightly with the crank and held with a locking pin. When the control knob or lever is released, the coil spring engages the ratchet with the flywheel to start the engine, Fig. 17-16.

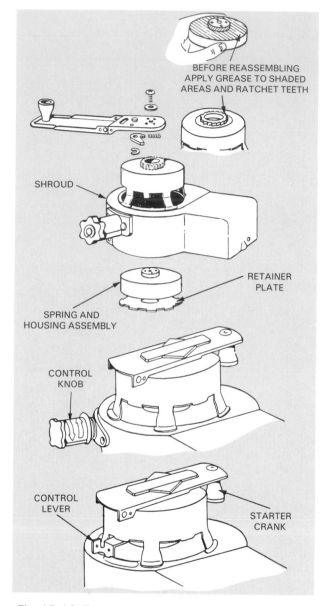

Fig. 17-16. The inertia starter is used by winding a coil spring with the starter crank. When ready to start the engine, turn the control knob or move the control lever. Spring will then rotate the engine flywheel for starting.

Electric starts are also available. The electric start mower has the added weight of a starter motor, switch, battery, and wiring, Fig. 17-17. A key is required to turn on the switch. A small battery charger is also needed to energize the battery after motor operation.

A recent innovation for convenience and safety is the "extended rope" starter. The advantage to this system is that the operator does not

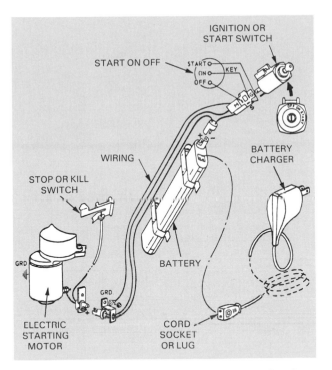

Fig. 17-17. An electric start system has an electric motor, battery, and switch, like an automobile. Some mowers have a hand start backup system in case the battery is discharged.

have to bend over as far and his or her feet are clear of the blade housing. See Fig. 17-18. Also, the blade brake must be released with one hand while the opposite hand pulls the rope.

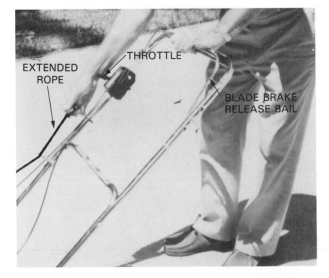

Fig. 17-18. Extended rope starting system helps keep hands and feet away from mower while starting.

BLADE BRAKES

One of the major safety features of every mower manufactured today is that the blade must automatically stop within three seconds after the operator's hands leave the handle. This requires a highly efficient braking system. There are basically two kinds of *blade brakes.*

In one system, a brake band wraps around and grips the flywheel to stop the engine and the blade, Fig. 17-19. The second system, as shown in Fig. 17-20, has a clutch release that allows the

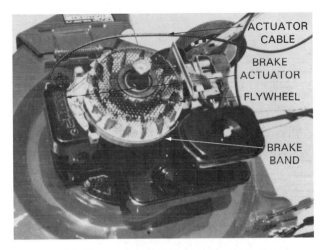

Fig. 17-19. To stop engine and blade within three seconds of release of handle, this engine has a flywheel brake. Cover has been removed for clarity to expose brake band around flywheel.

Fig. 17-20. This mower has a clutch that disengages the engine from the blade while the brake stops the blade. The engine can safely remain running when the operator releases the handle and bail.

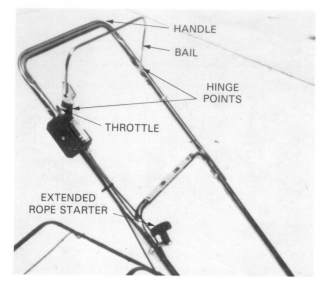

Fig. 17-21. Bail must be held against handle to release engine or blade brake.

Fig. 17-22. Check fuel level in tank before starting.

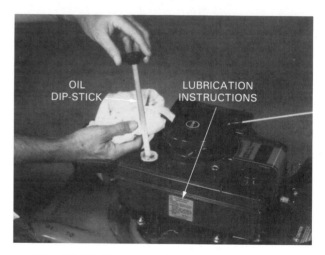

Fig. 17-23. This engine has an automotive-type dip-stick for checking oil level. Oil is added through dip-stick opening.

engine to remain running while a brake stops the blade. Both systems utilize a "bail" (hand lever) hinged to the handle. In the position shown in Fig. 17-21, the brake is engaged and the blade will not turn. When the bail is pulled into the handle, the blade begins to spin.

⚠ WARNING: In all cases, operators should thoroughly study the owner's manual and be familiar with all safety precautions and operating procedures before using a power mower.

PROCEDURE FOR STARTING AN ENGINE

The following is a basic procedure for starting a cold engine:

1. Fill fuel tank with proper fuel for engine type. Refer to Fig. 17-22.
2. Check oil level and condition of oil. Add or change if necessary, Fig. 17-23.
3. Prime engine with fuel, Fig. 17-24, or close choke.
4. Turn key on or advance throttle to START position, Fig. 17-25.
5. A–Vertical or Horizontal Rope Start: Place foot on mower deck and pull rope, Fig. 17-15.
   B–Extended Rope Start: Hold bail to handle and pull rope. Refer to Fig. 17-21.
   C–Electric Start: Turn key in switch past

Fig. 17-24. Depressing primer forces fuel into cylinder for quickly starting a cold engine. Engines without a primer pump utilize a choke.

RUN to START. Release key to RUN position as soon as engine starts, Fig. 17-26.

6. When engine starts, open choke and adjust throttle to operating speed (about 1/2 to 2/3 maximum speed). Refer to Fig. 17-25.

NOTE: For restarting a warm engine, priming or choking should not be necessary. A cold two cycle engine may need several prime pump strokes after the engine starts to keep it running.

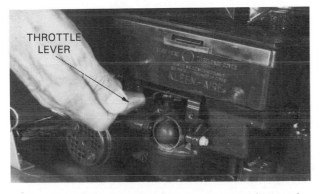

Fig. 17-25. Throttles have stop, start, and fast positions. Throttle lever is being moved to start position.

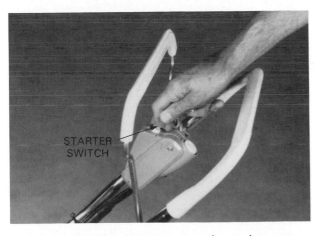

Fig. 17-26. Electric start mowers have a key-type switch. To start, insert key and turn to start position until engine begins to run. Release key immediately upon starting.

MINOR CHECKS

If engine idles too fast or too slow, adjust the idle speed screw on the carburetor with a screwdriver. The engine should idle slowly and smoothly. Refer to Fig. 17-27.

Fig. 17-27. An idle speed adjustment screw located on carburetor can be turned with a screwdriver to increase or decrease rpm.

Fig. 17-28. If engine will not start, remove spark plug. With spark plug wire about 3/16 inch from engine or mower, a blue spark should jump gap when engine is cranked. No spark indicates ignition system problem.

If the engine will not start, remove the spark plug as shown in Fig. 17-28. Test for spark as described in Chapter 13, Ignition System Service.

If a spark jumps the gap (3/16 in.) from the spark plug wire to the engine when the engine is cranked, replace the wire on the spark plug and ground the base of the plug against the engine (not close to gasoline tank), Fig. 17-29. Have someone hold the bail and crank the engine. Watch for spark between the electrodes. If there is no spark, service or replace the spark plug.

If the engine still does not start, refer to the troubleshooting chart near the end of this chapter.

Fig. 17-29. If a spark jumps the gap described in Fig. 17-28, connect wire to spark plug and ground plug to the engine. A bright blue spark should jump the electrode gap. If no spark is present, replace the spark plug.

GENERAL MAINTENANCE

As mentioned earlier, cleaning the mower after each use is an important maintenance procedure. Additional maintenance procedures will be described here.

BLADE SHARPENING

When mower blades become dull and nicked, it makes the engine work harder and the lawn is cut poorly. Blades can be sharpened by clamping them to a table or in a vise and filing. Always retain the same edge angle. Blades can also be sharpened on a grinding wheel, as in Fig. 17-30.

 WARNING: When grinding a blade, always wear safety goggles or glasses with side shields.

When sharpening the blade, balance it at the same time. Grind or file a small amount off of the heavy end until the blade balances horizontally. Test the blade by using a blade balancer, Fig. 17-31, or balance on a sharp edge, as in Fig. 17-32. An unbalanced lawn mower blade will create damaging engine vibration.

Hitting solid objects with the blade, such as large rocks, cement edges, pipes, etc., can be very damaging to a mower, Fig. 17-33. Besides damage to the blade, the engine crankshaft can be

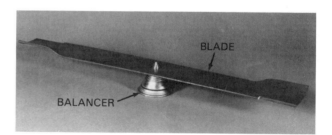

Fig. 17-31. This blade balancer is being used to test blade balance. An unbalanced blade can cause vibration and shorten engine life.

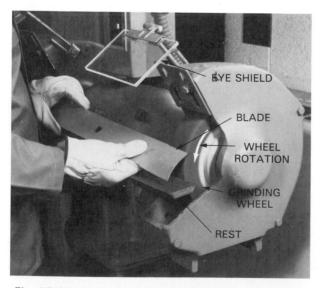

Fig. 17-30. Proper procedure for grinding cutting edges of blade. Note wheel rotation direction and position of blade. Always wear safety glasses with side shields.

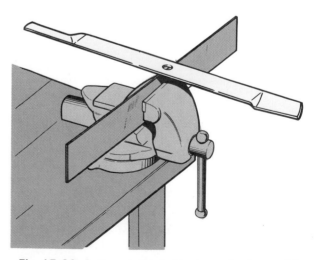

Fig. 17-32. A sharp-edged object can also be used for blade balancing. Center the blade hole over sharp edge.

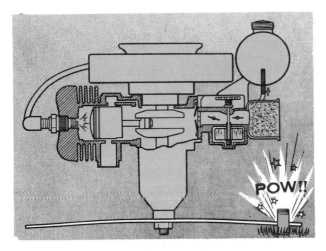

Fig. 17-33. Hitting solid objects with the blade can severely damage an engine.

bent, crankcase housings distorted, connecting rods bent, and flywheel keys sheared.

SPARK PLUG

A spark plug is normally easy to change. The spark plug should be carefully removed with a spark plug socket and ratchet wrench so the porcelain insulator is not damaged, Fig. 17-11.

Spark plugs can be cleaned and gapped as described in Chapter 13. When replacing the spark plug, clean the seat and do not overtighten the plug. Spark plugs should be serviced about every one-hundred hours of operation.

When replacing the spark plug, use only the kind specified for the engine with the proper reach (thread length) and heat range (insulator tip length).

AIR CLEANERS

Engines consume a tremendous amount of air, which passes through the air cleaner. When mowing, dust and dry grass materials are trapped by the filter. When too much debris collects in the air filter, it can restrict airflow. Therefore, the air cleaner element should be cleaned frequently, about every twenty-five hours of operation. Lawn mowers generally use the oil-foam type element, which is a spongy plastic saturated with oil. Refer to Figs. 17-9 and 17-10.

To clean the oil-foam filter:
1. Remove the filter cover.
2. Remove the foam element from the base.
3. Wash the element in kerosene or liquid detergent and water. Follow the steps in Fig. 17-34.
4. Wrap the foam in cloth and squeeze dry.
5. Saturate the foam with engine oil.
6. Squeeze out the excess oil.
7. Replace the foam and filter cover.
For other filter types, see Chapter 5.

CHANGING OIL

Four cycle engines should have the oil changed about every twenty-five hours of operation. To drain oil from lawn mower engines:
1. Place a pan under the mower housing.
2. Locate and remove the oil drain plug.
3. Drain all of the oil and replace the drain plug.
To replace oil:
1. Obtain the kind of oil recommended by the manufacturer of the engine.
2. Pour the oil into the oil-fill neck that is provided and labeled. Fill to the level indicated by the engine manufacturer. Some engines have dip-sticks. Fill to the "Full" line, as shown in Fig. 17-23.
3. Replace the filler cap.

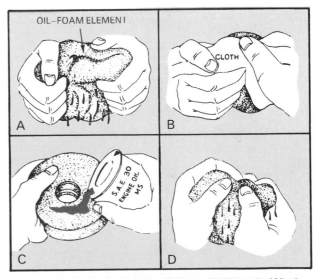

Fig. 17-34. Cleaning the oil-foam air filter. A–Wash foam element in kerosene or liquid detergent and water. B–Wrap foam in cloth and squeeze dry. C–Saturate foam with clean oil. D–Squeeze out excess oil.

MUFFLER REPLACEMENT

Engine mufflers get extremely hot and are subjected to the acidic products of combustion. They eventually need replacement, Fig. 17-35. There are many styles and shapes of mufflers.

Fig. 17-12 shows a rather large, extensive muffler that requires removal of the blade before it can be removed.

The mufflers in Fig. 17-35 simply require unscrewing the tapered pipe threads. Coat the threads with antiseize compound prior to replacement. Other mufflers are removed by loosening two bolts with a wrench.

Fig. 17-35. Most mufflers are inexpensive and easy to replace. Combustion products and heat cause mufflers to corrode.

V-BELTS

V-belts are a widely used means of power transmission, particularly on riding-type mowers. See Fig. 17-36. There may be four or five separate belts on a small tractor mower. V-belts may look very much alike, but there are many unseen differences that may affect their quality and life span.

Some belts have a fabric cover; others do not. The cords embedded inside the belt may be rayon, polyester, or Kevlar. Rayon will not last as long as the tremendously strong Kevlar. Polyester tends to shrink as it gets hot. Polyester could present a danger on a clutch, causing unintentional engagement due to shrinkage.

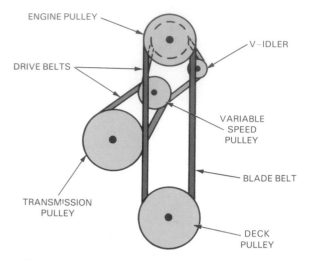

Fig. 17-36. This V-belt system for a riding mower illustrates multiple belting and pulleys.

Use only the type and size belt specified by the manufacturer. V-belt failures are usually a result of some other failure, such as those shown in Fig. 17-37.

CAUSE OF FAILURE	CORRECTION
1. Normal wear.	Replace belt.
2. Poor operating habits.	Do not engage and disengage clutch excessively.
3. Damaged or worn idler pulleys.	Replace idler, frozen bearings, and belt.
4. Incorrectly positioned belt guards.	Realign guards. Replace damaged guards.
5. Misaligned pulleys.	Replace pulleys.
6. Damaged or worn pulleys.	Align pulleys (except where an offset system is used with special pulleys).
7. Incorrect tensions.	Replace belt. Check idler spring tension. Lubricate idler brackets.
8. Oil and grease damage.	Replace belts. Eliminate oil leakage. Use oil resistant belts if possible.
9. Heat damage (140° or higher).	Use heat resistant belt. Avoid polyester belts. Shield belts from heat source.
10. Incorrect installation.	Install with care. Never use force. Recheck belts after 48 hours of use.

Fig. 17-37. Study typical causes and corrections for V-belt failure.

WARNING: Always keep hands and objects away from any exposed belts when engine is running. Remove spark plug wire when servicing belts. Always keep belt guards in place when operating implements.

STORING POWER LAWN MOWER

Several things can be done when preparing to store a lawn mower for extended periods. Proper preparation will help ensure long mower and engine life. It will also ensure easy starting the following season.

1. Clean the grass bag and hang it in a dry location.
2. Clean the mower of grass cuttings, mud, etc.
3. Avoid storing gasoline for long periods of time. Store only in approved "safety" containers. Never store fuel or mower in an enclosure where there is an open flame. If fuel must be stored, add a stabilizer to it. A fuel stabilizer is available from implement dealers.
4. Try to plan ahead and run the engine dry of fuel at the last use.
5. Drain the oil from the crankcase. Do not refill it now. See Chapter 11 for additional information on draining oil. Place a tag on the engine that says NO OIL.
6. Rotate the engine so the piston is at bottom of the cylinder. Remove the spark plug. Spray storage oil through the spark plug hole, or squirt about one tablespoon of clean motor oil through the spark plug hole with an oil can. Rotate the engine slowly several times to distribute the oil on the cylinder walls. Replace the spark plug.
7. Leave the spark plug lead disconnected. Using the pull rope, rotate the engine slowly until compression resistance is felt. Then rotate the engine an additional one-quarter turn to close off its ports. This seals the cylinder and prevents moisture entry.
8. Leave the throttle in the off position (closed) and close the choke.
9. Lubricate the mower, drive system, etc., as described by the manufacturer.

10. Coat the cutting blade with chassis grease to prevent rusting.
11. Store the mower in a dry, clean area.

REMOVING MOWER FROM STORAGE

1. Replace the grass bag.
2. Fill crankcase with new oil, or make new mixture for two cycle engine. See Chapter 11 for additional information on mixing oil and fuel.
3. Remove the spark plug. Using the pull rope or starter, spin the engine rapidly to remove excess oil from the cylinder. Clean or replace the spark plug.
4. Clean and oil the air filter if necessary.
5. Fill fuel tank.
6. Start the engine and idle until warm. Adjust the idle speed if necessary.
7. Increase the engine speed in normal manner.
8. Make a brief test run while listening to the engine and watching the condition of all parts.
9. If the engine does not start, review the troubleshooting chart in Fig. 17-38 and the other chapters of this text.

CHAIN SAWS

Small gasoline engine powered *chain saws*, Fig. 17-39, have become very popular for cutting firewood and trimming trees. Like any cutting tool, they work best when properly maintained. To avoid hazards, all safety devices should be in place and safe operating procedures must be carefully followed.

PURCHASING CONSIDERATIONS

When investing in a chain saw, there are a number of things to consider so that you will be satisfied with its performance and operation. Chain saws are manufactured in a number of sizes from about 10 inches to over 40 inches of blade length. The type of work to be done is the best indicator for size selection. A smaller saw works well for cutting branches, small trees, and fireplace logs. A large professional model would be for big trees and continuous rugged work.

TROUBLESHOOTING CHART

PROBLEM	CAUSE	REMEDY
1. Engine fails to start.	A. Blade control handle disengaged. B. Check fuel tank for gas. C. Spark plug lead wire disconnected. D. Throttle control lever not in the starting position. E. Faulty spark plug. F. Carburetor improperly adjusted, engine flooded. G. Old, stale gasoline. H. Engine brake engaged.	A. Engage blade control handle. B. Fill tank if empty. C. Connect lead wire. D. Move throttle lever to start position. E. Spark should jump gap between center electrode and side electrode. If spark does not jump, replace the spark plug. F. Remove spark plug. Dry the plug. Crank engine with plug removed, and throttle in off position. Replace spark plug and lead wire and resume starting procedures. G. Drain and refill with fresh gasoline. H. Follow starting procedure.
2. Hard starting or loss of power.	A. Spark plug wire loose. B. Carburetor improperly adjusted. C. Dirty air cleaner.	A. Connect and tighten spark plug wire. B. Adjust carburetor. See separate engine manual. C. Clean air cleaner as described in separate engine manual.
3. Operation erratic.	A. Dirt in gas tank. B. Dirty air cleaner. C. Water in fuel supply. D. Vent in gas cap plugged. E. Carburetor improperly adjusted.	A. Remove the dirt and fill tank with fresh gas. B. Clean air cleaner as described in separate engine manual. C. Drain contaminated fuel and fill tank with fresh gas. D. Clear vent or replace gas cap. E. Adjust carburetor. See separate engine manual.
4. Occasional skip (hesitates) at high speed.	A. Carburetor idle speed too slow. B. Spark plug gap too close. C. Carburetor idle mixture adjustment improperly set.	A. Adjust carburetor. See separate engine manual. B. Adjust to .030″. C. Adjust carburetor. See separate engine manual.
5. Idles poorly.	A. Spark plug fouled, faulty, or gap too wide. B. Carburetor improperly adjusted. C. Dirty air cleaner.	A. Reset gap to .030″ or replace spark plug. B. Adjust carburetor. See separate engine manual. C. Clean air cleaner as described in separate engine manual.
6. Engine overheats.	A. Carburetor not adjusted properly. B. Air flow restricted. C. Engine oil level low.	A. Adjust carburetor. See separate engine manual. B. Remove blower housing and clean as described in separate engine manual. C. Fill crankcase with the proper oil.
7. Excesssive vibration.	A. Cutting blade loose or unbalanced. B. Bent cutting blade.	A. Tighten blade and adapter. Balance blade. B. Replace blade.

Fig. 17-38. Study this engine troubleshooting chart.

The lighter a chain saw is, the easier it is to handle, control, and carry. Chain saws range in weight from less than 10 pounds to over 25 pounds. Weight is a consideration that should be planned according to the type of work to be done. When trimming tree branches from a ladder, a lighter saw will be less tiring.

Chain saws also come with different size engines. Rugged work and long continuous cutting require powerful engines.

Some states have laws that require certain safety devices to be used on every chain saw. Check which devices are required and be sure the saw is properly equipped.

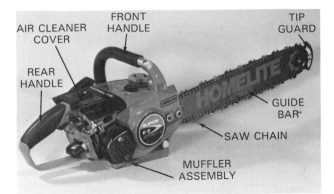

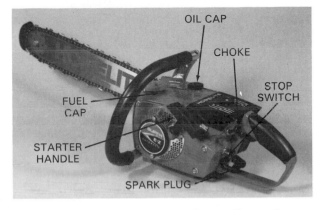

Fig. 17-39. Study the parts of a gasoline engine-powered chain saw.

SAFETY FEATURES

Safe operation of a chain saw comes from a good knowledge of correct operating procedure. However, there are a few safety features on many chain saws that are very important.

When the unshielded nose of a chain saw hits a solid surface, the spinning chain may cause the saw to fly back toward the operator. This is known as **kickback** and may be very dangerous. Some chain saws have a tip guard device that is attached to the end of the blade that helps prevent kickback. It can be installed or removed very quickly.

Another safety feature on many chain saws is the **quick-stop device.** Illustrated in Fig. 17-40, this device stops the chain to reduce the possibility of injury. With a sudden kickback, the operator's left hand moves forward to make contact with the front hand guard and the chain is stopped. The front hand guard is the quick-stop activating lever.

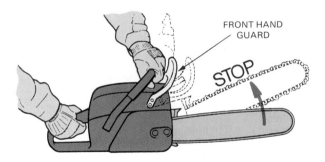

Fig. 17-40. When operator's left hand makes contact with front hand guard, it will activate quick-stop to stop chain and reduce risk of injury.

When carrying the chain saw by hand, the engine must be stopped and the saw must be in the proper position. Grip the front handle and place the muffler to the side away from the body. The chain guide bar should be behind you.

Some chain saws come with a case, Fig. 17-41, that protects the saw during transportation and storage.

Another safety feature is the **chain guard** (scabbard) shown in Fig. 17-42. The chain guard protects the operator from the sharp blade when not in use and protects the blade from moisture and rust.

Hold the chain saw with both hands when cutting. See Fig. 17-43. You should also wear heavy gloves to protect your hands and dampen vibration.

Another safety feature required by some states is the **spark arrestor.** The spark arrestor is built

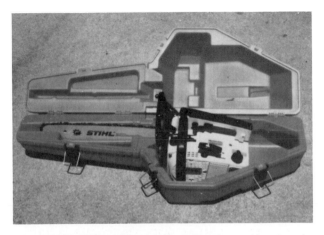

Fig. 17-41. This chain saw is well protected in a tough, plastic carrying case.

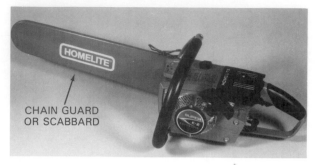

Fig. 17-42. A chain guard (scabbard) protects saw blade during transportation and storage.

Fig. 17-43. Correct position of hands when operating chain saw. Note use of heavy gloves for protection.

into the exhaust system to prevent sparks from the system from catching dry grass or wood chips on fire. These devices are sometimes known as fire arrestor screens.

RULES FOR SAFE OPERATION

The following rules for the operation of a chain saw should be followed carefully.

1. Never operate a chain saw when you are tired.
2. Use safety footwear; snug-fitting clothing; and eye, hearing, and head protection devices.
3. Always use caution when handling fuel. Move the chain saw at least 10 feet (3 m) from the fueling point before starting the engine.
4. Do not allow other persons near the chain saw when starting or cutting. Keep bystanders and animals out of the work area.

5. Never start cutting until you have a clear work area, secure footing, and a planned retreat path from the falling tree.
6. Always hold the chain saw firmly with both hands when the engine is running. Use a firm grip with thumb and fingers encircling the chain saw handles.
7. Keep all parts of your body away from the saw chain when the engine is running.
8. Before you start the engine, make sure the saw chain is not contacting anything.
9. Always carry the chain saw with the engine stopped, with the guide bar and saw chain to the rear, and the muffler away from your body.
10. Never operate a chain saw that is damaged, improperly adjusted, or is not securely assembled. Be sure that the saw chain stops moving when the throttle control trigger is released.
11. Always shut off the engine before setting the chain saw down.
12. Use extreme caution when cutting small brush and saplings because slender material may catch the saw chain. This could fling the saw toward you or pull you off balance.
13. When cutting a limb that is under tension, be alert for springback so that you will not be struck when the tension in the wood fibers is released.
14. Keep the handles dry, clean, and free of oil or fuel mixture.
15. Do not operate the chain saw with a deteriorated or removed muffler system. Fire-preventing mufflers (fire arrestor screen types) should be used in dry areas.
16. Operate the chain saw only in well ventilated areas.
17. Do not operate a chain saw in a tree unless you are specially trained to do so.
18. Guard against kickback. Kickback can lead to severe injuries.

TO AVOID KICKBACK:

1. Hold the chain saw firmly with both hands. Do not reach too far.

2. Do not let the nose of the guide bar contact a log, branch, ground, or any other obstruction.
3. Cut at high engine speeds.
4. Do not cut above shoulder height.
5. Follow manufacturer's sharpening and maintenance instructions for the saw chain.

CHAIN SAW MAINTENANCE

Careful maintenance of the chain saw engine and chain provide for long-lasting service life and safe use. Never operate a chain saw that is damaged, improperly adjusted, or not completely assembled.

⚠ WARNING: Always stop the engine and be sure that the chain is stopped before doing any maintenance or repair work on the saw.

FUEL AND CARBURETOR

Always use the correct gasoline and oil mixture for a two-cycle engine as recommended by the manufacturer. Before refueling, carefully clean the filler cap and the area around it to ensure that no dirt falls into the tank. See Fig. 17-44.

Do not adjust the carburetor unless it is necessary. The high speed and low speed carburetor adjustments on a chain saw engine are very critical. Incorrect settings of these speeds can cause serious damage to the engine. If adjustments become necessary, follow the manufacturer's recommendations.

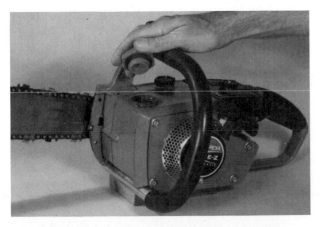

Fig. 17-44. Clean area around filler cap before refueling engine.

If the engine stops while idling, the exhaust smokes, or engine does not run smoothly, try adjusting the carburetor, Fig. 17-45. Normally, you must turn the adjusting screw clockwise when the idle setting is too lean and counterclockwise when the setting is too rich.

Apart from minor adjustments, carburetor repairs should be made by a trained technician who has all the necessary service tools and equipment.

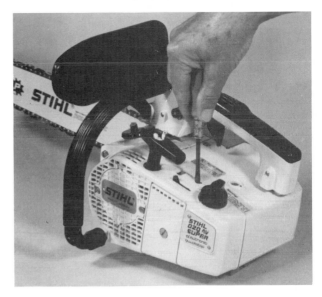

Fig. 17-45. Turn idle speed adjusting screw with a screwdriver to obtain correct idle.

CYLINDER FINS

Check the cylinder fins periodically since clogged fins result in poor engine cooling. Remove the dirt and dust from between the fins to allow cooling air to pass freely. This can be done with a small stick, brush, or compressed air. See Fig. 17-46.

AIR CLEANER (FILTER)

The function of the air filter is to catch dust and dirt in the inlet air and reduce wear on engine components. Clogged air cleaners cut down on engine power, increase fuel consumption, and make starting more difficult. The air cleaner should be cleaned every day in very dusty operating conditions.

Fig. 17-46. Cylinder fins can be cleaned with a small wooden stick.

Before removing the air cleaner, close the choke valve so that no dirt can get into the carburetor, Fig. 17-47. Unscrew the air cleaner cover and remove the element, Fig. 17-48. Lightly brush off dust or wash cleaner element in cleaning solvent if extremely dirty. It should be dried completely before replacing it on the engine.

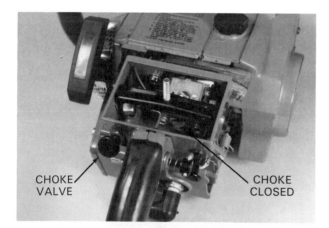

Fig. 17-47. Move choke into closed position so dirt cannot enter carburetor.

FUEL FILTER

Check the fuel filter periodically. A clogged fuel filter will cause engine trouble, such as hard starting. Remove the fuel filler cap and fish for the flexible pick-up tube with a hook. Pull the fuel pick-up out, Fig. 17-49. Remove the old filter sleeve and slide a new sleeve in place. Return the flexible pick-up tube to the gas tank.

LUBRICATION

The saw chain and guide bar must be continuously lubricated during operation to protect them from excessive wear. This is provided for by the automatic chain oiling system. Clean the lubricating oil supply hole daily. It is located at the

Fig. 17-48. This air filter cover is removed and air filter is ready for cleaning.

Fig. 17-49. Flexible fuel pick-up tube is removed through fuel filler hole.

Fig. 17-50. Removing oil tank cap to fill tank with oil.

base of the blade. Always fill the oil tank with chain oil each time the engine is refueled, Fig. 17-50. NEVER use waste oil for this purpose.

MUFFLER

The muffler should be kept clean and open. Do not run a chain saw without the muffler. If regulations require the use of a spark arrestor, check its condition periodically. Carbon deposits in the muffler and cylinder exhaust port will cause lower engine power output and sparking from the muffler. If needed, remove the muffler and clean the cylinder exhaust port, Fig. 17-51. Use a wood stick or dowel to protect the metal surfaces.

Fig. 17-52. Set engine stop-start switch to "off" position when servicing.

Fig. 17-51. Clean cylinder exhaust port with a wood stick or dowel.

SPARK PLUG

If the engine does not start, it may be due to a wet, fouled, or faulty spark plug. Check the spark plug periodically and clean or install a new one as necessary. Adjust the spark gap if it is wider or narrower than the standard gap. Be sure the stop-start switch is in the "off" position when checking the spark plug, Fig. 17-52.

GUIDE BAR

Clean the *guide bar* daily or before each use of the chain saw. Remove any burrs that may be found along the bar rails. On most chain saws, the roller nose bearing must be lubricated. Place the chain saw on its side so that the bar nose is

firmly supported. Clean the grease hole and pump grease in as shown in Fig. 17-53.

STORAGE AFTER USE

Inspect and make adjustments of every part of the chain saw before storage. Clean all parts and apply a thin coat of oil to all metal surfaces to prevent rust. Remove the chain and the guide bar. Apply a sufficient oil coat and wrap them in a plastic bag. Drain the fuel tank and pull the starter a few times to drain the carburetor.

Pour a small amount of oil in the spark plug hole and replace the spark plug. Slowly pull the starter to crank the engine a couple of revolutions. Place the saw in its case and store it in a dry, dust-free area.

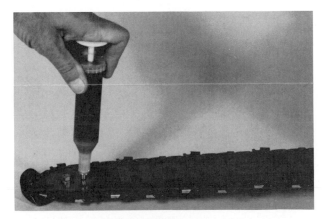

Fig. 17-53. Lubricate roller nose bearing with a grease pump as shown.

EDGER/TRIMMERS

The combination *edger/trimmers* powered by small gasoline engines are versatile units for lawn care maintenance, Fig. 17-54. Since the edger has an exposed, fast-spinning blade, it is very important that safety devices designed for the unit are always in place. Use proper operating procedures to assure safe, dependable service.

Fig. 17-54. This is a combination edger/trimmer set for edging operation.

PURCHASING CONSIDERATIONS

When purchasing a combination edger/trimmer, Fig. 17-55, a number of factors should be considered. Edgers come in a variety of sizes with respect to the construction and power of the unit. Small engines in the range of two to five horsepower are generally used to drive a belt to the blade.

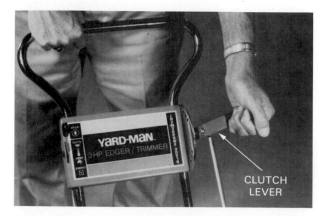

Fig. 17-55. Disengage clutch lever when you are not trimming or edging.

The most popular edgers are the small units designed for home lawn care and weed cutting. Commercial edgers for heavy-duty work are built to be more rugged. They incorporate a larger, more powerful engine and are considerably more expensive.

Some edgers are single units designed for edging along sidewalks and driveways. Others are designed so the blade can be tilted from vertical to horizontal for trimming long grass or weeds under fences and close to buildings. Consider the type of work the edger will be expected to perform when planning a purchase.

SAFETY FEATURES AND ADJUSTMENTS

The blade of a gasoline engine-driven edger is usually belt driven from the engine. The blade clutch should always be disengaged, Fig. 17-55, when starting the engine or when doing maintenance work on the unit. This loosens the belt drive to the blade, Fig. 17-56, and allows the engine shaft to turn freely.

⚠ WARNING: Never attempt to make any adjustments on an edger/trimmer while the engine is running. Serious injury could result.

The guide wheel can normally be adjusted horizontally by loosening the lever on the front of the frame and moving the wheel to either side of the frame as needed. See Fig. 17-57. The lever should then be tightened to hold the wheel firmly in place.

Fig. 17-56. This blade drive belt is in loosened position.

Fig. 17-57. Locking front guide wheel in a newly adjusted position.

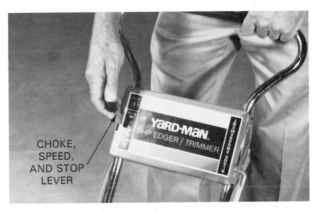

CHOKE, SPEED, AND STOP LEVER

Fig. 17-59. Adjusting choke, speed, and stop lever.

With some designs, the depth of the edger blade can be controlled by raising and lowering the guide wheel. The height of the wheel is adjusted by placing the lever on the right of the frame in the desired notch, Fig. 17-58.

One lever on the handle of the edger usually operates the choke, regulates engine speed from slow to fast, and stops the engine, Fig. 17-59. When starting the engine, place this lever in the choke position and pull the starter handle rapidly, Fig. 17-60. Grip the handle firmly and place your foot behind the rear wheel so that the unit will not move during starting.

The *blade guard* should always be in place during use. The cam lever releases the blade guard, Fig. 17-61, so the guard can be rotated to cover the blade in any edging or trimming position.

Fig. 17-60. Pulling starter handle to start engine. Note belt is in loose position.

Fig. 17-58. Raising front guide wheel by placing adjusting lever in correct notch.

Fig. 17-61. Releasing cam lever so blade guard can be set at a new position.

Tighten the cam lever securely before operating the edger.

For edging along a sidewalk or driveway, the blade is in the vertical position. When the unit is to be used for trimming, the blade is set horizontally, as shown in Fig. 17-62. It can also be set at an angle for special edging jobs, Fig. 17-63. This adjustment is made by loosening the belt with the blade clutch lever, releasing the lever on the pivot bracket, and rotating the spindle housing. See Fig. 17-64. The notches in the bracket will hold the blade firmly in any position.

RULES FOR SAFE OPERATION

1. Thoroughly inspect the area where the equipment is to be used and remove all stones, sticks, wire, bones, and other foreign objects.
2. Do not operate equipment when barefoot or when wearing open sandals. Always wear substantial footwear.
3. Check the fuel before starting the engine. Do not fill the gasoline tank indoors, when the engine is running, or while the engine is still hot. Wipe off any spilled gasoline before starting the engine.
4. Disengage the blade clutch before starting the engine.
5. Never attempt to make a wheel adjustment while the engine is running.
6. Never operate the equipment in wet grass. Always be sure of your footing; keep a firm hold on the handle.
7. Do not change the engine governor settings or overspeed the engine.
8. Do not put hands or feet near or under rotating parts. Keep clear of the discharge opening at all times.

Fig. 17-62. Blade is set horizontally for trimming.

Fig. 17-63. Blade is set at an angle for special trimming jobs.

Fig. 17-64. Lever on pivot bracket is placed in correct notch for trimming.

9. Stop the blade when crossing a gravel drive, walk, or road.

10. After striking a foreign object, stop the engine. Remove the wire from the spark plug. Thoroughly inspect the edger for any damage and repair the damage before restarting and operating the edger.

11. If the equipment should start to vibrate abnormally, stop the engine and check for the cause. Vibration is generally a warning of trouble.

12. Stop the engine whenever you leave the equipment, before cleaning the guard assembly, and when making any repairs or inspections.

13. When cleaning, repairing, or inspecting, make certain the blade and all moving parts have stopped. Disconnect the spark plug wire, and keep the wire away from the plug to prevent accidental starting.

14. Do not run the engine indoors.

15. Shut the engine off and wait until the blade comes to a complete stop before unclogging guard assembly.

16. Safety glasses or other eye protection should always be used when operating an edger.

EDGER/TRIMMER MAINTENANCE

The standard edger blade, shown in Fig. 17-65, is 10 inches long and is notched on the ends. Since the blade scrapes the edges of driveways or sidewalks during operation, it wears down quickly and should be replaced as needed.

To change the blade, raise the front wheel and loosen the nut on the drive shaft, Fig. 17-66. Remove the old blade, and replace it with a new one. Be sure the blade nut is tightened properly. Al-

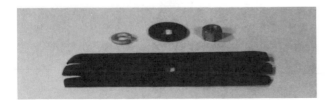

Fig. 17-65. This is a standard edger blade with lock washer, plain washer, and nut.

Fig. 17-66. Loosening blade nut on drive shaft. Hold blade with a heavy glove to protect hand.

ways wear a glove to hold the blade to prevent injury to your hand.

Edger blades are never sharpened. They are just replaced when they become too short to make good contact with a surface being edged.

To replace a worn belt, remove the belt guard on the engine pulley and the belt guard on the spindle housing. Remove the belt. Replace with a proper size V-belt and secure the belt guards.

Check the engine oil level before starting the engine and after every six hours of use. Add oil as necessary to keep level on full. Before removing the filler plug, clean the area around the plug to prevent dirt from entering the oil fill opening.

Change the oil after 30 hours of operation by draining oil through the lower oil drain plug. Refill with the correct amount and weight of fresh oil.

Lubricate all moving parts of the edger with engine oil periodically. Check and clean or replace the spark plug each operating season. Remove and clean the air filter at recommended intervals.

EDGER/TRIMMER STORAGE

The following steps should be taken to prepare edgers and trimmers for storage:

1. Clean and lubricate the unit thoroughly.

2. Loosen the belt so it will not stretch during storage.

3. Coat the cutting blade with oil to prevent rusting.
4. Remove the spark plug and pour a tablespoon of clean engine oil into the spark plug hole. Rotate the crankshaft a few times and replace the spark plug.
5. Check the blade and engine mounting bolts for proper tightness.
6. Never store the edger with gasoline in the tank when inside of a building where fumes may reach a spark or open flame.
7. Store the edger in a dry, clean area.

SUMMARY

Many considerations should be made when purchasing a power lawn mower. For the average yard, a rotary-type mower with a 22 in. diameter blade is satisfactory. Selecting a push-type or self-propelled mower is a matter of personal preference. Both two cycle and four cycle engines are reliable.

There are several mechanical methods for starting small engines including the recoil starter, inertia starter, and electric starter. Every mower manufactured today must be equipped with a blade brake.

After each use, the mower should be cleaned. Blades can be sharpened when they become dull or nicked. Spark plugs should be cleaned and gapped as recommended by the manufacturer. Air filters should be cleaned and oil should be changed after every twenty-five hours of operation.

Always follow safe operating procedures when using a gasoline powered chain saw. When purchasing a saw, make sure it is suited for the type of work to be done. Safe operation of a chain saw comes from a thorough knowledge of correct operating procedures.

Always stop the engine and make sure the chain is stopped before doing maintenance work on a chain saw. Follow all manufacturers maintenance instructions.

Always keep safety devices in place when using a gas-powered edger/trimmer. Never attempt to make adjustments on the edger/trimmer when the engine is running.

Common maintenance procedures on an edger/trimmer include replacing worn blades and belts, changing oil, and lubricating moving parts.

Special precautions should be taken when storing any small engine powered implement.

KNOW THESE TERMS

Rotary mowers, Reel-type mowers, Discharge opening, Dethatcher blade, Independent rope start, Recoil starter, Inertia starter, Electric starter, Blade brakes, Chain saw, Kickback, Tip guard, Quick stop device, Chain guard, Spark arrestor, Guide bar, Edger/trimmer, Blade guard.

REVIEW QUESTIONS–Chapter 17

1. Name five safety features used on power lawn mowers.
2. List eight considerations for purchasing a power lawn mower.
3. List five safe operating practices when using a power lawn mower.
4. Name seven lawn mower conveniences.
5. List the procedures for preparing an engine before starting.
6. List the cold engine starting procedure for an extended rope type lawn mower with blade brake.
7. If the engine idles too fast, you can adjust the idle speed _____ on the _____.
8. If the engine will not start, though proper procedures have been used, the first trouble check should be the _____ _____.
9. List the five maintenance steps that will keep the power lawn mower in proper working condition.
10. When the unshielded nose of the chain saw hits a solid surface, it may jump toward the operator. This is called _____.
11. The device often built into the muffler of a chain saw to prevent sparks from causing a fire is called a _____ _____.
12. The _____ _____ should always be in place when operating an edger.

13. A standard edger blade is _____ long.
14. A small chain saw works well for cutting _____, small _____, and _____ _____.

SUGGESTED ACTIVITIES

1. Change the oil in a power lawn mower.
2. Service an air filter.
3. Clean and gap a used spark plug.
4. Sharpen and balance a lawn mower blade.
5. Demonstrate engine starting procedures.
6. Adjust engine idle setting.
7. Replace a burned out muffler.
8. Go to a local implement store and compare lawn mowers as though you were planning to purchase one.
9. From literature obtained from dealers, make a list of chain saws produced by different manufacturers. Describe the advantages and disadvantages of each in terms of safety.
10. Replace a worn out blade on an edger.
11. Make a list of maintenance features to be done before placing a chain saw in storage.
12. Check and change, if necessary, the gas line filter in a chain saw fuel tank.
13. Check the muffler of a chain saw to see if it has a spark arrestor device.

Cutaway view of a rotary lawn mower equipped with a one-cylinder, two cycle engine. Can you locate the piston, crankshaft, flywheel, and air filter? (Jacobsen Mfg. Co.)

Many small gas engine-powered implements, such as the lawn mower pictured above, have become extremely complex. As a result, there is a growing demand for qualified people to design, manufacture, and service these products. (Deere & Co.)

CHAPTER 18

CAREER OPPORTUNITIES

After studying this chapter, you will be able to:
- [ ] Identify several career opportunities in the small gas engine field.
- [ ] List qualities that are essential for anyone pursuing a career in small engines.
- [ ] List the advantages and disadvantages of entrepreneurship

The small gas engine field offers career opportunities for men and women in three different areas: manufacturing, sales, and service. Through training, study, and work experience, you can become an engine mechanic, service manager, sales manager, or general manager of a small engine service center. You can also be a manufacturer's technician, service representative, or engineer.

ENGINE MECHANIC

Implement sales facilities and equipment rental centers need *engine mechanics* to do tuneups, service equipment, and make repairs. Often, the quality of workmanship and reliability of the mechanic who services the customers' equipment directly affects the reputation and sales volume of the business.

Small gas engine mechanics must be able to diagnose engine troubles and make appropriate repairs and/or part replacements, Fig. 18-1. They

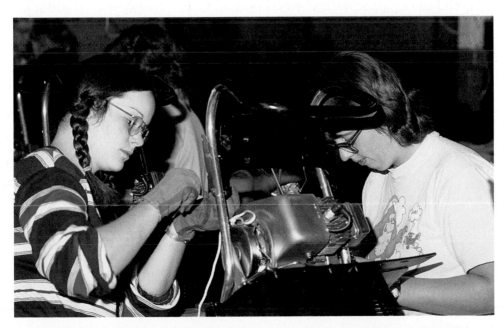

Fig. 18-1. A small gas engine mechanic must know engine construction and principles of operation. Mechanics must be proficient in troubleshooting, maintenance, service, and repair.

should be able to analyze the mechanical condition and performance of an engine and make proper recommendations to the owner. A mechanic should also be competent in the use of test equipment and thoroughly familiar with manufacturers' manuals.

Good mechanics keep their tools and equipment in first-class condition and organized for convenient use. They must have specialty tools (pullers, drivers, etc.), which are available from the engine manufacturers or tool manufacturers. Small gas engine mechanics must know how to use micrometers and dial indicators. Basic machining and welding experience is also desirable. Engine mechanics generally receive their knowledge and skill at technical career centers or vocational schools. Some engine manufacturers have their own training programs for service personnel.

SERVICE MANAGER

Men and women are needed as service managers in small gas engine shops with more than one mechanic. *Service managers* are responsible for quality workmanship and satisfactory shop operation. They must plan and supervise the activities of all service department employees.

Service managers discuss service problems with customers, make recommendations, write job tickets, and assign work to the mechanics. They handle customer complaints, are responsible for the training of apprentices, and, generally, inspect all finished repair work. They report directly to the general manager or the owner of the service facility.

SALES MANAGER

Sales managers are needed to sell or rent implements and vehicles that utilize small gasoline engines. Some sales managers handle a variety of products, such as yard and farm equipment, marine and sports vehicles, construction equipment, and emergency repair or rescue equipment. Others specialize in one field or sell a few closely related products such as motorcycles, all-terrain vehicles (ATVs), and snowmobiles.

The geographical location of a sales or rental business often determines the type of vehicle or equipment that is most in demand by the consumer. Gasoline engine applications are numerous, and new uses are being developed each year.

GENERAL MANAGER

Managing a successful small engine sales and service business requires experience and education. Keeping accurate sales and service records, budgeting, promoting sales, and maintaining adequate tools, parts, supplies, and accessories are just a few of the responsibilities of the *general manager.*

The general manager must have a sincere, personal commitment to provide fair, quality service to customers. Often, a mechanic or sales person works up to this position through years of work experience and post high school courses in business and management.

TECHNICIAN

Small gas engine manufacturers need men and women to develop prototype engines or engine parts and test new design theories. *Technicians* need to be skilled in the use of tools, materials, and machine processes in order to produce a special part or engine unit. They are usually required to run exhaustive tests using dynamometers and other specialized testing equipment, observing and recording test results.

Technicians are generally involved with experiments, tests, and analysis of various engine systems and designs in the plant and in actual field use. When testing is completed, they present the test results and recommend changes to engineers and others involved in the project. Their observations and recommendations may be presented orally or in writing. Therefore, they must be able to communicate in clear, technical language. This requires adequate communicative skills combined with technical talents.

Many colleges, technical institutes, and universities offer programs for technicians. Engine mechanics can become technicians if they have the desire to further their education through evening courses and service training programs.

SERVICE REPRESENTATIVE

Small gas engine manufacturers may train certain employees with broad service experience to become *service representatives.* These representatives are required to work closely with service managers and mechanics in the field to catch and correct chronic service problems. In some cases, service representatives write and distribute service bulletins concerning these problems. They also meet with and report findings to company engineers involved in engine design.

ENGINEER

Manufacturers need *engineers* to design engines that will perform satisfactorily under specific environmental conditions. For example, a small gas engine designed for use in a garden tractor is quite different from one intended for use in a chain saw. Engineers must use their knowledge of scientific principles to design and create engines that will meet all of the specified operating requirements.

Engineers usually have college engineering degrees. Sometimes, however, an engineer's license can be obtained by passing special examinations. Engineers must have a strong background in science, mathematics, and many specialized technical subjects. They must be analytical and creative, with a practical knowledge of manufacturing processes and materials.

EXECUTIVE

Any of the careers outlined in this chapter can serve as stepping stones to high-level management positions in the small gas engine field. Most successful *executives* in this field began their careers in engine production, design, sales, or service. Almost invariably, the key to their success is learning to do the job at hand to the best of their ability.

ENTREPRENEUR

Many people in the small engine field start their own business. These people are called *entrepreneurs.* Entrepreneurs must have a total understanding of the managerial, financial, and technical aspects of the small engine business.

There are several advantages associated with owning a business. As the owner, you have total control over the way a business grows and develops. You have the opportunity to hire and train people as you desire. Additionally, your income is only limited by the success of your business.

On the other hand, owning a business can be extremely difficult. Entrepreneurs work very long hours trying to establish and maintain a profitable business. They also take many risks to get the new business started. Most entrepreneurs spend years repaying loans that were taken to go into business. Entrepreneurs are responsible for these loans even if their business fails.

SUMMARY

The small gas engine field offers career opportunities in several areas. The engine mechanic diagnoses engine trouble and makes appropriate repairs. Service managers plan and supervise the activities of all service department employees. Sales managers sell implements and vehicles that utilize small gas engines. General managers oversee both the service and sales aspects of the business.

Small engine technicians develop prototype engines and test new design theories. Manufacturers often train employees with broad service experience to become service representatives. Service representatives work to solve chronic service problems. Manufacturers also need engineers to design new engines that will perform satisfactorily under specific conditions.

Any career in small engines can lead to high-level management positions. Most successful executives began their careers in production, design, sales, or service. Many individuals in the small engine field start their own business. These people are entrepreneurs.

KNOW THESE TERMS

Engine mechanic, Service manager, Sales manager, General manager, Technician, Service representative, Engineer, Executive, Entrepreneur.

REVIEW QUESTIONS–CHAPTER 18

1. Small engine mechanics generally receive their training at _____ _____.
2. The service manager _____ the activities of all service department employees.
3. General managers are often required to keep accurate _____ and _____ records.
4. Small engine technicians must have good oral and written communications skills. True or False?
5. The _____ _____ works closely with service managers and mechanics to catch and correct chronic service problems.
6. Many successful executives in the small engine field began their careers in production, sales, or service.
7. An entrepreneur is an individual who starts his or her own business. True or False?

USEFUL INFORMATION

CHECK SHEET FOR 4 CYCLE ENGINE RECONDITIONING

ENGINE: MAKE _____ MODEL NO. _____
         HP _____ SERIAL NO. _____

CHECK LIST FOR DISASSEMBLY

_____ 1. Remove all gasoline from engine.
_____ 2. Inspect engine for broken or missing parts.
_____ 3. Record all important data on an information sheet.
_____ 4. Remove spark plug. Check gap and condition of electrodes.
_____ 5. Take compression reading. Record on information sheet.
_____ 6. Check ignition output with spark tester.
_____ 7. Drain oil from crankcase.
_____ 8. Disconnect all linkage from remote throttle assembly to engine.
_____ 9. Remove engine from mountings.
_____ 10. Clean engine housing or mounting area.
_____ 11. Remove blower housing from engine.
_____ 12. Remove carburetor and carburetor linkage.
_____ 13. Remove governor air vane and governor linkage.
_____ 14. Remove muffler.
_____ 15. Remove valve chamber cover.
_____ 16. Remove cylinder head and head gasket.
_____ 17. Measure bore and stroke. Record measurements.
_____ 18. Check and record valve clearance.
_____ 19. Remove air deflector shields.
_____ 20. Check and record armature air gap.
_____ 21. Remove starter clutch and flywheel nut.
_____ 22. Remove flywheel.
_____ 23. Remove ignition breaker point cover.
_____ 24. Check and record ignition point gap setting.
_____ 25. Remove ignition points, condenser, and ignition cam (or solid state ignition unit).
_____ 26. Remove magneto assembly.
_____ 27. Remove all rust and burrs from end of crank shaft.
_____ 28. Remove mounting flange, if any.
_____ 29. Check timing marks.
_____ 30. Remove camshaft and oil pump.
_____ 31. Remove valve tappets.
_____ 32. Remove piston and rod assembly. Note: Mark rod first.
_____ 33. Remove piston rings from piston.
_____ 34. Check and record ring end gap.
_____ 35. Clean ring grooves in piston.
_____ 36. Check and record piston ring-to-land clearance.
_____ 37. Remove crankshaft and inspect it.
_____ 38. Remove intake and exhaust valves.
_____ 39. Wash and clean all parts that will not be damaged by solvent.
_____ 40. Inspect engine block for scores or imperfections.
_____ 41. Check all bearings and oil seals for possible damage.

_____ 42. Recondition or replace necessary engine components.

CHECK LIST FOR REASSEMBLY

_____ 1. Clean valve seats with wire wheel and brush.
_____ 2. Have instructor check valve parts after they are cleaned.
_____ 3. Lap valves against valve seats, using lapping compound.
_____ 4. Install valve assemblies.
_____ 5. Install crankshaft.
_____ 6. Fit rings on piston in proper order.
_____ 7. Oil cylinder wall. Install piston and rod assembly in proper direction.
_____ 8. Torque rod bolts to specifications. Bend up retainer clips.
_____ 9. Install tappets, camshaft, and oil pump.
_____ 10. Align timing marks on camshaft and crankshaft.
_____ 11. Bolt mounting flange (crankcase cover) on engine. Check for proper fit.
_____ 12. Check and record valve clearance measurements.
_____ 13. Assemble valve cover and breather and bolt to engine block.
_____ 14. Install cylinder head and head gasket. Torque to specifications.
_____ 15. Reassemble and install ignition system. If necessary, set point gap to specifications.
_____ 16. Install ignition point cover (if necessary).
_____ 17. Install flywheel key and flywheel. Torque flywheel nut.
_____ 18. Set armature air gap to specifications.
_____ 19. Fasten governor air vane to engine block.
_____ 20. Mount carburetor on engine block.
_____ 21. Install muffler.
_____ 22. Install blower housing.
_____ 23. Connect fuel lines and valve breather tube.
_____ 24. Mount engine in implement or equipment.
_____ 25. Connect engine to drive train.
_____ 26. Connect all linkage between remote throttle and engine.
_____ 27. Tighten oil plug and fill engine with proper oil.
_____ 28. Clean and install air cleaner.
_____ 29. Check engine compression.
_____ 30. Clean and set gap of spark plug electrodes. Install spark plug.
_____ 31. Check to be sure all components are tight and properly adjusted.
_____ 32. Fill fuel tank with clean gasoline.
_____ 33. If indoors, turn on exhaust fan and wear goggles.
_____ 34. Engage carburetor choke and start engine.
_____ 35. Adjust carburetor.

Note: Keep a record on file of all reconditioning repairs, part replacements, and engine identification information.

GENERAL TORQUE SPECIFICATIONS IN CONSIDERATION OF FASTENER QUALITY

THE FOLLOWING RULES APPLY TO THE CHART:

1. CONSULT MANUFACTURERS SPECIFIC RECOMMENDATIONS WHEN AVAILABLE.
2. THE CHART MAY BE USED DIRECTLY WHEN ANY OF THE FOLLOWING LUBRICANTS ARE USED:
   a. NEVER-SEEZ COMPOUND, MOLYKOTE, FEL-PRO C-5, GRAPHITE AND OIL OR SIMILAR MIXTURES.
3. INCREASE THE TORQUE BY 20% WHEN USING ENGINE OIL OR CHASSIS GREASE AS A LUBRICANT.
   (THESE LUBRICANTS ARE NOT GENERALLY RECOMMENDED FOR FASTENERS)
4. REDUCE TORQUE BY 20% WHEN PLATED BOLTS ARE USED.
5. INCREASE TORQUE BY 20% WHEN MULTIPLE TAPERED TOOTH LOCKWASHERS ARE USED.

	MOST USED	MUCH USED	USED AT TIMES	USED AT TIMES	RECOMMENDED FOR COMPETITION AND CRITICAL USE
CURRENT AUTOMOTIVE USAGE					
MIN. TENSILE STRENGTH	64,000 P.S.I.	105,000 P.S.I.	133,000 P.S.I.	150,000 P.S.I.	160,000 P.S.I.
MATERIAL	LOW CARBON STEEL	MEDIUM CARBON STEEL TEMPERED	MEDIUM CARBON STEEL QUENCHED/TEMP.	MED. CARBON ALLOY STEEL QUENCHED/TEMP.	SPECIAL ALLOY STEEL QUENCHED/TEMP.
DEFINITION	INDETERMINATE QUALITY	MINIMUM COMMERCIAL QUALITY	MEDIUM COMMERCIAL QUALITY	BEST COMMERCIAL QUALITY	BEST QUALITY SUPERTANIUM
GRADE MARKINGS	(hex)	(hex)	(hex)	(hex)	(hex)
BOLT SIZE	S.A.E. GRADE 1 OR 2	S.A.E. GRADE 5	S.A.E. GRADE 6	S.A.E. GRADE 8	EXCEEDS ALL S.A.E. GRADES
1/4	5	7	10	10.5	11
5/16	9	14	19	22	24
3/8	15	25	34	37	40
7/16	24	40	55	60	65
1/2	37	60	85	92	97
9/16	53	88	120	132	141
5/8	74	120	167	180	192
3/4	120	200	280	296	316
7/8	190	302	440	473	503
1	282	466	660	714	771

NOTE: TO CONVERT FROM FOOT-POUNDS TO NEWTON-METRES, MULTIPLY FOOT-POUNDS BY .7376. (ACTUALLY POUND FEET)

NOTE: THE TORQUE SPECIFICATIONS ARE GIVEN IN FOOT-POUNDS. INCH-POUND EQUIVALENT MAY BE OBTAINED BY MULTIPLYING BY 12.

(P. A. Sturtevant)

The Metric System of Measurement

Measures of Length

1 Millimetre (mm) = ...0.03937079 inch, or about 1/25 inch

10 Millimetres = 1 Centimetre (cm) = ...0.3937079 inch

10 Centimetres = 1 Decimetre (dm) = ...3.3937079 inch

10 Decimetres = 1 Metre (m) = ...39.37079 inches, 3.2808992 feet, or 1.09361 yards

10 Metres = 1 Dekametre (dam) = ...32.808992 feet

10 Dekametres = 1 Hectometre (hm) = ...19.927817 rods

10 Hectometres = 1 Kilometre (km) = ...1093.61 yards, or 0.6213824 mile

10 Kilometres = 1 Myriametre (mym) = ...6.213824 miles

1 inch = 2.54 cm, 1 foot = 0.3048 m, 1 yard = 0.9144 m, 1 rod = 0.5029 m,
1 mile = 1.6093 km

Measures of Weight

1 Gram (g) = 15.4324874 gr. Troy, or 0.03215 oz. Troy, or ..0.03527398 oz. avoirdupois

10 Grams = 1 Dekagram (dag) = ...0.3527398 oz. avoirdupois

10 Dekagrams = 1 Hectogram (hg) = ...3.527398 oz. avoirdupois

10 Hectograms = 1 Kilogram (kg) = ...2.20462125 lbs.

1000 Kilograms = 1 Ton (t) = 2204.62125 lbs., or 1.1023 tons of 2000 lbs., or 0.9842 ton
of 2240 lbs., or 19.68 cwts.

1 grain = 0.0648 g, 1 oz. avoirdupois = 28.35 g, 1 lb. = 0.4536 kg,
1 ton (2000 lbs.) = 0.9072 t, 1 ton (2240 lbs.) = 1.016 t, or 1016 kg

Measures of Capacity

1 Litre (L) = 1 cubic decimetre = 61.0270515 cubic in., or 0.03531 cu. ft., or 1.0567 liquid
qts., or 0.908 dry qt., or 0.26417 Amer. gal.

10 Litres = 1 Decalitre (dal) = 2.6417 gal, or 1.135 pk.

10 Decalitres = 1 Hectolitre (hl) = 2.8375 bu.

10 Hectolitres = 1 Kilolitre (kl) = 61027.0515 cu. in., or 28.375 bu.

1 cu. foot = 28.317 Litres, 1 gallon (American) = 3.785 Litres,
1 gallon (British) = 4.543 Litres

Metric Conversion Table

Millimetres	×	.03937	= Inches
Millimetres	×	25.400	× Inches
Metres	×	3.2809	= Feet
Metres	×	.3048	× Feet
Kilometres	×	.621377	= Miles
Kilometres	×	1.6093	× Miles
Square centimetres	×	.15500	= Square inches
Square centimetres	×	6.4515	× Square inches
Square metres	×	10.76410	= Square feet
Square metres	×	.09290	× Square feet
Square kilometres	×	247.1098	= Acres
Square kilometres	×	.00405	× Acres
Hectares	×	2.471	= Acres
Hectares	×	.4047	× Acres
Cubic centimetres	×	.061025	= Cubic inches
Cubic centimetres	×	16.3866	× Cubic inches
Cubic metres	×	35.3156	= Cubic feet
Cubic metres	×	.02832	× Cubic feet
Cubic metres	×	1.308	= Cubic yards
Cubic metres	×	.765	× Cubic yards
Litres	×	61.023	= Cubic inches
Litres	×	.01639	× Cubic inches
Litres	×	.26418	= U.S. gallons
Litres	×	3.7854	× U.S. gallons
Grams	×	15.4324	= Grains
Grams	×	.0648	× Grains
Grams	×	.03527	= Ounces, avoirdupois
Grams	×	28.3495	× Ounces, avoirdupois
Kilograms	×	2.2046	= Pounds
Kilograms	×	.4536	× Pounds
Kilograms per square centimetre	×	14.2231	= Pounds per square inch
Kilograms per square centimetre	×	.0703	× Pounds per square inch
Newton-metres	×	1.3558	= Pound feet
Newton-metres	×	.7376	× Pound feet
Metric tons (1,000 kilograms)	×	1.1023	= Tons (2,000 pounds)
Metric tons	×	.9072	× Tons (2,000 pounds)
Kilowatts	×	1.3405	= Horsepower
Kilowatts	×	.746	× Horsepower
Calories	×	3.9683	= B.T. units
Calories	×	.2520	× B.T. units

(L. S. Starrett)

Decimal Equivalents of 8ths, 16ths, 32nds, 64ths

8ths	16ths	32nds	64ths	64ths
1/8 = .125	1/16 = .0625	1/32 = .03125	1/64 = .015625	33/64 = .515625
1/4 = .250	3/16 = .1875	3/32 = .09375	3/64 = .046875	35/64 = .546875
3/8 = .375	5/16 = .3125	5/32 = .15625	5/64 = .078125	37/64 = .578125
1/2 = .500	7/16 = .4375	7/32 = .21875	7/64 = .109375	39/64 = .609375
5/8 = .625	9/16 = .5625	9/32 = .28125	9/64 = .140625	41/64 = .640625
3/4 = .750	11/16 = .6875	11/32 = .34375	11/64 = .171875	43/64 = .671875
7/8 = .875	13/16 = .8125	13/32 = .40625	13/64 = .203125	45/64 = .703125
	15/16 = .9375	15/32 = .46875	15/64 = .234375	47/64 = .734375
		17/32 = .53125	17/64 = .265625	49/64 = .765625
		19/32 = .59375	19/64 = .296875	51/64 = .796875
		21/32 = .65625	21/64 = .328125	53/64 = .828125
		23/32 = .71875	23/64 = .359375	55/64 = .859375
		25/32 = .78125	25/64 = .390625	57/64 = .890625
		27/32 = .84375	27/64 = .421875	59/64 = .921875
		29/32 = .90625	29/64 = .453125	61/64 = .953125
		31/32 = .96875	31/64 = .484375	63/64 = .984375

Decimal Equivalents of Millimetres

mm	Inches	mm	Inches	mm	Inches	mm	Inches	mm	Inches
.01	.00039	.41	.01614	.81	.03189	21	.82677	61	2.40157
.02	.00079	.42	.01654	.82	.03228	22	.86614	62	2.44094
.03	.00118	.43	.01693	.83	.03268	23	.90551	63	2.48031
.04	.00157	.44	.01732	.84	.03307	24	.94488	64	2.51968
.05	.00197	.45	.01772	.85	.03346	25	.98425	65	2.55905
.06	.00236	.46	.01811	.86	.03386	26	1.02362	66	2.59842
.07	.00276	.47	.01850	.87	.03425	27	1.06299	67	2.63779
.08	.00315	.48	.01890	.88	.03465	28	1.10236	68	2.67716
.09	.00354	.49	.01929	.89	.03504	29	1.14173	69	2.71653
.10	.00394	.50	.01969	.90	.03543	30	1.18110	70	2.75590
.11	.00433	.51	.02008	.91	.03583	31	1.22047	71	2.79527
.12	.00472	.52	.02047	.92	.03622	32	1.25984	72	2.83464
.13	.00512	.53	.02087	.93	.03661	33	1.29921	73	2.87401
.14	.00551	.54	.02126	.94	.03701	34	1.33858	74	2.91338
.15	.00591	.55	.02165	.95	.03740	35	1.37795	75	2.95275
.16	.00630	.56	.02205	.96	.03780	36	1.41732	76	2.99212
.17	.00669	.57	.02244	.97	.03819	37	1.45669	77	3.03149
.18	.00709	.58	.02283	.98	.03858	38	1.49606	78	3.07086
.19	.00748	.59	.02323	.99	.03898	39	1.53543	79	3.11023
.20	.00787	.60	.02362	1.00	.03937	40	1.57480	80	3.14960
.21	.00827	.61	.02402	1	.03937	41	1.61417	81	3.18897
.22	.00866	.62	.02441	2	.07874	42	1.65354	82	3.22834
.23	.00906	.63	.02480	3	.11811	43	1.69291	83	3.26771
.24	.00945	.64	.02520	4	.15748	44	1.73228	84	3.30708
.25	.00984	.65	.02559	5	.19685	45	1.77165	85	3.34645
.26	.01024	.66	.02598	6	.23622	46	1.81102	86	3.38582
.27	.01063	.67	.02638	7	.27559	47	1.85039	87	3.42519
.28	.01102	.68	.02677	8	.31496	48	1.88976	88	3.46456
.29	.01142	.69	.02717	9	.35433	49	1.92913	89	3.50393
.30	.01181	.70	.02756	10	.39370	50	1.96850	90	3.54330
.31	.01220	.71	.02795	11	.43307	51	2.00787	91	3.58267
.32	.01260	.72	.02835	12	.47244	52	2.04724	92	3.62204
.33	.01299	.73	.02874	13	.51181	53	2.08661	93	3.66141
.34	.01339	.74	.02913	14	.55118	54	2.12598	94	3.70078
.35	.01378	.75	.02953	15	.59055	55	2.16535	95	3.74015
.36	.01417	.76	.02992	16	.62992	56	2.20472	96	3.77952
.37	.01457	.77	.03032	17	.66929	57	2.24409	97	3.81889
.38	.01496	.78	.03071	18	.70866	58	2.28346	98	3.85826
.39	.01535	.79	.03110	19	.74803	59	2.32283	99	3.89763
.40	.01575	.80	.03150	20	.78740	60	2.36220	100	3.93700

Rules Relative to the Circle

To Find Circumference—
Multiply diameter by 3.1416.....................Or divide diameter by 0.3183

To Find Diameter—
Multiply circumference by 0.3183.................Or divide circumference by 3.1416

To Find Radius—
Multiply circumference by 0.15915...............Or divide circumference by 6.28318

To Find Side of an Inscribed Square—
Multiply diameter by 0.7071
Or multiply circumference by 0.2251.............Or divide circumference by 4.4428

To Find Side of an Equal Square—
Multiply diameter by 0.8862
Or multiply circumference by 0.2821.............Or divide diameter by 1.1284
Or divide circumference by 3.545

Square—
A side multiplied by 1.4142 equals diameter of its circumscribing circle.
A side multiplied by 4.443 equals circumference of its circumscribing circle.
A side multiplied by 1.128 equals diameter of an equal circle.
A side multiplied by 3.547 equals circumference of an equal circle.
Square inches multiplied by 1.273 equals circle inches of an equal circle.

To Find the Area of a Circle—
Multiply circumference by one-quarter of the diameter.
Or multiply the square of diameter by 0.7854
Or multiply the square of circumference by .07958
Or multiply the square of ½ diameter by 3.1416

(L. S. Starrett)

Tap Drill Sizes

For Machine Screw Threads
75% Depth of Thread

A bolt inserted in an ordinary nut, which has only one-half of a full depth of thread, will break before stripping the thread. Also a full depth of thread, while very difficult to obtain, is only about 5% stronger than a 75% depth.

These tables give the exact size of the hole, expressed in decimals, that will produce a 75% depth of thread, and also the nearest regular stock drill to this size. Holes produced by these drills are considered close enough for any commercial tapping.

$$\text{Diameter of Tap, Minus } \frac{.974}{\text{No. threads per Inch}} = \text{Diameter of Hole}$$

Tap Size	Threads per Inch	Diameter Hole	Drill
0	80	.048	3/64
1	72	.060	53
1	64	.058	53
2	64	.071	50
2	56	.069	50
3	56	.082	45
3	48	.079	47
4	48	.092	42
4	40	.088	43
4	36	.085	44
5	44	.103	37
5	40	.101	38
5	36	.098	40
6	40	.114	33
6	36	.111	34
6	32	.108	36
7	36	.124	1/8
7	32	.121	31
7	30	.119	31
8	36	.137	29
8	32	.134	29
8	30	.132	30
9	32	.147	26
9	30	.145	27
9	24	.136	29
10	32	.160	21
10	30	.158	22
10	28	.155	23
10	24	.149	25
12	28	.181	14
12	24	.175	16
14	24	.201	7
14	20	.193	10
16	22	.224	2
16	20	.219	7/32
16	18	.214	3
18	20	.245	D
18	18	.240	B
20	20	.271	I
20	18	.266	17/64
22	18	.292	L
22	16	.285	9/32
24	18	.318	O
24	16	.311	5/16
26	16	.337	R
26	14	.328	21/64
28	16	.353	23/64
28	14	.354	T
30	16	.389	25/64
30	14	.380	V

Tap Drill Sizes *For Fractional Size Threads*
75% Depth Thread
AMERICAN NATIONAL THREAD FORM

Tap Size	Threads per Inch	Diam. Hole	Drill
1/16	72	.049	3/64
1/16	64	.047	3/64
1/16	60	.046	56
5/64	72	.065	52
5/64	64	.063	1/16
5/64	60	.062	1/16
5/64	56	.061	53
3/32	60	.077	5/64
3/32	56	.076	48
3/32	50	.074	49
3/32	48	.073	49
7/64	56	.092	42
7/64	50	.090	43
7/64	48	.089	43
1/8	48	.105	36
1/8	40	.101	38
1/8	36	.098	40
1/8	32	.095	3/32
9/64	40	.116	32
9/64	36	.114	33
9/64	32	.110	35
5/32	40	.132	30
5/32	36	.129	1/8
5/32	32	.126	27
11/64	36	.145	27
11/64	32	.141	20
3/16	32	.157	22
3/16	30	.155	23
3/16	24	.147	26
13/64	32	.173	17
13/64	30	.171	11/64
13/64	24	.163	20
7/32	32	.188	12
7/32	28	.184	13
7/32	24	.178	16
15/64	32	.204	6
15/64	28	.200	8
15/64	24	.194	10
1/4	32	.220	7/32
1/4	28	.215	3
1/4	27	.214	3
1/4	24	.209	4
1/4	20	.201	7
5/16	32	.282	9/32
5/16	27	.276	J
5/16	24	.272	I
5/16	20	.264	F
5/16	18	.258	F
3/8	27	.339	R
3/8	24	.334	Q
3/8	20	.326	21/64
3/8	16	.314	X
7/16	27	.401	Y
7/16	24	.397	Z
7/16	20	.389	25/64
7/16	14	.368	U
1/2	27	.464	15/32
1/2	24	.460	29/64
1/2	20	.451	29/64
1/2	13	.425	27/64
1/2	12	.419	27/64
9/16	27	.526	17/32
9/16	18	.508	33/64
9/16	12	.481	31/64
5/8	27	.589	19/32
5/8	18	.571	37/64
5/8	12	.544	35/64
5/8	11	.536	17/32
11/16	16	.627	5/8
11/16	11	.599	19/32
3/4	27	.714	23/32
3/4	16	.689	11/16
3/4	12	.669	43/64
3/4	10	.653	21/32
13/16	12	.731	47/64
13/16	10	.715	23/32
7/8	27	.839	27/32
7/8	18	.821	53/64
7/8	14	.805	13/16
7/8	12	.794	51/64
7/8	9	.767	49/64
15/16	12	.856	55/64
15/16	9	.829	53/64
1	27	.964	31/32
1	14	.930	15/16
1	12	.919	59/64
1	8	.878	7/8
1 1/16	8	.941	15/16
1 1/8	12	1.044	1 3/64
1 1/8	7	.986	63/64
1 3/16	7	1.048	1 3/64
1 1/4	12	1.169	1 11/64
1 1/4	7	1.111	1 7/64
1 3/16	12	1.173	1 11/64
1 3/8	12	1.294	1 19/64
1 3/8	6	1.213	1 7/32
1 1/2	12	1.338	1 11/32
1 1/2	6	1.448	1 15/32
1 5/8	5 1/2	1.555	1 9/16
1 3/4	5	1.680	1 11/16
1 7/8	5	1.783	1 25/32
2	4 1/2	1.909	1 29/32
2 1/8	4 1/2	2.034	2 1/32
2 1/4	4 1/2	2.131	2 1/8
2 3/8	4	2.256	2 1/4
2 1/2	4	2.381	2 3/8
2 5/8	3 1/2	2.506	2 1/2
2 3/4	3 1/2	2.597	2 19/32
2 7/8	3 1/2	2.722	2 23/32
3	3 1/4	2.847	2 27/32
3 1/8	3 1/4	2.972	2 31/32
3 1/4	3 1/4	3.075	3 1/16
3 3/8	3 1/4	3.200	3 3/16
3 1/2	3 1/4	3.325	3 5/16
3 3/4	3	3.425	3 7/16
4	3	3.675	3 11/16

(L. S. Starrett)

TYPICAL IGNITION COMPONENT TESTS

The following pages illustrate typical ignition component tests using three common testers. These examples are for illustrative purposes only and may not apply to all makes and models. Always refer to an appropriate service manual for specific hookup information before performing ignition component tests.

CONDENSER CONTINUITY TEST USING A MERC-O-TRONIC MODEL 79 TESTER:

1. Insert the test leads into the "Test Leads" jacks.
2. Connect the black test clip to the base of the condenser.
3. Connect the red test clip to the condenser lead.
4. Turn the selector switch to the "Cont./Cond. Test" position.
5. Press the "Cont./Cond. Test" push switch. The L.E.D. will glow brightly and then slowly dim as the condenser becomes charged.
6. If the L.E.D. remains bright, the condenser is defective (due to leakage or short).

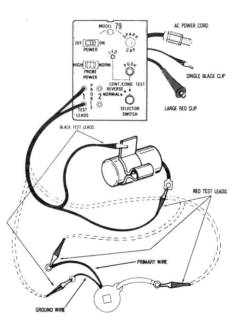

COIL CONTINUITY TEST USING A MERC-O-TRONIC MODEL 79 TESTER:

1. Insert the test leads into the "Test Leads" jacks. Turn the selector switch to "Cont./Cond. Test."
2. By connecting the red and black test clips to the ignition coil primary or secondary wires (or across any wire), continuity can be checked.
3. Press the "Cont./Cond. Test" push switch. The L.E.D. will light if there is continuity.

SOLID STATE PULSE TRANSFORMER TEST USING A MERC-O-TRONIC MODEL 79 TESTER:

1. Insert the test leads into the "Test Leads" jacks.
2. Connect the black test lead to the coil primary ground.
3. Connect the red test lead to the coil positive primary wire.
4. Connect the large red test lead to the high tension wire (use adapter if needed).
5. Connect the single black test lead from the analyzer to the coil primary ground wire. Normally, both black test leads will be connected in the same place.
6. Place the tester's selector switch in the "Normal" position.
7. Turn the power on.
8. Push the probe power switch to "Normal."
9. A strong, steady spark should occur across the spark gap.
10. If the spark is faint, intermittent, or does not occur at all, the coil is defective and must be replaced.

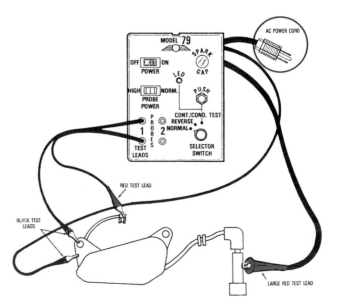

CONDENSER CONTINUITY TEST USING A GRAHAM-LEE MODEL 31-SM (31-SM X-H) TESTER:

Note: This is a test for continuity only–Microfarad values are not used here. The condenser may be tested on the stator, but it must be disconnected at the terminals. "Short" the condenser lead to the condenser body to remove any charge.

1. Plug the tester into a 110/120 volt outlet.
2. Turn the toggle "Main Switch" to the "On" position. Allow the tester to warm up for a minute.
3. Connect the black lead (alligator clip) of the tester to the condenser body.
4. Connect the red lead (alligator clip) of the tester to the condenser terminal.
5. The continuity light will glow briefly if the condenser is good. If the continuity light remains on, the condenser is leaking.
6. To repeat this test, disconnect the red and black leads, ground the condenser to the terminal lead, and repeat the steps to achieve the results in step 5.

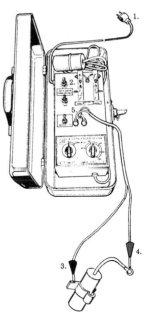

COIL FIRING TEST USING A GRAHAM-LEE MODEL 31-SM (31-SM X-H) TESTER:

Note: This coil test can be made on the engine, but the condenser and the coil leads must be disconnected at the terminal. (If the condenser and coil leads are clamped together, they must be separated.)

1. Plug the tester into a 110/120 volt outlet.
2. Turn the toggle "Main Switch" to the "On" position. Allow the tester to warm up for a minute.
3. Connect the black lead (alligator clip) of the tester to the ground lead of the coil.
4. Connect the red lead (alligator clip) of the tester to the primary lead of the coil.
5. Connect a lead from the spark plug gap labeled "Normal" to the secondary terminal (high tension lead) of coil.
6. Turn the "Coil Index" knob counterclockwise as far as it will go.
7. Turn the "Coil Index" knob clockwise until a spark arcs across the air gap. The setting should not exceed 70.

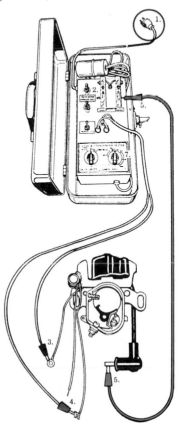

SOLID STATE IGNITION SYSTEM TEST (IGNITION UNITS) USING A GRAHAM-LEE MODEL 31-SM (31-SM X-H) TESTER:

Note: This solid state unit must be tested off of the engine.

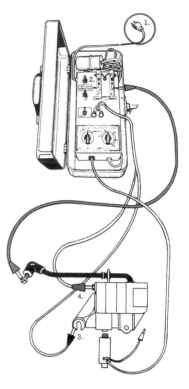

1. Plug the tester into a 110/120 volt outlet.
2. Turn the toggle "Main Switch" to the "On" position. Allow the tester to warm up.
3. Connect the black lead (alligator clip) to the ignition unit ground (frame) or the trigger pack ground lead.
4. Connect the red lead (alligator clip) to the ignition unit terminal.
5. Connect the spark plug terminal to the spark gap labeled "Normal" on the tester.
6. Set the "Coil Index" to 80. Hold the "Coil Tests" switch to the "On" position.
7. With the red end of the trigger coil up to the trigger magnet on the ignition unit, a low frequency arcing should occur. When the trigger pack is pulled away, firing should stop.
8. Release the spring switch while the trigger pack is pulled away from the ignition unit and disconnect the red lead.
9. Operate the spring switch and bring the trigger coil red end to the trigger magnet on the ignition unit as before. A single spark should jump the spark gap, indicating that the capacitor in the ignition unit is holding a charge.

CONDENSER CAPACITY TEST USING A MERC-O-TRONIC MODEL 9800 (98, 98A) TESTER:

Note: To determine if a condenser meets its rated value, check the appropriate specification table. See the example below. A condenser can be tested on the stator but must be disconnected at the terminals.

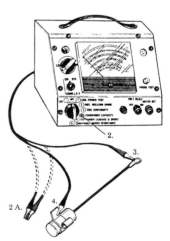

1. If a Merc-O-Tronic tester 98 or 98A is used, plug the tester into a 110/120 volt outlet. (Model 9800 does not require 110/120 volts.)
2. Position the "Selector Switch" to position 4, "Condenser Capacity."
   To Re-calibrate Meter:
   a. Clip the black and red test leads together.
   b. For Merc-O-Tronic testers 98 and 98A, depress the red button to set the meter to the top of the scale. For a 9800 model, turn the "Meter Set" knob to set line on scale 4.
   c. Unclip the test leads.
3. Connect the red test lead (alligator clip) to the condenser terminal.
4. Connect the black test lead (alligator clip) to the body of the condenser.
5. Read the value on scale 4. (On 98 and 98A models, the red button must be depressed.)
6. If the value found in the test is not the same as listed in the specification table, replace the condenser.

CONDENSERS	
29164	.16 - .20
29177	.15 - .18
30548A	.16 - .18
30548B	.16 - .18
610370	.15 - .19
610467	.12 - .16
610588	.18 - .22
610642	.16 - .18
610767A	.16 - .18

COIL POWER TEST USING A MERC-O-TRONIC MODEL 9800 (98, 98A) TESTER:

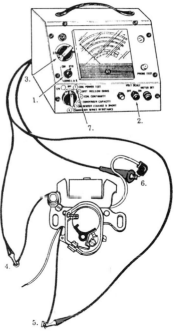

Note: Coil test can be made on the engine. Isolation of the coil leads can be made by placing a piece of cardboard between the points or by separating the coil primary lead, ground lead, and condenser lead as shown.
WARNING: Never perform this test without the spark plug high tension lead attached to the tester's large red lead.

1. The ignition selector switch on the 9800 tester must be in the "Std." position. (If 98 or 98A testers are used, a 55-980 adapter must be used.)
2. The "Volt Scale" must be in the "Off" position.
3. The "Lo-Hi" control knob must be in the lowest possible position.
4. Connect the black test lead (alligator clip) to the coil primary ground lead or the stator plate.
5. Connect the red test lead (alligator clip) to the primary coil lead.
6. Connect the large red test lead (alligator clip) to the terminal of the spark plug wire.
7. Turn the selector switch to coil power test No. 1.
8. Slowly turn the "Lo-Hi" knob clockwise and note the value on scale 1. When the meter reads the operating amperage for a particular winding (refer to coil manufacturer's specifications), stop turning the knob. The 5 mm gap should fire steadily.

COIL	OPERATING AMPERAGE	COIL	OPERATING AMPERAGE
27947	2.8	610371	2.8
29176	2.8	610466	2.8
29632	2.3	610477	2.8
30546	2.7	610586	2.8
30560A	1.6	610633	1.8
30560	1.6	610657	2.25
34431	1.6	610678	2.25

SOLID STATE IGNITION SYSTEM TEST USING A MERC-O-TRONIC MODEL 9800 (98, 98A) TESTER AND A CD 55-700 ADAPTER:

If using a model 9800 tester, turn the ignition selector switch to the "Std." position.

1. Attach a lead from the ignition coil terminal to the terminal on the ignition unit.
2. Attach a jumper lead from the ignition mounting plate to the ignition coil mounting plate.
3. Connect the black test lead (alligator clip) to the black terminal on the CD 55-700 unit.
4. Connect the red test lead (alligator clip) to the red terminal on the CD 55-700 unit.
5. Connect the large red test lead from the tester to the spark plug terminal on the high tension lead.
6. Attach the CD 55-700 adapter red lead to the ground cut-off terminal on the ignition unit (not pictured, located on the backside of the ignition unit).
7. Attach the CD 55-700 adapter black lead to the ignition unit mounting plate.
8. Place the electro-magnetic triggering device near the trigger terminal of the ignition unit.
9. Position the selector switch to "Coil Power Test."
10. Slowly turn the "Lo-Hi" current knob clockwise until the ignition unit fires when viewed through the spark gap window.

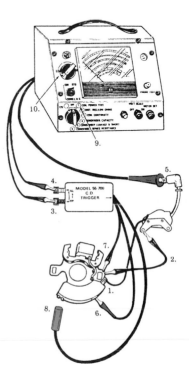

Note: While the test is in progress, it may be necessary to move the triggering device to achieve proper alignment with the trigger on the ignition unit.

(Ignition component tests courtesy of Tecumseh Products Co.)

SPARK PLUG CONDITION CHART

NORMAL APPEARANCE
A spark plug in a sound engine operating at the proper temperature will have deposits that range from tan to gray in color. If LP-Gas is used, the deposits will be brown. Under normal conditions, the electrode should wear slightly, but there should be no evidence of burning.

CARBON FOULING
Carbon fouling (dry, black, sooty carbon deposits) can be caused by plugs that are too cold for the engine, an over-rich fuel mixture, a clogged air cleaner, a faulty choke, or sticking valves. Installing a hotter plug will temporarily solve this problem.

OIL FOULING
Oil fouling (wet, black deposits) is caused by excessive oil in the combustion chamber. Worn rings, valve guides, valve seals, and cylinder walls can cause oil fouling. Switching to a hotter spark plug may temporarily relieve the symptoms, but will not correct the problem.

ASH FOULING
Ash fouling is caused by the buildup of heavy combustion deposits. These deposits are formed by burning oil and/or fuel additives. Although ash fouling is not conductive, excessive deposits can cause a spark plug to misfire.

SPLASHED FOULING
Splashed fouling can occur after a long-delayed tune-up. When a new plug is installed in an engine with excessive piston and combustion chamber deposits, the plug will restore regular firing impulses and raise the combustion temperature. When this occurs, accumulated engine deposits may flake off and stick to the hot plug insulator.

GAP BRIDGING
Gap bridging (combustion deposit bridging the center and ground electrodes) is caused by a sudden burst of high speed operation following excessive idling. It can also be caused by improper fuel additives, obstructed exhaust ports (two cycle engines), and excessive carbon in the cylinder.

HIGH SPEED GLAZING
High speed glazing (hard, shiny, electrically conductive deposits) can be caused by a sudden increase in plug temperature during hard acceleration or loading. High speed glazing can cause the engine to misfire at high speeds. If high speed glazing recurs, a cooler plug should be used.

PREIGNITION
Preignition (fuel charge ignited by a glowing combustion chamber deposit or a hot valve edge before the spark plug fires) can cause extensive plug damage. When plugs show evidence of preignition, check the heat range of the plugs, the condition of the plug wires, and the condition of the cooling system.

DETONATION
Detonation can cause the insulator nose of a spark plug to fracture and chip away. The explosions that occur during heavy detonation produce extreme pressure in the cylinder. Detonation can be caused by low octane fuel, advanced ignition timing, or an excessively lean fuel mixture.

OVERHEATING
Overheating (dull, white insulator and eroded electrodes) can occur when a spark plug is too hot for the engine. Advanced ignition timing, cooling system problems, detonation, sticking valves, and excessive high-speed operation can also cause spark plug overheating.

MECHANICAL DAMAGE
Mechanical damage can be caused by a foreign object in the combustion chamber. It can also occur if the piston hits the firing tip of a spark plug with improper reach. When working on an engine, keep spark plug hole(s) and carburetor throat covered to prevent foreign objects from entering the combustion chamber.

WORN OUT
Extended use will cause the spark plug's center electrode to erode. When the electrode is too worn to be filed flat, the plug must be replaced. Typical symptoms of worn spark plugs include excessive fuel consumption and poor engine performance.

DICTIONARY OF TERMS

A

ABRASION: Wearing or rubbing away.

ADDITIVE: A material that is added to the oil to give it certain properties. Example: an additive is blended with engine oil to lessen its tendency to congeal or thicken at low temperatures.

AEA: Automotive Electric Association.

AIR: A gas containing approximately 4/5 nitrogen, 1/5 oxygen and some carbonic gas.

AIR BLEED: A tube in a carburetor through which air can pass into fuel moving through a fuel passage.

AIR CLEANER: A device for filtering, cleaning, and removing dust from the air admitted to an engine.

AIR-COOLED ENGINE: An engine cooled by air.

AIR-FUEL RATIO: Ratio, by weight, of fuel as compared to air in carburetor mixture.

AIR GAP: The space between the spark plug electrodes.

AIR HORN: Part of air passage in carburetor that is on the atmospheric side of the venturi. The choke valve is located in the air horn.

AIR LOCK: A bubble of air trapped in a fluid circuit that interferes with normal circulation of the fluid.

ALIGNMENT: An adjustment to bring related components into a line.

ALLEN WRENCH: A hexagonal (six-sided) wrench that fits into a recessed hexagonal hole. Used commonly with set screws.

ALLOY: A mixture of different metals. For example, solder is an alloy of lead and tin.

ALTERNATING CURRENT: An electrical current alternating flow back and forth in a circuit.

ALTERNATOR: A generator that produces alternating current.

ALUMINUM: A metal noted for its lightness. Aluminum is often alloyed with small quantities of other metals.

AMMETER: An instrument for measuring the flow of electric current.

AMPERE: Unit of measurement for flow of electric current.

AMPERE-HOUR CAPACITY: A term to indicate the capacity of a storage battery.

ANTIFREEZE: A material, such as ethylene glycol, that is added to water to lower its freezing point.

ANTIFRICTION BEARING: A bearing constructed with balls or rollers between journal and bearing surfaces to provide rolling instead of sliding friction.

APEX SEALS: The spring-loaded seals located at the apexes (points or tips) of the triangular rotor in the Wankel engine.

API: American Petroleum Institute.

ARC: A discharge of electric current across a gap, such as between electrodes.

ARMATURE: Part of an electrical motor or generator that includes the main current-carrying winding.

ASME: American Society of Mechanical Engineers.

ATMOSPHERIC PRESSURE: The weight of air at sea level (about 14.7 psi; less at higher altitudes).

B

BABBITT: An alloy of tin, copper, and antimony having good antifriction properties. Used as a facing for bearings.

BACKFIRE: Ignition of the mixture in the intake manifold by a flame from the cylinder, such as might occur from a leaking or open intake valve.

BACKLASH: The clearance or "play" between two parts, such as meshed gears.

BACK PRESSURE: A resistance to free flow, such as a restriction in the exhaust line.

BAFFLE OR BAFFLE PLATE: An obstruction for checking or deflecting the flow of gases or sound.

BALL BEARING: An anti-friction bearing consisting of a hardened inner and outer race with hardened steel balls set between the two races.

BATTERY: Any number of complete electrical cells assembled in one housing or case.

BCI: Battery Council International.

BDC: Bottom dead center.

BEARING: A part in which a journal, pivot, or similar object turns or moves.

BEVEL: The angle that one surface makes with another when they are not at right angles.

BHP: Brake horsepower. A measurement of power developed by an engine in actual operation.

BLOW-BY: A leakage or loss of pressure, often used in reference to leakage of compression past piston ring between piston and cylinder.

BOILING POINT: The temperature at atmospheric pressure at which bubbles or vapors rise to the surface of a liquid and escape.

BORE: The diameter of a hole, such as a cylinder; also to enlarge a hole as opposed to making a hole with a drill.

BOSS: An extension or strengthened section that supports piston pin or piston pin bushings.

BOUNCE: As applied to engine valves, a condition where the valve is not held tightly to its seat when the cam is not lifting it. In an ignition distributor, a condition where breaker points make and break contact when they should remain closed.

BREAKER ARM: Movable part of a pair of contact points in an ignition distributor or magneto.

BREAKER POINTS: Two separable points that interrupt primary circuit in a distributor or magneto for the purpose of inducing a high tension current in the ignition system.

Breaker points are usually faced with silver, platinum, or tungsten.

BREAK-IN: The process of wearing into a desirable fit between surfaces of two new or reconditioned parts.

BRUSHES: The bars of carbon or other conducting material that contact the commutator of an electric motor or generator.

BTU (British Thermal Unit): A measurement of the amount of heat required to raise the temperature of 1 lb. of water, 1°F.

BUSHING: A removable liner for a bearing.

BYPASS: An alternate path for a flowing substance.

C

CALIBRATE: To determine or adjust the graduation or scale of any instrument giving quantitative measurements.

CAM OR BREAKER CAM: The lobed cam rotating in the ignition system that interrupts the primary circuit to induce a high tension spark for ignition.

CAM ANGLE: The number of degrees of distributor shaft rotation during which contact points are closed.

CAM-GROUND PISTON: A piston ground to a slightly oval shape, which, under the heat of operation, becomes round.

CAMSHAFT: Shaft containing lobes or cams that operate engine valves.

CAPACITANCE: The property that opposes any change in voltage in all electrical circuits.

CAPACITOR: A device that possesses capacitance (stores electricity). In simple state, consists of two metal plates separated by an insulator.

CARBON: A common nonmetallic element that is an excellent conductor of electricity. It also forms in the combustion chamber of an engine during the burning of fuel and oil.

CARBON MONOXIDE: Gas formed by incomplete combustion. Colorless, odorless, and very poisonous.

CARBURETOR: A device for automatically mixing fuel in proper proportion with air to produce a combustible gas.

CARBURETOR "ICING": A term used to describe formation of ice on a carburetor throttle plate during certain atmospheric conditions.

CASE HARDEN: To harden the surface of steel.

CDI: Capacitive discharge ignition.

CELL: Part of a battery containing a group of positive and negative plates along with electrolyte.

CELL CONNECTOR: Lead bar or link connecting the pole of one battery cell to the pole of another.

CELSIUS: A measurement of temperature. Zero on Celsius scale is 32° on Fahrenheit scale.

CENTRIFUGAL FORCE: A force that tends to move a body away from its center of rotation. Example: a whirling weight attached to a string.

CHAMFER: A bevel or taper at the edge of a hole.

CHARGE (or Recharge): Passing an electrical current through a battery to restore it to activity.

CHECK VALVE: A gate or valve that allows passage of gas or fluid in one direction only.

CHOKE: A reduced passage, such as a valve placed in a carburetor air inlet to restrict the volume of air admitted.

CIRCUIT: The path of electrical current, fluids or gases. Examples: for electricity, a wire; for fluids and gases, a pipe.

CLEARANCE: The space allowed between two parts, such as between a journal and a bearing.

CLOCKWISE ROTATION: Rotation in same direction as the hands of a clock.

COEFFICIENT OF FRICTION: A measurement of the amount of friction developed between two surfaces that are rubbed together.

COIL: An electrical device made up of a number of coils in spiral form to provide electrical resistance.

COMBUSTION: The process of burning.

COMBUSTION SPACE OR CHAMBER: Volume of cylinder above piston with piston at top center.

COMMUTATOR: A ring of copper bars in a generator or electric motor providing connections between armature coils and brushes.

COMPOUND: A mixture of two or more ingredients.

COMPRESSION: The reduction in volume or the "squeezing" of a gas. As applied to metal, such as a coil spring, compression is the opposite of tension.

COMPRESSION RATIO: Volume of combustion chamber at end of compression stroke as compared to volume of cylinder and chamber with piston on bottom center.

COMPONENTS: The parts that constitute a whole.

CONCENTRIC: Two or more circles having a common center.

CONDENSATION: The process of vapor becoming a liquid; the reverse of evaporation.

CONDENSER: A device for temporarily collecting and storing a surge of electrical current for later discharge.

CONDUCTION: The flow of electricity through a conducting body.

CONDUCTOR: A material along or through which electricity will flow with slight resistance; silver, copper, and carbon are good conductors.

CONNECTING ROD: Rod that connects the piston to the crankshaft.

CONTACT BREAKER: A device for interrupting an electrical circuit; often automatic and may be known as a "circuit breaker," "interrupter," "cut-out," or "relay."

CONTACT POINTS: Also called breaker points. Two separable points usually faced with silver, platinum, or tungsten, which interrupt the primary circuit in the distributor or magneto for the purpose of inducing a high tension current in the ignition system.

CONTRACTION: A reduction in mass or dimension; the opposite of expansion.

CONVECTION: A transfer of heat by circulating heated air.

CONVERTER: As used in connection with liquefied petroleum gas, it is a device that converts or changes LP-Gas from liquid to vapor state for use in the engine.

CORRODE: To eat away gradually, especially by chemical action.

COUNTERCLOCKWISE ROTATION: Rotating in the opposite direction of the hands on a clock. Anti-clockwise rotation.

COUPLING: A connecting means for transferring movement from one part to another; may be mechanical, hydraulic, or electrical.

CRANKCASE: The housing for the crankshaft and other related internal parts.

CRANKCASE DILUTION: Under certain conditions of operation, unburned fuel finds its way past the piston rings and into the crankcase and oil reservoir, where it dilutes or "thins" the engine lubricating oil.

CRANKSHAFT: The main shaft of an engine that, working with the connecting rods, changes the reciprocating motion of the pistons into rotary motion.

CRANKSHAFT COUNTER-BALANCE: A series of weights attached to or forged integrally with the crankshaft and placed to offset the reciprocating weight of each piston and rod assembly.

CRUDE OIL: Oil as it comes from the ground. Unrefined.

CU. IN.: Cubic inch.

CURRENT: The flow of electricity.

CYLINDER: A round chamber of some depth bored to receive a piston; also sometimes referred to as "bore" or "barrel."

CYLINDER BLOCK: The largest single part of an engine. The basic or main mass of metal in which the cylinders are bored or placed.

CYLINDER HEAD: A detachable portion of an engine fastened securely to the top of the cylinder block that contains all or a part of the combustion chamber.

CYLINDER SLEEVE: A liner or tube placed between the piston and the cylinder wall or cylinder block to provide a readily renewable wearing surface of the cylinder.

D

DEAD CENTER: The extreme upper or lower position of the crankshaft throw at which the piston is not moving in either direction.

DEGREE: Abbreviated deg. or indicated by a "°" placed alongside of a figure; may be used to designate temperature readings or angularity, one degree being 1/360 part of a circle.

DEMAGNETIZE: To remove the magnetization of a pole that has previously been magnetized.

DENSITY: Compactness; relative mass of matter in a given volume.

DETERGENT: A compound used in engine oil to remove engine deposits and hold them in suspension in the oil.

DETONATION: A excessively rapid burning or explosion of the mixture in the engine cylinders. It becomes audible through a vibration of the combustion chamber walls and is sometimes confused with a "ping" or spark knock.

DIAPHRAGM: A flexible partition or wall separating two cavities.

DIE CASTING: An accurate and smooth casting made by pouring molten metal or composition into a metal mold or die under pressure.

DILUENT: A fluid that thins or weakens another fluid. Example: gasoline dilutes oil.

DILUTION: See Crankcase Dilution.

DIRECT CURRENT: Electric current that flows continuously in one direction. Example: a storage battery.

DISCHARGE: The flow of electric current from the battery; the opposite of charge.

DISPLACEMENT: See Piston Displacement.

DISTORTION: A warpage or change in form from the original shape.

DOWEL PIN: A pin inserted in matching holes in two parts to maintain those parts in fixed relation to each other.

DOWN-DRAFT: Used to describe a carburetor type in which the mixture flows downward to the engine.

DROP FORGING: A piece of steel shaped between dies while hot.

DWELL PERIOD: The number of degrees the breaker cam rotates from the time the breaker points close until they open again. Also known as cam angle.

DYNAMOMETER: A machine for measuring the actual power produced by an internal combustion engine.

E

ECCENTRIC: One circle within another circle, each having a different center. Example: a cam on a camshaft.

ELECTRODE: Usually refers to the insulated center rod of a spark plug. It is also sometimes used to refer to the rods attached to the shell of the spark plug.

ELECTROLYTE: A mixture of sulphuric acid and distilled water used in wet-type storage batteries.

ELEMENT: One set of positive plates and one set of negative plates complete with separators assembled together.

EMF (Electromotive Force or Voltage): See Electrical Section.

ENERGY: The capacity for doing work.

ENGINE: The term applies to the prime source of power generation.

ENGINE DISPLACEMENT: The sum of the displacement of all the engine cylinders. See Piston Displacement.

EPITROCHOID: A geometric path followed by a specific point located in a generating circle, which is rolled around the periphery of a base circle.

ETHYL GASOLINE: Gasoline to which Ethyl fluid has been added. Ethyl fluid is a compound of tetraethyl lead, ethylene dibromide, and ethylene dichloride. The purpose of the material is to slow down and control the burning rate of the fuel in the cylinder to produce an expansive force rather than an explosive force. Thus, detonation or "knocking" is reduced.

EVAPORATION: The process of changing from a liquid to a vapor, such as boiling water to produce steam. Evaporation is the opposite of condensation.

EXHAUST: The spent fuel after combustion takes place in an internal combustion engine.

EXHAUST PIPE: The pipe connecting the engine to the muffler to conduct the exhausted or spent gases away from the engine.

EXPANSION: An increase in size. Example: when a metal rod is heated, it increases in length and diameter. Expansion is the opposite of contraction.

F

FAHRENHEIT (F): A scale of temperature measurement ordinarily used in English-speaking countries. The boiling point of water is 212° Fahrenheit, as compared to 100° Celsius.

FEELER GAGE: A metal strip or blade finished accurately with regard to thickness and used for measuring the clearance between two parts. Feeler gages ordinarily come in a set of different blades graduated in thickness by increments of .001 in.

FERROUS METAL: Metals that contain iron or steel and are, therefore, subject to rust.

FIELD: In a generator or electric motor, the area in which magnetic flow occurs.

FIELD COIL: A coil of insulated wire surrounding the field pole.

FILLET: A rounded filling, such as a weld between two parts joined at an angle.

FILTER (Oil, Water, Gasoline, etc.): A unit containing an element, such as a screen of varying degrees of fineness. The screen or filtering element is made of various materials depending upon the size of the foreign particles to be eliminated from the fluid being filtered.

FIT: A kind of contact between two machined surfaces.

FLANGE: A projecting rim or collar on an object for keeping it in place.

FLASHOVER: Tendency of current to travel down the outside of a spark plug instead of through the center electrode.

FLASH POINT: The temperature at which an oil will flash and burn.

FLOAT: A hollow tank that is lighter than the fluid in which it rests. A float is ordinarily used to automatically operate a valve controlling the entrance of a fluid.

FLOATING PISTON PIN: A piston pin that is not locked in the connecting rod or the piston, but is free to turn or oscillate in both the connecting rod and the piston.

FLOAT LEVEL: The predetermined height of the fuel in the carburetor bowl, usually regulated by means of a suitable valve.

"FLUTTER" OR "BOUNCE": A condition arising from a valve not being held tightly on its seat when the cam is not lifting it.

FLYWEIGHTS: Special weights that react to centrifugal force to provide automatic control of other mechanisms, such as accelerators or valves.

FLYWHEEL: A heavy wheel in which energy is absorbed and stored by means of momentum.

FOOT POUND (or ft.-lbs.): This is a measure of the amount of energy or work required to lift 1 lb. 1 ft.

FORCE-FIT: Also known as a press-fit, interference-fit or drive-fit. This term is used when a shaft is slightly larger than a hole and must be forced in place.

FORGE: To shape plastic or hot metal by hammering.

FOUR CYCLE ENGINE: Also known as Otto cycle. In a four cycle engine, an explosion occurs every other revolution of the crankshaft. A cycle is considered as a half revolution of the crankshaft. The cycles (strokes) are (1) intake; (2) compression; (3) power and; (4) exhaust stroke.

FRICTION: Resistance to motion created when one surface rubs against another.

FUEL KNOCK: Same as detonation.

FULCRUM: The support on which a lever turns in moving a body.

G

GAS: A substance that can be changed in volume and shape according to the temperature and pressure applied to it. Example: air is a gas that can be compressed into smaller volume and into any shape desired by pressure. It can also be expanded by the application of heat.

GASKET: Anything used as a packing, such as a nonmetallic substance placed between two metal surfaces to act as a seal.

GASSING: The bubbling of the battery electrolyte that occurs during the process of charging a battery.

GEAR RATIO: The number of revolutions made by a driving gear as compared to the number of revolutions made by a driven gear of a different size. Example: if one gear makes three revolutions while the other gear makes one revolution, the gear ratio would be 3 to 1.

GENERATOR: A device consisting of an armature, field coils, and other parts, which, when rotated, generates electricity. A generator is usually driven by a belt from the engine crankshaft.

GLAZE: An extremely smooth or glossy surface finish, such as a cylinder wall that is highly polished over a long period of time by the friction of the piston rings.

GLAZE BREAKER: A tool for removing the glossy surface finish in an engine cylinder.

GLOW PLUG: A device with a fine wire connected in series to an electrical circuit for the purpose of creating enough resistance and heat to ignite fuel in a combustion chamber. Used in place of spark plugs on some engines.

GOVERNOR: A mechanical, hydraulic, or electrical device that controls and regulates speed.

GRID: The metal framework of an individual battery plate in which the active material is placed.

GRIND: To finish or polish a surface by means of an abrasive material.

GUM: Oxidized petroleum products that accumulate in the fuel system, carburetor, or engine parts.

H

HEAT TREATMENT: A combination of heating and cooling operations timed and applied to a metal in a solid state in a way that will produce desired properties.

HIGH TENSION: The secondary or induced high-voltage electrical current. High tension is also used in reference to the wiring from the distributor cap to the coil and the spark plugs.

HONE: An abrasive tool for correcting small irregularities or differences in the diameter of a cylinder.

HOT SPOT: Refers to a comparatively thin section or area of the wall between the inlet and exhaust manifold of an engine, the purpose being to allow the hot exhaust gases to heat the comparatively cool incoming mixture. Also used to designate local areas of the cooling system that have above average temperatures.

HP: Horsepower. The energy required to lift 550 lbs. 1 ft. in 1 second.

HYDROCARBON: Any compound composed entirely of carbon and hydrogen, such as petroleum products.

HYDROCARBON ENGINE: An engine using petroleum products, such as gas, liquefied gas, gasoline, kerosene, or fuel oil as a fuel.

HYDROMETER: An instrument for determining the state of charge in a battery by finding the specific gravity of the electrolyte.

I

ICEI: Internal Combustion Engine Institute, Inc.

ID: Inside Diameter.

IDLE: Refers to the engine operating at its slowest practical speed.

IGNITION DISTRIBUTOR: An electrical unit containing the circuit breaker for the primary circuit and providing a means for conveying the secondary or high tension current to the spark plug wires as required.

IGNITION SYSTEM: The means for igniting the fuel in the cylinders; includes spark plugs, wiring, ignition distributor, ignition coil, and source of electrical current supply.

IHP: Indicated horsepower developed by an engine. A measurement of the pressure of the explosion within the cylinder expressed in pounds per square inch.

IMPELLER: A rotor or wheel with blades to pump water or propel objects through water or other fluids.

INDUCTION: The influence of different strength magnetic fields that are not electrically connected to one another.

INDUCTION COIL: Essentially a transformer that, through the action of induction, creates a high tension current by means of an increase in voltage.

INERTIA: A physical law that tends to keep a motionless body at rest and also tends to keep a moving body in motion. Therefore, effort required to start a mass moving or to retard or stop it once it is in motion.

INHIBITOR: A material to restrain or hinder some unwanted action, such as a rust inhibitor, which is a chemical added to a cooling system to retard rust formation.

INLET VALVE OR INTAKE VALVE: A valve that permits a fluid or gas to enter a chamber and seals against exit.

INSULATION: Any material that does not conduct electricity; used to prevent the flow or leakage of current from a conductor. Also, used to describe a material that does not conduct heat readily.

INSULATOR: An electrical conductor covered or shielded with a nonconducting material, such as a copper wire within a rubber tube.

INTAKE MANIFOLD OR INLET PIPE: The tube used to conduct the gasoline and air mixture from the carburetor to the engine cylinders.

INTEGRAL: Formed as a unit with another part.

INTENSIFY: To increase or concentrate, such as to increase the voltage of an electrical current.

INTERFERENCE FIT: Difference in angle between mating surfaces of a valve and a valve seat.

INTERMITTENT: Motion or action that is not constant but occurs at intervals.

INTERNAL COMBUSTION: The burning of a fuel within an enclosed space.

J

JOURNAL: The part of a shaft or crank that rotates inside of a bearing.

JUMP SPARK: A high tension electrical current that jumps through the air from one terminal to another.

K

KEY: A small block inserted between a shaft and hub to fasten a pulley or gear to the shaft.

KEYWAY OR KEYSEAT: A groove or slot cut to permit the insertion of a key.

KNURL: To indent or roughen a finished surface.

KNOCK: A general term used to describe various noises occurring in an engine; may be used to describe noises made by loose or worn mechanical parts, preignition, or detonation.

L

LACQUER: A solution of solids in solvents that evaporate with great rapidity.

LAMINATE: To build up or construct out of a number of thin sheets. Example: the laminated core in a magneto coil.

LAPPING: The process of fitting one surface to another by rubbing them together with an abrasive material between the two surfaces.

LEAD: A short connecting wire that makes electrical contact between two points.

LINER: Usually a thin section placed between two parts, such as a replaceable cylinder liner in an engine.

LINKAGE: Any series of rods, yokes, and levers, etc., used to transmit motion from one unit to another.

LIQUID: Any substance that assumes the shape of the vessel in which it is placed without changing volume.

LOBE: An off center or eccentric enlargement on a shaft that converts rotary motion to reciprocating motion. Also called a cam.

LOST MOTION: Motion between a driving part and a driven part that does not move the driven part. Also see Backlash.

LP-GAS: Made usable as a fuel for internal combustion engines by compressing volatile petroleum gases to liquid form. When so used, must be kept under pressure or at low temperature in order to remain in liquid form.

M

MAGNET (Permanent): A piece of hard steel that can be charged with and retain magnetic power. Magnets are often bent into a "U" shape so as to have opposite poles.

MAGNETIC FIELD: The flow of magnetic force or magnetism between the opposite poles of a magnet.

MAGNETO: An electrical device that generates current when rotated by an outside source of power; may be used for the generation of either low tension or high tension current.

MANIFOLD: A pipe with multiple openings used to connect various cylinders to one inlet or outlet.

MECHANICAL EFFICIENCY: The ratio between the indicated horsepower and the brake horsepower of an engine.

MEMA: Motor and Equipment Manufacturer's Association.

MEWA: Motor and Equipment Wholesaler's Association.

MICROMETER: A measuring instrument for either external or internal measurement in thousandths or ten thousandths of an inch.

MILL: To cut or machine with rotating, toothed cutters.

MISFIRING: Failure of an explosion to occur in one or more cylinders while the engine is running; may be continuous or intermittent failure.

MODULE: A packaged functional assembly of wired electronic components for use with other such assemblies.

MOTOR: This term should be used in connection with an electric motor and should not be used when referring to the engine.

MUFFLER: A chamber attached to the end of the exhaust pipe that allows the exhaust gases to expand and cool. It is usually fitted with baffles or porous plates and serves to subdue much of the noise created by the exhaust.

N

NEEDLE BEARING: An anti-friction bearing using a great number of small, cylindrical rollers; also known as a quill-type bearing.

NEGATIVE POLE: The point from which an electrical current flows through the circuit. It is designated by a minus sign (-).

NEON TUBE: An electric "bulb" or tube filled with a rare gas. Neon tubes are often used on ignition test instruments.

NONFERROUS METALS: This designation includes practically all metals that do not contain iron (or contain very little iron) and, therefore, are not subject to rusting.

NORTH POLE: The pole of a magnet where the lines of force start; the opposite of south pole.

NSPA: National Standard Parts Association.

O

OD: Outside diameter.

OHM: A measurement of the resistance to the flow of an electrical current through a conductor.

OIL PUMPING: A term used to describe an engine that is using an excessive amount of lubricating oil.

OPEN CIRCUIT: A break or opening in an electrical circuit that interferes with the passage of the current.

OSCILLATE: To swing back and forth like a pendulum.

OTTO CYCLE: Also called four-stroke cycle. Named after the man who adopted the principle of four cycles of operation for each explosion in an engine cylinder. The four cycles of operation are: (1) intake, (2) compression, (3) power, (4) exhaust.

OXIDIZE: To combine an element with oxygen or convert it into its oxide. The process is often accomplished by a combination. For example, when carbon burns, it combines with oxygen to form carbon dioxide or carbon monoxide. When iron rusts, the iron has combined with oxygen from the air to form the rust (oxide of iron).

P

PETROLEUM: A group of liquid and gaseous compounds composed of carbon and hydrogen, which are removed from the earth.

PHILLIPS SCREW: A screw head having a cross instead of a slot for a corresponding type of screwdriver.

PISTON: A cylindrical part that is closed at one end and connected to the crankshaft by the connecting rod. The force of the explosion in the cylinder is exerted against the closed end of the piston, causing the connecting rod to move the crankshaft.

PISTON COLLAPSE: An abnormal reduction in the diameter of the piston skirt due to heat or stress.

PISTON DISPLACEMENT: The volume of air moved or displaced by moving the piston from one end of its stroke to the other.

PISTON HEAD: The part of the piston above the rings.

PISTON LANDS: Those parts of a piston between the piston rings.

PISTON PIN (Wrist Pin): The journal for the bearing in the small end of an engine connecting rod. The piston pin also passes through the piston walls.

PISTON RING: An expanding ring placed in the grooves of the piston to seal it against the passage of fluid or gas.

PISTON RING EXPANDER: A spring placed behind the piston ring in the groove to increase the pressure of the ring against the cylinder wall.

PISTON RING GAP: The clearance between the ends of the piston ring.

PISTON RING GROOVE: The channel or slots in the piston in which the piston rings are placed.

PISTON SKIRT: The part of the piston below the rings and the bosses.

PISTON SLAP: Rocking of loose fitting piston in a cylinder, making a hollow bell-like sound.

PIVOT: A pin or short shaft upon which another part rests, turns, rotates, or oscillates.

PLATINUM: An expensive metal having an extremely high melting point and good electrical conductivity. Often used in magneto breaker points.

POLARITY: The positive or negative terminal of a battery or an electric circuit; also the north or south poles of a magnet.

POLARIZE: To give polarity to an electric circuit so current will flow in the proper direction.

POPPET VALVE: A valve structure consisting of a circular head with an elongated stem attached in the center. A poppet valve is designed to open and close a circular hole or port.

PORCELAIN: General term applied to the material or element used for insulating the center electrode of a spark plug.

PORT: The openings in the cylinder block for valves, exhaust and inlet pipes, or water connections. In two cycle engines, the openings for inlet and exhaust purposes.

POSITIVE POLE: The point to which an electrical current returns after passing through the circuit. This is designated by a plus sign (+).

POST: The heavy circular part to which the group of plates is attached. The post extends through the cell cover to provide a means of attachment to the adjacent cell or battery cable.

POTENTIAL: An indication of the amount of energy available.

POTENTIAL DIFFERENCE: A difference of electrical pressure that sets up a flow of electric current.

PREIGNITION: Ignition occurring earlier than intended. Example: an explosive mixture could be fired in a cylinder by a flake of incandescent carbon before the electric spark occurs.

PRESS-FIT: Also known as a force-fit or drive-fit. Fit accomplished by forcing a shaft into a hole that is slightly smaller in diameter than the shaft.

PRIMARY WINDING: In an ignition coil or magneto armature, a wire that conducts the low tension current, which is to be transformed by induction into high tension current in the secondary winding.

PRIMARY WIRES: The wiring circuit used for conducting

the low tension or primary current to the points, where it is to be used.

PRONY BRAKE: A machine for testing the power of an engine while it is running against a friction brake.

PROPANE: A petroleum hydrocarbon compound that has a boiling point about 44°F and is used as an engine fuel. It is loosely referred to as LP-Gas and is often combined with Butane.

PSI: A pressure measurement in pounds per square inch.

PUSHROD: A connecting link in an operating mechanism, such as the rod interposed between the valve lifter and rocker arm on an overhead valve engine.

R

RATIO: The relation or proportion that one number bears to another.

R.C. ENGINE: Rotating combustion engine. Same as rotary engine.

REAM: To finish a hole accurately with a rotating, fluted tool.

RECIPROCATING: A back and forth movement, such as the action of a piston in a cylinder.

RECTIFIER: Device used to convert alternating current to pulsating direct current.

REED VALVE: A flat, springy valve covering the ports between the carburetor and the crank chamber in a two cycle engine.

REGULATOR: An automatic pressure reducing or regulating valve.

RESISTANCE: Quality of an electric circuit, or any component in it, to oppose the flow of electrical current.

RETARD: To cause the spark to occur at a later time in the cycle of engine operation. Opposite of spark advance.

ROCKER ARM: Device used in overhead valve system to transfer the upward motion of the pushrod to a downward force on the valve.

ROLLER BEARING: An inner and outer race upon which hardened steel rollers operate.

ROTARY VALVE: A valve construction in which ported holes move in and out of register with each other to allow fluids or gases to enter and exit.

ROTARY ENGINE: An engine that transmits energy from burning fuel directly to a rotating mechanical member.

ROTOR: A rotating valve or conductor for carrying fluid or electrical current from a central source to the individual outlets as required. In Wankel rotary engines, the triangular rotating member that transmits energy from burning fuel to an eccentric shaft is the rotor.

RPM: Revolutions per minute.

RUBBER: An elastic, vibration-absorbing material of either natural or synthetic origin.

RUNNING-FIT: Sufficient clearance allowed between a shaft and journal to allow free running without overheating.

S

SAE: Society of Automotive Engineers.

SCALE: A flaky deposit occurring on steel and iron or the mineral and metal build-up in a cooling system.

SCORE: A scratch, ridge, or groove marring a finished surface.

SEAT: A surface, usually machined, upon which another part rests or seats. Example: the surface upon which a valve face rests.

SECONDARY WINDING: In an ignition coil or magneto armature, a wire in which a secondary or high tension current is created by induction due to the interruption of current in the adjacent primary winding.

SEDIMENT: Active material of the battery plates, which is gradually shed and accumulates in a space provided below the plates.

SEIZE: When one surface moving on another binds and sticks. Example: a piston will seize in a cylinder because of lack of lubrication or excessive heat-related expansion.

SEMICONDUCTOR: A substance that shares the characteristics of both a conductor and an insulator.

SEPARATORS: Sheets of rubber or wood inserted between the positive and negative plates of a cell to keep them out of contact with each other.

SHIM: Thin sheets used as spacers between two parts, such as the two halves of a journal bearing.

SHORT CIRCUIT: To provide a shorter path; often used to indicate an accidental ground in an electrical device or conductor.

SHRINK-FIT: An exceptionally tight fit achieved by the heating and/or cooling of parts. The outer part is heated above its normal operating temperature or the inner part chilled below its normal operating temperature and assembled in this condition.

SHROUD: A light cover over the flywheel that shields the flywheel and helps direct airflow over the engine to carry away heat.

SHUNT: To bypass or turn aside. In electrical apparatus, an alternate path for the current.

SHUNT WINDING: An electric winding or coil of wire that forms a bypass or alternate path for electric current. When applied to electric generators or motors, each end of the field winding is connected to an armature brush.

SIDE SEALS: The spring-loaded seals located on the sides of the triangular rotor in the Wankel engine.

SILLMENT SEAL: Compacted powder that helps ensure permanent assembly of a spark plug and eliminates compression leakage under operating conditions.

SLIDING-FIT: Clearance between a shaft and journal that is sufficient to allow free running without overheating.

SLUDGE: A composition of oxidized petroleum products along with an emulsion formed by the mixture of oil and water. This forms a pasty substance, clogs oil lines and passages, and interferes with engine lubrication.

SOLENOID: An iron core surrounded by a coil of wire that moves due to magnetic attraction when an electrical current is fed to the coil. Solenoids are often used to actuate mechanisms by electrical means.

SOLVENT: A solution that dissolves some other material. Example: water is a solvent for sugar.

SOUTH POLE: The pole of a magnet to which the lines of force flow; the opposite of north pole.

SPARK: An electrical current possessing sufficient pressure to jump an air gap from one conductor to another.

SPARK ADVANCE: Causing the spark to occur at an earlier time in the cycle of engine operation; opposite of retard.

SPARK GAP: The space between the electrodes of a spark plug across which the spark jumps. Also, a safety device in a magneto to provide an alternate path for the current when it exceeds a safe value.

SPARK KNOCK: See Preignition.

SPARK PLUG: A device inserted into the combustion chamber of an engine. Spark plugs contain an insulated central electrode for conducting the high tension current from the ignition distributor or magneto. This insulated electrode is spaced a predetermined distance from the shell or side electrode in order to control the dimensions of the gap that the spark must jump.

SPECIFIC GRAVITY: The relative weight of a substance as compared to water. Example: if a cubic inch of acid weighs twice as much as a cubic inch of water, the specific gravity of the acid would be 2.

SPIRAL BEVEL GEARS: A gear and pinion wherein the mating teeth are curved and placed at an angle with the pinion shaft.

SPLINE: A long keyway.

SPURT-HOLE: A hole drilled through a connecting rod and bearing that allows oil under pressure to be squirted out of the bearing for additional lubrication of the cylinder walls.

SQ. FT.: Square foot.

SQ. IN.: Square inch.

STAMPING: A piece of sheet metal cut and formed into the desired shape with the use of dies.

STARTER: An electric motor attached by gearing to an engine to provide power to turn it over for starting.

STATOR: Stationary coils of an alternating current generator.

STRESS: The force or strain to which a material is subjected.

STROBOSCOPE: A term applied to an ignition timing light, which, by being attached to the distributor points, gives the effect of making a mark on a rapidly rotating wheel, such as a flywheel, appear to stand still for observation.

STROKE: The distance moved by the piston.

STUD: A rod with threads cut on both ends, such as a cylinder stud, which screws into the cylinder block on one end and has a nut placed on the other end to hold the cylinder head in place.

SUCTION: Suction exists in a vessel when the pressure is lower than the atmospheric pressure. Also see Vacuum.

SULFATED: When a battery is improperly charged or allowed to remain in a discharged condition for some length of time, an abnormal amount of lead sulfate will collect on the plates. The battery is then said to be "sulfated."

SUMP: The part of the block in a small four-stroke engine that holds and collects the lubricating oil.

SUPERCHARGER: A device for increasing the volume of the air charge for an internal combustion engine.

SYNCHRONIZE: To cause two events to occur in unison or at the same time.

T

TACHOMETER: A device for measuring and indicating the rotational speed of an engine.

TAP: To cut threads in a hole with a tapered, fluted, threaded tool.

TAPPET: The adjusting screw for varying the clearance between the valve stem and the cam. May be built into the valve lifter or into the rocker arm on an overhead valve engine.

TCI: Transistor controlled ignition.

TDC: Top dead center.

TENSION: Effort devoted towards elongation or "stretching" of a material.

TERMINAL: In electrical work, a junction point where connections are made, such as the terminal fitting on the end of a wire.

THERMAL EFFICIENCY: A gallon of fuel contains a certain amount of potential energy in the form of heat when burned in the combustion chamber. Some of this heat is lost and some is converted into power. The thermal efficiency is the ratio of work accomplished compared to the total quantity of heat contained in the fuel.

THERMOSTAT: A heat-controlled valve used in the cooling system of an engine to regulate the flow of water or used in the electrical circuit to control the current.

THROW: With reference to an engine, usually the distance from the center of the crankshaft main bearing to the center of the connecting rod journal.

TIMING CHAIN: Chain used to drive camshaft and accessory shafts of an engine.

TIMING GEARS: Any group of gears that are driven from the engine crankshaft to cause the valves, ignition, and other engine-driven apparatus to operate at the desired time during the engine cycle.

TOLERANCE: A permissible variation between the two extremes of specified dimensions.

TORQUE: An effort devoted toward twisting or turning.

TORQUE WRENCH: A special wrench with a built-in indicator to measure the applied turning force.

TRANSFER PORT: In two cycle engines, an opening in the cylinder wall permitting fuel to enter from the crankcase.

TRANSISTOR: A semiconductor device that is often used for switching applications. A transistor is used in place of breaker points in a TCI system.

TROUBLESHOOTING: Refers to a process of diagnosing or determining the source of trouble or troubles from observation and testing.

TUNE-UP: The process of accurate and careful adjustments to obtain maximum engine performance.

TURBULENCE: A disturbed or disordered, irregular motion of fluids or gases.

TWO CYCLE ENGINE: An engine design permitting one power stroke for each revolution of the crankshaft.

U

UP-DRAFT: Used to describe a carburetor type where the mixture flows upward to the engine.

UPPER CYLINDER LUBRICATION: A method of introducing a lubricant into the fuel or intake manifold in order to permit lubrication of the upper cylinder, valve guides, and other parts.

V

VACUUM: A perfect vacuum has not been created as this would involve an absolute lack of pressure. The term is ordinarily used to describe a partial vacuum, that is, a pressure less than atmospheric pressure. See Suction.

VACUUM GAGE: An instrument designed to measure the amount of vacuum existing in a chamber.

VALVE: A device for alternately opening and sealing an aperture.

VALVE CLEARANCE: Air gap allowed between end of valve stem and valve lifter or rocker arm to compensate for heat expansion.

VALVE FACE: Part of a valve that mates with and rests upon a seating surface.

VALVE GRINDING: A process of mating the valve seat and valve face performed with the aid of an abrasive.

VALVE HEAD: The portion of the valve upon which the valve face is machined.

VALVE KEY OR VALVE LOCK: The key, washer, or other device that holds the valve spring cup or washer in place on the valve stem.

VALVE LIFTER: A rod or plunger that transfers motion from the cam and the other valve train components. A lifter is often adjustable to vary the length of the unit.

VALVE MARGIN: The space or rim between the surface of the head and the surface of the valve face.

VALVE OVERLAP: An interval expressed in degrees where both valves of an engine cylinder are open at the same time.

VALVE SEAT: The matched surface upon which the valve face rests.

VALVE SPRING: A spring attached to a valve to return it to the seat after it has been released from the lifting or opening operation.

VALVE STEM: That portion of a valve that rests within a guide.

VALVE STEM GUIDE: A bushing or hole in which the valve stem is placed. A valve stem guide allows only two-way motion.

VANES: Any plate or blade attached to an axis and moved by air or liquid.

VAPORIZER: A device for transforming or helping to transform a liquid into a vapor; often includes the application of heat.

VAPOR LOCK: A condition wherein fuel boils in the fuel system, forming bubbles that retard or stop the flow of liquid fuel to the carburetor.

VENTURI: Two tapering streamlined tubes joined at their small ends so as to reduce the internal diameter.

VIBRATION DAMPER: A device to reduce the torsional or twisting vibration that occurs along the length of the crankshaft used in multiple cylinder engines; also known as a harmonic balancer.

VISCOSITY: The resistance to flow or adhesive characteristics of an oil.

VOLATILITY: The tendency for a fluid to evaporate rapidly or pass off in the form of a vapor. Example: gasoline is more volatile than kerosene because it evaporates at a lower temperature.

VOLT: A unit of electrical force that will cause a current of one ampere to flow through a resistance of one ohm.

VOLTAGE REGULATOR: An electrical device for controlling or regulating voltage.

VOLTMETER: An instrument for measuring the voltage in an electrical circuit.

VOLUME: The measure of space expressed as cubic inches, cubic feet, or other units of linear measure.

VOLUMETRIC EFFICIENCY: A combination between the ideal and actual efficiency of an internal combustion engine. If the engine completely filled each cylinder on each intake stroke, the volumetric efficiency of the engine would be 100%. In actual operation, however, volumetric efficiency is lowered by the inertia of the gases, the friction between the gases and the manifolds, the temperature of the gases, and the pressure of the air entering the carburetor. Volumetric efficiency is ordinarily increased by the use of large valves, ports, and manifolds. It can be further increased with the aid of a supercharger.

W

WANKEL, FELIX: German inventor of the Wankel rotary engine.

WATER COLUMN: A reference term used in connection with a manometer.

WATT: A measuring unit of electrical power obtained by multiplying amperes by volts.

WIRING DIAGRAM: A detailed drawing of all wiring, connections, and components in an electrical circuit.

WRIST PIN: See Piston Pin.

INDEX